Lecture Notes in Computational Science and Engineering **132**

Editors:

More information about this series at http://www.springer.com/series/3527

Harald van Brummelen • Alessandro Corsini •
Simona Perotto • Gianluigi Rozza
Editors

Numerical Methods for Flows

FEF 2017 Selected Contributions

Editors

Harald van Brummelen
Department of Mechanical Engineering
Eindhoven University of Technology
Eindhoven
Noord-Brabant, The Netherlands

Alessandro Corsini
Department of Mechanical and Aerospace Engineering
Sapienza University of Rome
Roma, Italy

Simona Perotto
MOX - Department of Mathematics
Politecnico di Milano
Milano, Italy

Gianluigi Rozza
SISSA mathLab, Mathematics Area
International School for Advanced Studies
Trieste, Italy

ISSN 1439-7358 ISSN 2197-7100 (electronic)
Lecture Notes in Computational Science and Engineering
ISBN 978-3-030-30707-3 ISBN 978-3-030-30705-9 (eBook)
https://doi.org/10.1007/978-3-030-30705-9

Cover illustration: Velocity 1st POD mode field for the problem of the flow around a circular cylinder. There are no parameters considered in this case where POD is performed on a set of snapshots acquired in different time instants. Reynolds number in this case is equal to 1e4. Courtesy of S. Hijazi (SISSA mathLab).

This Springer imprint is published by the registered company Springer Nature Switzerland AG.
The registered company address is: Gewerbestrasse 11, 6330 Cham, Switzerland

Preface

This special volume contains selected contributions from the Finite Elements in Flow (FEF) conference held in Rome in April 2017: http://congress.cimne.com/fef2017/.

The overarching objective of the FEF 2017 conference, in a similar vein to the previous editions, was to provide a forum for the exchange of ideas and recent results in finite element-type methods for applications in fluid dynamics and related areas. Both the methodological and the applicative goals of the FEF series have broadened over recent editions, extending beyond traditional finite element methods and traditional fluid mechanics. Indeed, FEF 2017 attracted many participants using numerical techniques other than finite element methods or considering applications other than fluid dynamics, often with multiphysics couplings.

The volume "Numerical Methods for Flows" brings together up-to-date contributions in applied mathematics, numerical analysis, numerical simulation, and scientific computing related to fluid mechanics problems. The authors are world-leading scientists who participated in the FEF 2017 conference and accepted our invitation to contribute a chapter. All papers were selected after anonymous peer review. Comprising 30 chapters, the book presents the state of the art in topics relating to numerical simulation for flows and provides very interesting insights and perspectives regarding current and future methodological and numerical developments in computational science. The contributions are organised from the most methodological ones to the most application-oriented ones in computational fluid dynamics, and the book will meet the needs of both researchers and graduate students.

We thank the contributors and the reviewers for their outstanding work and would like to express our gratitude to the International Association for Computational Mechanics for granting us the privilege of organizing the 19th edition of the FEF conference series. We also cordially thank the administration of Sapienza University of Rome for providing the splendid venue, as well as Professor Maurizio Falcone for his help and contacts with the Department of Mathematics "Guido Castelnuovo" at the university. We furthermore gratefully acknowledge the administrative support of CIMNE and, in particular, Mr. Alessio Bazzanella. Finally, we would like to thank the editorial board of Lecture Notes in Computational Science and Engineering for hosting this special volume in the series and offer special thanks to Francesca Bonadei and Francesca Ferrari of Springer Milan for their constant support during the preparation of the volume.

Eindhoven, The Netherlands — Harald van Brummelen
Rome, Italy — Alessandro Corsini
Milan, Italy — Simona Perotto
Triste, Italy — Gianluigi Rozza
March 2019

Contents

About the Editors

Harald van Brummelen holds the chair of Multiscale Engineering Fluid Dynamics at Eindhoven University of Technology in the Netherlands. He received an MSc in Mechanical Engineering from Twente University in 1997 and a PhD in Numerical Analysis from the University of Amsterdam in 2002. He is the current Director of the Dutch National Graduate School for Engineering Mechanics and Secretary General of the European Community on Computational Methods in Applied Sciences (ECCOMAS). Professor van Brummelen's research is concerned with advanced numerical methods for multiscale problems in fluid dynamics and, in particular, adaptive finite element techniques, fluid-structure interaction, rarefied-gas dynamics and Boltzmann's equation, and phase-field models. He has (co-)authored more than 50 journal papers. He is the local program director of the industrial-private partnership program FIP (Fundamental Fluid-Dynamics Challenges in Inkjet Printing) and principal investigator of the project RareTrans (Transport in Rarefied Gases in Next Generation Photo Lithography Machines).

Alessandro Corsini is Professor of Energy Conversion and Fluid Machinery in the Department of Mechanical and Aerospace Engineering at Sapienza University of Rome. He received his PhD from the Sapienza University of Rome in 1996, and since 1997 has lectured on Thermal Machines and Energy Systems Courses at the university. He has been involved in the development of CFD methods for the prediction of turbulent, incompressible internal flows, with specific emphasis on the development of stabilized formulations for turbomachinery CFD. Professor Corsini has been collaborating with the Team for Advanced Flow Simulation and Modeling TAFSM at Rice University, USA, since 2004. He is active in the field of industrial turbomachinery and energy systems, working on CFD-based design concepts, noise reduction technologies, stall dynamics, and control and fault monitoring. Alessandro Corsini is Associate Editor of the IMechE Journal of Power and Energy and a member of the editorial boards of the International Journal of Rotating Machinery and Periodica Polytechnica. He is a reviewer for international journals in the fields of Energy, Turbomachinery, and Computational Mechanics. He is the author of more than 150 scientific publications and holds nine international patents.

Simona Perotto is Associate Professor in Numerical Analysis at the Department of Mathematics of Politecnico di Milano, Italy. She gained her Master's degree in Mathematics from the University of Torino in 1995 and her PhD in Computational Mathematics and Operations Research from the University of Milan in 1999. She then spent 2 years as a postdoc in Scientific Computing and Mathematical Modeling at the EPFL, Lausanne. Her primary research fields cover anisotropic mesh adaptation for computational fluid dynamics problems and adaptive model reduction in the framework of a finite element approximation of partial differential equations. She is the author of more than 50 scientific publications in peer-reviewed journals. Professor Perotto has supervised more than 50 Master's theses in Engineering at Politecnico di Milano and seven PhD students. She has been Co-PI of the project FIRB2008 on "Advanced Statistical and Numerical Methods for the Analysis of High Dimensional Functional Data in Life Sciences and Engineering", and of the NSF Project, DMS 1419060, on "Model Reduction Techniques for Incompressible Fluid-Dynamics and Fluid-Structure Interaction Problems" (July 2014 to June 2018). Finally, she has organized several international and national minisymposia, workshops, and conferences, including the 8th edition of the International Conference on Adaptive Modeling and Simulation and the 19th International Conference on Finite Elements in Flow Problems.

Gianluigi Rozza is Professor in Numerical Analysis and Scientific Computing at the International School for Advanced Studies (SISSA), Trieste, Italy. He gained an MSc in Aerospace Engineering from Politecnico di Milano in 2002 and a PhD in Applied Mathematics from EPFL in 2005; he was subsequently a post-doc at MIT. At SISSA Professor Rozza is coordinator of the SISSA doctoral program in Mathematical Analysis, Modelling, and Applications, as well as a lecturer on the Master's degrees in High Performance Computing, Mathematics, and Data Science and Scientific Computing. He is SISSA Director's delegate for Valorisation, Innovation, Technology Transfer, and Industrial Cooperation. His research is mostly focused on numerical analysis and scientific computing, developing reduced order methods. He is the author of more than 100 scientific publications, the editor of five books, and the author of two books. He has been Principal Investigator of the European Research Council Consolidator Grant (H2020) AROMA-CFD and for the project FARE-AROMA-CFD, funded by the Italian Government. Within SISSA mathLab he is responsible for the UBE (Under Water Blue Efficiency), SOPHYA (Seakeeping Of Planing Hull for YAchts) and PRELICA projects within the regional maritime technology cluster MARE FVG, and coordinator of industrial projects with several companies, such as Danieli, Electrolux, and Fincantieri. He is an associate editor of SIAM SINUM, SIAM/ASA JUQ, Computing and Visualisation in Science, and Mathematics in Engineering. He is a member of the Applied Mathematics Committee of the European Mathematical Society.

Simulation of Complex High Reynolds Flows with a VMS Method and Adaptive Meshing

Luisa Silva, David Chalet, Thierry Coupez, Audrey Durin, Tristan Launay, and Christelle Ratajczack

Abstract The whistling noise phenomenon, which is related to vortexes appearance in high Reynolds air flows in ducts, implies a very precise description of the flow at the very small scales, especially near the solid walls, on which boundary layer division may occur. In this work, the Variational Multiscale method has been coupled to automatic anisotropic adaptive meshing, allowing the capture of very complex flows at high Reynolds number. The adaptive procedure is based on the error evaluation on several chosen quantities (phase location, velocity, velocity direction changes) and it provides the capture of very thin flow motions, even close to the walls or boundaries. Simulations of flows on resonator-like geometries have been performed, reputed to whistle for certain flow rates. A method to qualitatively discriminate whistling from non-whistling flow rates has been implemented, based on the appearance of certain vortexes on the obtained flow patterns.

Keywords Anisotropic adaptive meshing · Computational fluid dynamics · VMS solver · Finite elements · Parallel computing

---

L. Silva (✉) · T. Coupez
High Performance Computing Institute (ICI), École Centrale de Nantes, Nantes, France
e-mail: luisa.rocha-da-silva@ec-nantes.fr; thierry.coupez@ec-nantes.fr

D. Chalet · A. Durin
Hydrodynamics, Energetics & Atmospheric Environment Laboratory (LHEEA), École Centrale de Nantes, Nantes, France
e-mail: david.chalet@ec-nantes.fr; audrey.durin@ec-nantes.fr

T. Launay · C. Ratajczack
Mann+Hummel, Research Center of Laval, Laval, France
e-mail: tristan.launay@mann-hummel.com; christelle.ratajczack@mann-hummel.com

H. van Brummelen et al. (eds.), *Numerical Methods for Flows*,
Lecture Notes in Computational Science and Engineering 132,
https://doi.org/10.1007/978-3-030-30705-9_1

1 Introduction

Resonator type devices are used in inlet and exhaust systems of internal combustion engines to reduce the running noises. They are constituted of chambers connected to the pipe through openings which allow reducing the pressure (and thus the noise), but are paradoxically subject to whistling issues.

Generally, these unwanted sounds are caused by vortex shedding [1]. When a fluid flows around a solid object, a fluid boundary layer is formed around it. Pushed by the main flow, this boundary layer will create a rotating vortex. Behind the object, two vortexes are released alternatively and a vortex street is formed. Vortexes generate aerodynamic sound, and can be described as acoustic sources and integrated in an acoustical study through an equivalent force field, as described by Lighthill [2], where the general problem discussed was how to estimate the sound emitted from such a given fluctuating flow. The authors consider two major assumptions: the first one is that the propagation of fluctuations in the flow is not considered; the second is that the reaction of produced sound on the flow itself is neglected. It is worth noticing that this reaction of the sound on flow is expected when a resonator (a chamber) is close to the flow, but this theory is not applicable to supersonic flows. In reality, the acoustic perturbation will interact with the flow and potentially existing solids. It can be thus amplified or dissipated, or create new vortexes interacting with unstable shear layers, leading to complex coupled cases.

For example, in classical open cavity or side-branch cases [3–5], a thin shear layer is created when air is blown above the orifice. If there is an existing acoustic perturbation, it will interact with the shear layer, generating vortexes that will in return interact with the acoustic field, absorbing the acoustical energy, or amplifying it by a feed-back loop phenomenon in the cavity. To obtain whistle, the velocity at which the vortexes will cross the orifice have to match the resonance frequency of the cavity so that the latter will whistle at certain air flow rates. The final frequency of the noise is given by the resonator, which is the cavity itself. Recent studies of silencer whistling [5, 6] have shown that the same phenomenon occurs with the shear layers above the resonator holes. In an experimental study of a perforated straight-through pipe type muffler, other authors [7] have visualized the flow, showing that the air is flowing out the tube into the chamber through almost all the holes, except the last one, through which the air was strongly flowing back from the chamber into the tube. They have also found that the fluctuating velocities near the last row of holes show peaks at frequencies corresponding to the noise generated, and have concluded that the whistling noise generated by the perforated straight-through pipe type muffler is caused by the turbulent air flow passing through the last row of holes, and that it is strongly correlated with the resonance of the tail pipe and/or the cavity. A turbulent flow is characterized by chaotic fluctuations of velocity and pressure appearing at high Reynolds number. Whereas Reynolds number increase, a bifurcation is created and the flow develops a turbulent behaviour, irregular both in space and time. There is no analytic solution of a turbulent flow. Efforts are focusing on the numerical research. A direct solution would be to solve Navier–

Stokes equation with appropriate boundary and initial conditions. That is called Direct Numerical Simulation (DNS). Nevertheless, the extent of the scales that must be solved in time and space is very large and beyond the power domain of today's computers. Because of that, the DNS method is always limited to moderate Reynolds numbers. There are two alternative solutions, the Large Eddy Simulation methods (LES) and the Reynolds Averaged Navier–Stokes equations (RANS). The first of these approaches consists in filtering the flow scales in order to fully solve the large scale structures while the effect of the small scales are modelled and are not explicitly computed (unlike in the VMS method, see below). Nevertheless, this method remains very costly. In the other hand, the RANS method averages the equations in time and space, suppressing totally the fluctuations and leading to a set of equations less costly. Nevertheless, as the RANS method does not solve the small scales, closure models are required to take into account the effects of turbulence.

In this study, a Variational Multiscale Method [8] (VMS, see below) was used and coupled with adaptive meshing [9]. In this method, the scales under the mesh size are described through stabilization terms. These ones are implicitly computed, unlike in the LES method where the physical models taking into account the small scale are explicitly computed. Combining the VMS method with adaptive meshing allows a reduction of the "approximated" part of the small scale in the computation (reducing the mesh size), which brings it close to a DNS method. The full methodology is based on an implicit boundary approach [10], established within a massively parallel framework [11, 12].

2 Whistling Study Through CFD

2.1 *Numerical Resolution of the Navier–Stokes Equations*

The fluid motion is described by the non-linear Navier–Stokes equations. The discrete Galerkin problem implies solving the mixed problem: find the discrete velocity-pressure pair $(\mathbf{v}_h, p_h) \in (\mathscr{V}_h, \mathscr{Q}_h)$ for $(\mathbf{w}_h, q_h) \in (\mathscr{V}_{h,0}, \mathscr{Q}_h)$ such that

$$\begin{aligned}
&(\rho\partial_t \mathbf{v}_h, \mathbf{w}_h) + (\rho \mathbf{v}_h \cdot \nabla \mathbf{v}_h, \mathbf{w}_h) + (2\mu\varepsilon(\mathbf{v}_h) : \varepsilon(\mathbf{w}_h)) - (p_h, \nabla \cdot \mathbf{w}_h) = \\
&\qquad (\mathbf{f}_h, \mathbf{w}_h) + (h_N, \mathbf{w}_h)_{\Gamma_N} \\
&(\nabla \cdot \mathbf{v}_h, q_h) = 0
\end{aligned} \tag{1}$$

where Γ_N is the part of the boundary where boundary conditions are imposed. To avoid numerical instabilities, the Variational MultiScale (VMS) method [8] is used, by applying an orthogonal decomposition of the functional spaces, to get $\tilde{\mathscr{V}} = \mathscr{V}_h \oplus \mathscr{V}'$ and $\tilde{\mathscr{Q}} = \mathscr{Q}_h \oplus \mathscr{Q}'$. Taking $\tilde{\mathbf{v}} = \mathbf{v}_h + \mathbf{v}' \in \tilde{\mathscr{V}}$ and $\tilde{p} = p_h + p' \in \tilde{\mathscr{Q}}$, we decompose the velocity and the pressure fields respectively into two scales: the resolvable coarse scale and the unresolvable fine one, which we model to provide additional stabilization. After simplification, one gets

$$
\begin{aligned}
&\underbrace{(\rho(\partial_t \mathbf{v}_h + \mathbf{v}_h \cdot \nabla \mathbf{v}_h), \mathbf{w}_h)_\Omega + (2\mu\varepsilon(\mathbf{v}_h) : \varepsilon(\mathbf{w}_h))_\Omega - (p_h, \nabla \cdot \mathbf{w}_h)_\Omega - (\mathbf{f}_h, \mathbf{w}_h)_\Omega}_{\text{Galerkin}} \\
&\underbrace{+ \sum_{K \in \Omega} \tau_K (\rho(\partial_t \mathbf{v}_h + \mathbf{v}_h \cdot \nabla \mathbf{v}_h) + \nabla p_h - \mathbf{f}_h, \rho \mathbf{v}_h \cdot \nabla \mathbf{w}_h)_K}_{\text{Upwind stabilization}} \\
&\underbrace{+ \sum_{K \in \Omega} \tau_K (\rho(\partial_t \mathbf{v}_h + \mathbf{v}_h \cdot \nabla \mathbf{v}_h) + \nabla p_h - \mathbf{f}_h, \rho \mathbf{v}_h \cdot \nabla q_h)_K}_{\text{Pressure stabilization}} \\
&\underbrace{+ \sum_{K \in \Omega} (\tau_c \nabla \cdot \mathbf{v}_h, \nabla \cdot \mathbf{w}_h)_K}_{\text{Grad-div stabilization}} = 0
\end{aligned}
\tag{2}
$$

Stabilisation parameters computation has been detailed in [9]. One important point is that in the definitions of such parameters one needs to compute the local mesh size, h_K, which usually refers to the element diameter. This choice is not optimal when we use an anisotropic mesh and, for example, in convection dominated problems, we will use the element diameter computed in the flow direction.

2.2 *Immersed Boundaries and Automatic Anisotropic Adaptation*

One unique mesh is defined for both fluid and solid domains [10], and only one set of the Navier–Stokes equations is used to solve the problem. The geometrical mesh built to solve it through the previously described VMS method is iteratively adapted by using the interpolation error theory [13]. In this way, the mesh discretization takes into account, the velocity direction variations, the velocity magnitude and phase functions describing the boundaries. With this methodology, even the small vortexes developed by the solution will be captured and boundary layers at the fluid solid interface. In this context, the solid domain is also discretized (and has its own phase function), supposing that it is a rigid body, with zero velocity. This is achieved by weakly imposing this condition in the formulation.

3 Whistling Simulations

3.1 *Vortex Detection*

In the following examples, appearance of a vortex will indicate under which criteria whistling will occur. Several criteria exist in the literature allowing the location of a vortex core in a flow, but we have chosen to compute λ_2, the second eigenvalue of the summation of the squares of the symmetric and antisymmetric components of the velocity gradient. This criterion or vortex detection is based on the fact that, in a vortex, pressure tends to have a local minimum in the axe of the circulatory motion of the vortex when the centrifugal force is balanced by the pressure force (cyclotrophic balance). In this case, the vortex zones are the ones where λ_2 is large and to well identify it, one plots the isovalue 0 of λ_2 a line in 2D or as a surface in 3D. More details are given in [14].

3.2 *Validation in a Simple Air Flow Benchmark*

Simulations of flows in straight pipes (2D and 3D) at high Reynolds number have been performed to check the accuracy of our method. Theoretical values used for comparison are presented in [15]. A similar validation for adaptive finite element approximations has been performed in [16].

The pipe measures 78 cm long and has a 3.5 cm diameter, and its section is centered in $y = 0$. Numerical probes are placed in the radial direction. We simulate the flow with an imposed bulk velocity of 48 m/s (Re $= 1054$, considering the pipe diameter as the length scale), which corresponds to the 155kg/h flow rate measured during the whistling experiments on the silencer prototypes, which will be described hereafter. Density of the air is $1.255\,\text{kg/m}^3$ and its viscosity is $1.85e^{-5}$ Pa s.

Expressions for the velocity profile across the section are different for the sublayers close to the boundary. A central layer represents 80% of the thickness, where the axial velocity is not so much affected by the conditions at the wall. The internal layer (inner region) is defined as the thickness where the later is much affected by the wall conditions. One may distinguish three zones in this internal layer: the viscous sub-layer, near the wall, where the flow is laminar. Its thickness grows in the flow direction until reaching an instability, in a buffer zone, and then one attains a third region, the logarithmic sublayer.

Figure 1 shows the computed and theoretical axial velocity profiles, as a function of the radial position, in 2D (left) and in 3D (right). Probes are located at $x = 65$ cm, $z = 0$ in 3D and at y values that include all the described layers. The obtained mean values in time are computed on 100 times steps with $dt = 10^{-4}$ s.

In 2D, one observes that the velocity error is maximal near the buffer zone in which there is no “real” theoretical value, and where we thus cannot estimate the error. Out of the buffer zone, the error does not exceed 10% except on two probes:

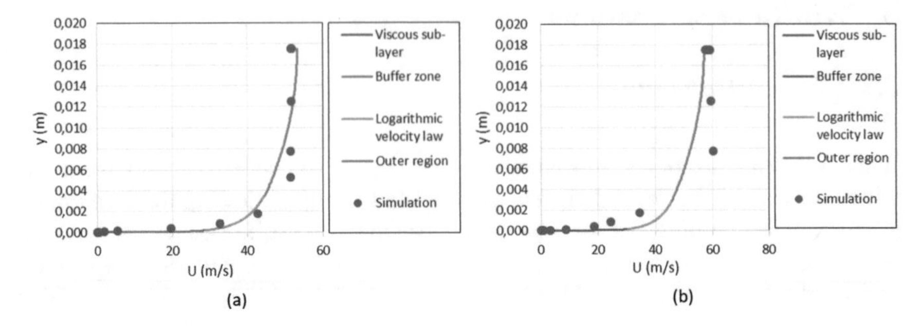

Fig. 1 Pipe flow: theoretical and computed mean velocities in 2D (left, **a**) and in 3D (right, **b**)

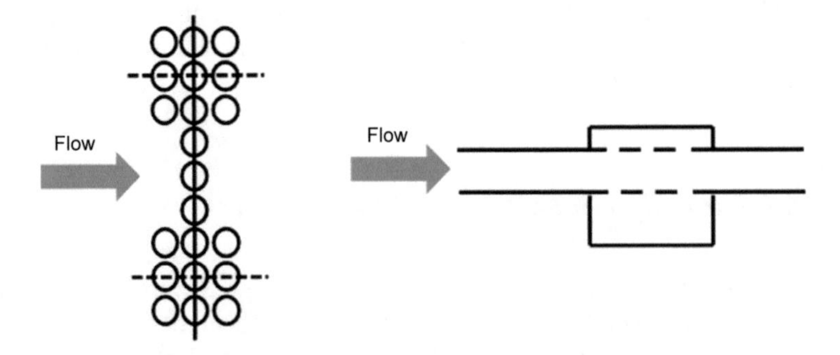

Fig. 2 Holes repartition in the tube (left) with a visualization of the eccentric chamber (right)

in the center of the flow for the pressure (15%) and in the logarithmic zone for the velocity (27%). In 3D, the velocity error is maximal in the buffer zone but also in the logarithmic zone (15–33%). In this zone, the velocity gradient is less strong than in the viscous sub-layer, leading to bigger mesh sizes, but it is still strong enough to lead to numerical errors. Nevertheless, overall results remain satisfying.

3.3 Industrial Application

Whistling simulations have been performed on a prototype reproducing the typical conditions in silencers. In fact, it is a simple pipe connected to a chamber through holes. The holes repartition can be used to describe the geometric characteristics of the silencer in a simple way. It is worth noticing that the chamber is not symmetric (due to the holes repartition) as described in Fig. 2. If there is no specification, the 2D visualization below are performed doing a slice of the geometry along the flow direction, passing through the second and the eighth horizontal row of holes (Fig. 2, on the left).

The tube is characterized by a 30.5 mm input diameter and 35 mm output diameter. As depicted in Fig. 2, the eccentric chamber is connected to the pipe by using different holes. Flow simulations were performed on these complex geometries for a 100 kg/h and a 155 kg/h rate flows for which it is supposed not to whistle and to whistle, respectively. The Reynolds number are equal to 7.10^4 and 10^5, depending on the mass flow rate. According to the analytical models described above for 3D tubes, the characteristic thickness of the shear layer near the walls is comprised between 7.10^{-6} and 10^{-5} m. As a consequence, the minimal mesh size is set to 10^{-7} m and the maximal mesh size is set to 5.10^{-3} m. The number of mesh nodes needed to achieve the computations varies from one simulation to another. The maximal number we have reached is 8×10^6, and it has necessitated 792 cores to run. The computations generally last about 1 day, without considering the pre-processing (geometry format conversion, immersion, etc.).

In the following paragraphs, the computation time step is 10^{-4} s. A uniform velocity is imposed on the pipe input and a zero pressure is imposed on the pipe output. This velocity is gradually increased to reach 100 kg/h or 155 kg/h. It is worth noticing that the boundary conditions (velocities, pressure, zero velocity in the solid) are imposed through a strong imposition method i.e. the values are imposed at the nodes. On the following pictures, the air direction is from the left to the right. In both cases, we do not find the pattern described previously. As described in Fig. 3 (λ_2 plots), there are no strong vortexes in the flows. Nevertheless, at 155 kg/h, before the flow is stabilized, there is an important perturbation occurring just after the last row of holes, as depicted in Fig. 4. We can visualize it both with the λ_2 and pressure criteria. This observation shows similarities with the phenomena experimentally observed by Kojima et al. [7]. To sum up, a vortex zone was found at 155 kg/h, but only during the transitory flow before stabilization.

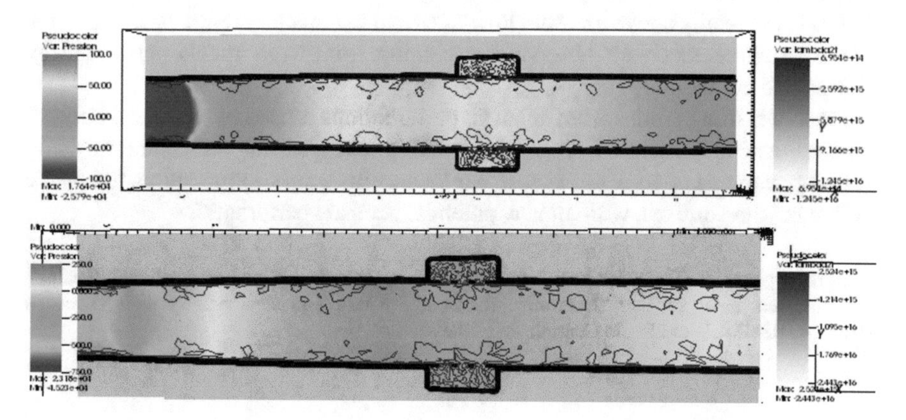

Fig. 3 $\lambda_2 = -1000$ isovalue contours and colored pressure values for 100 kg/h (top) and 155 kg/h (bottom)

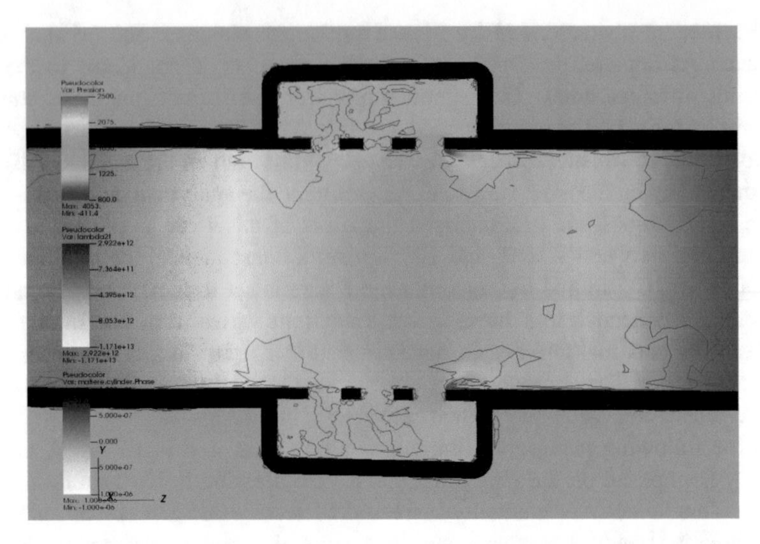

Fig. 4 $\lambda_2 = -1000$ isovalue contours and colored pressure values for 155 kg/h (transient period)

4 Conclusions and Perspectives

Simulations of turbulent air flow in different geometries at several flow rates have been performed, using a parallel scientific computation library, based on an adaptive monolithic finite element approach.

In particular, an industrial resonator prototype has been studied, and we have noticed that there is an area just below the last row of holes that seems to generate vortexes, especially above a certain flow rates and in transitory regimes, information that is in agreement with observations coming from experiments performed by Kojima and co-authors [7].

Since whistling comes from acoustic perturbations caused by vortexes, but also from acoustic amplification, we can try to avoid it suppressing the geometric cause of the appearance of vortexes. Future work concerns shape optimization, to be able to "break" the vortexes, without new potential acoustic perturbations.

Acknowledgements The work in this article has been done in a joined International Teaching and Research Chair entitled "Innovative Intake and Thermo-management Systems" between MANN+HUMMEL and Ecole Centrale de Nantes.

References

1. Powell, A.: Theory of vortex sound. J. Acoust. Soc. Am. **36**(1), 177–195 (1964)
2. Lighthill, M.J.: On sound generated aerodynamically. I. Gen. Theory Proc. R. Soc. Lond. A: Math. Phys. Eng. Sci. **211**, 564–587 (1952)
3. Lafon, P., Caillaud, S., Devos, J.P., Lambert, C.: Aeroacoustical coupling in a ducted shallow cavity and fluid/structure effects on a steam line. J. Fl. Struct. **18**(6), 695–713 (2003)
4. Ziada, S.: Flow-excited acoustic resonance in industry. J. Press. Vessel. Tech. **132**(1), 1–9 (2010)
5. Karlsson, M., Knutsson, M., Abom, M.: Predicting Fluid Driven Whistles in Automotive Intake and Exhaust Systems. SAE International, Warrendale (2016)
6. Du, L., Abom M., Karlsson, M., Knutsson, M.: Modelling of Acoustic Resonators Using the Linearized Navier Stokes Equations. SAE International, Warrendale (2016)
7. Kojima, N., Nakamura, Y., Fukuda, M.: A study on the correlation between fluctuating velocity in a muffler and air flow noise: heat transfer, power, combustion, thermophysical properties. JSME Int. J. **30**(265), 1113–1120 (1987)
8. Hughes, T.J.R.: Multiscale phenomena: green's functions, the Dirichlet-to-Neumann formulation, subgrid scale models, bubbles and the origins of stabilized methods. Comp. Meth. Appl. Mech. Eng. **127**, 387–401 (1995)
9. Hachem, E., Rivaux, B., Kloczko, T., Digonnet, H., Coupez, T.: Stabilized finite element method for incompressible flows with high Reynolds number. J. Comp. Phys. **229**(23), 8643–8665 (2010)
10. Coupez, T., Silva, L., Hachem, E.: Implicit boundary and adaptive anisotropic meshing. In: Perotto, S., Formaggia, L. (eds.) New Challenges in Grid Generation and Adaptivity for Scientific Computing. Springer, Berlin (2015)
11. Silva, L., Coupez, T., Digonnet, H.: Massively parallel mesh adaptation and linear system solution for multiphase flows. Int. J. Comp. Fl. Dyn. **30**(6), 431–436 (2016)
12. Digonnet, H., Coupez, T., Laure, P., Silva, L.: Massively parallel anisotropic mesh adaptation. Int. J. High Perf. Comp. Appl. (2017). https://doi.org/10.1007/s001090000086
13. Coupez, T.: Metric construction by length distribution tensor and edge based error for anisotropic adaptive meshing. J. Comp. Phys. **230**, 2391–2405 (2011)
14. Durin, A., Chalet, D., Coupez, T., Launay, T., Ratajczack, C., Silva, L.: Understanding whistling and acoustical perturbations through adaptive computational fluid dynamics simulations. Eng. Comp. (2018, submitted)
15. Padet, J.: Fluides en écoulement: Méthodes et modèle. Elsevier Masson, Paris (2008)
16. Hoffman, J., Jansson, J., Jansson, N., Vilela de Abreu, R.: Towards parameter-free method for high Reynolds number turbulent flow simulation based on adaptive finite-element approximation. Comp. Meth. Appl. Mech. Engrg. **288**, 60–74 (2015)

Comparison of Coupled and Decoupled Solvers for Incompressible Navier–Stokes Equations Solved by Isogeometric Analysis

Bohumír Bastl, Marek Brandner, Jiří Egermaier, Hana Horníková, Kristýna Michálková, Jan Šourek, and Eva Turnerová

Abstract This paper is devoted to the problem of solving the steady incompressible Navier–Stokes equations discretized by the Galerkin method on the spaces generated by the B-spline/NURBS basis functions, which is called isogeometric analysis. Two pressure-correction methods are presented for the solution of the incompressible flow in the benchmark backward facing step and also in Kaplan water turbine as conforming multipatch domains. The velocity and pressure under-relaxation is considered and the computational examples are compared with the coupled approach.

Keywords Incompressible Navier–Stokes equations · Isogeometric analysis · B-spline/NURBS objects · Pressure-correction methods · Coupled method · Decoupled methods

1 Introduction

The development of numerical methods for simulating fluid flows is fundamental in practice. Classical numerical methods, which are used to find approximate solutions of incompressible fluid flow, are finite difference methods, finite volume methods, finite element methods, spectral methods etc. Recently, a modification of the FEM based on B-spline/NURBS objects appeared (cf. [5]). This approach is known as isogeometric analysis and it is similar to FEM, only triangular/tetrahedral meshes typically used in FEM are replaced by meshes composed of parts of B-spline/NURBS surfaces/volumes representing a computational domain.

B. Bastl (✉) · M. Brandner · J. Egermaier · H. Horníková · K. Michálková · J. Šourek · E. Turnerová
NTIS – New Technologies for the Information Society, Faculty of Applied Sciences, University of West Bohemia, Plzeň, Czech Republic
e-mail: bastl@kma.zcu.cz

H. van Brummelen et al. (eds.), *Numerical Methods for Flows*, Lecture Notes in Computational Science and Engineering 132,
https://doi.org/10.1007/978-3-030-30705-9_2

This is more suitable for consequent computations, because it allows to avoid the time-consuming step of generating triangular/tetrahedral meshes and performs computations directly. Moreover, since the discretization of a computational domain is always exact, this approach reduces errors in the computational analysis and is more suitable for the formulation of automatic shape optimization algorithms.

The mathematical model for incompressible fluid flow simulation is based on the Navier–Stokes equations, two approaches are compared—coupled and decoupled. The "coupled" scheme means that one large linear system, arising from the discretization of the weak formulation via continuous Galerkin (and discontinuous Galerkin for non-conforming meshes) method, is solved for all components of velocity and pressure together. In practical examples, this usually leads to large linear systems which are difficult to solve. Direct solvers are time- and memory-consuming, iterative solvers without suitable preconditioners are very time-consuming. One possible approach to overcome this obstacle is to use the "decoupled" scheme which allows to compute components of velocity and pressure independently by solving smaller linear systems. However, this approach is not suitable in cases, when periodic (cyclic) boundary conditions with respect to rotation are specified, or when terms representing rotation of a computational domain need to be added.

2 NURBS Objects

NURBS objects are standard objects in geometric modelling for shape representation of curves, surfaces or volumes. A NURBS volume of degree (p, q, r) is determined by a control net of $(m + 1) \times (n + 1) \times (l + 1)$ control points $\mathbf{P}_{ijk}$, with weights w_{ijk}, $i = 0, \ldots, m$, $j = 0, \ldots, n$, $k = 0, \ldots, l$ and three knot vectors $\mathbf{U} = (u_0, \ldots, u_{m+p+1})$, $\mathbf{V} = (v_0, \ldots, v_{n+q+1})$, $\mathbf{W} = (w_0, \ldots, w_{l+r+1})$. The parameterization is then

$$\mathbf{v}(u, v, w) = \frac{\sum_{i=0}^{m} \sum_{j=0}^{n} \sum_{k=0}^{l} N_{i,p}(u) N_{j,q}(v) N_{k,r}(w) w_{ijk} \mathbf{P}_{ijk}}{\sum_{i=0}^{m} \sum_{j=0}^{n} \sum_{k=0}^{l} N_{i,p}(u) N_{j,q}(v) N_{k,r}(w) w_{ijk}}, \tag{1}$$

where $N_{i,p}(u)$, $N_{j,q}(v)$ and $N_{k,r}(w)$ are B-spline basis functions (see [5]) of degrees p, q and r corresponding to the knot vectors $\mathbf{U}$, $\mathbf{V}$ and $\mathbf{W}$, respectively. In the case of B-spline volumes all weights w_{ijk} of control points $\mathbf{P}_{ijk}$ are equal.

In isogeometric analysis, the B-spline/NURBS basis which generates the geometry of the computational domain is refined (if needed) and used as basis for the solution space.

3 Mathematical Model

The mathematical simulation of viscous incompressible Newtonian fluid flow is based on the Navier–Stokes equations. Let $\Omega \subset \mathbf{R}^d$ be a bounded domain, d the number of spatial dimensions, with boundary $\partial\Omega$ consisting of two disjoint parts $\partial\Omega_D$ and $\partial\Omega_N$. The steady incompressible Navier–Stokes problem can be written as

$$
\begin{aligned}
(\mathbf{u}\cdot\nabla)\mathbf{u} - \nu\Delta\mathbf{u} + \nabla p &= \mathbf{f} && \text{in } \Omega,\\
\nabla\cdot\mathbf{u} &= 0 && \text{in } \Omega,\\
\mathbf{u} &= \mathbf{g}_D && \text{on } \partial\Omega_D,\\
\nu\frac{\partial\mathbf{u}}{\partial\mathbf{n}} - \mathbf{n}p &= \mathbf{g}_N && \text{on } \partial\Omega_N,
\end{aligned}
\tag{2}
$$

where $\mathbf{u}$ is the flow velocity, p is the kinematic pressure, ν is the kinematic viscosity and $\mathbf{f}$ is a source function.

3.1 *Weak Formulation*

In order to derive the weak formulation of the problem, we define the solution space V and the test function space V_0 as follows

$$
V = \{\mathbf{u} \in H^1(\Omega)^d \mid \mathbf{u} = \mathbf{g}_D \text{ on } \partial\Omega_D\}, \qquad V_0 = \{\mathbf{v} \in H^1(\Omega)^d \mid \mathbf{v} = \mathbf{0} \text{ on } \partial\Omega_D\}.
\tag{3}
$$

By multiplying the first equation of (2) by a test function $\mathbf{v} \in V_0$ and the second equation by a test function $q \in L_2(\Omega)$ and using Green's theorem we obtain the weak formulation: find $\mathbf{u} \in V$ and $p \in L_2(\Omega)$ such that

$$
\begin{aligned}
\nu\int_\Omega \nabla\mathbf{u} : \nabla\mathbf{v} + \int_\Omega(\mathbf{u}\cdot\nabla\mathbf{u})\mathbf{v} - \int_\Omega p\nabla\cdot\mathbf{v} &= \int_\Omega \mathbf{f}\cdot\mathbf{v} + \nu\int_{\partial\Omega_N}\mathbf{g}_N\cdot\mathbf{v}, && \forall\mathbf{v}\in V_0,\\
\int_\Omega q\nabla\cdot\mathbf{u} &= 0, && \forall q \in L_2(\Omega).
\end{aligned}
\tag{4}
$$

Let us assume that the boundary integral in (4) is equal to zero in the rest of the text. To treat the non-linearity in the convective term we employ Picard's method, where the problem is solved iteratively and the non-linear term is linearized using the solution from the previous step. In the k-th iteration we look for $\mathbf{u}^{k+1} \in V$ and $p^{k+1} \in L_2(\Omega)$ such that

$$
\begin{aligned}
\nu\int_\Omega \nabla\mathbf{u}^{k+1} : \nabla\mathbf{v} + \int_\Omega(\mathbf{u}^k\cdot\nabla\mathbf{u}^{k+1})\mathbf{v} - \int_\Omega p^{k+1}\nabla\cdot\mathbf{v} &= \int_\Omega \mathbf{f}\cdot\mathbf{v}, && \forall\mathbf{v}\in V_0,\\
\int_\Omega q\nabla\cdot\mathbf{u}^{k+1} &= 0, && \forall q \in L_2(\Omega).
\end{aligned}
\tag{5}
$$

When we choose the initial velocity $\mathbf{u}^0$ to be zero, then we obtain the solution of the Stokes problem in the first iteration.

3.2 Solution Methods

In this section we describe two approaches we use to find the approximate solution of the discrete problem.

3.2.1 Coupled Approach

A straightforward way is to assemble the whole linear system and find its solution using some direct or iterative method for solving linear systems. As already mentioned, direct solvers are rather inapplicable for large problems because of very high time and memory requirements. On the other hand, iterative methods with standard available preconditioners are not very efficient, hence, the implementation of special preconditioners for Navier–Stokes equations is required. This is a topic for future work. For the time being, we are left with direct solvers.

3.2.2 Decoupled Approach

An alternative to solving the original stationary system, is to search for a steady-state solution of a time-dependent system of equations in the semi-discrete form (discrete in time) on which we apply the *pressure-correction* method on the continuous level. In this method, velocity and pressure fields are computed separately. Thanks to decoupling of the system we solve several smaller systems instead of one large system in each time step. Of course, this method can also be used for unsteady computations with sufficiently small time step. In the following we summarize two versions of the pressure-correction algorithm which we have implemented.

Pressure-Correction Method 1
In each iteration we perform the following steps:

1. We search for an intermediate velocity field $\mathbf{u}^*$ using the pressure from the previous iteration

$$\frac{\mathbf{u}^* - \mathbf{u}^n}{\Delta t} + (\mathbf{u}^n \cdot \nabla)\mathbf{u}^* - \nu \Delta \mathbf{u}^* + \nabla p^n = \mathbf{f}. \tag{6}$$

   The intermediate velocity field does not satisfy the condition $\nabla \cdot \mathbf{u}^* = 0$.

2. We compute a pressure correction p' from

$$\Delta p' = \frac{1}{\Delta t}\nabla \cdot \mathbf{u}^*. \tag{7}$$

   Additional boundary conditions for the pressure have to be introduced.
3. We compute a velocity correction $\mathbf{u}'$ as

$$\mathbf{u}' = -\Delta t\, \nabla p'. \tag{8}$$

4. We update the intermediate velocity field and the pressure field with the computed corrections

$$\mathbf{u}^{n+1} = \mathbf{u}^* + \mathbf{u}', \qquad p^{n+1} = p^n + p'. \tag{9}$$

   The new velocity field satisfies the condition $\nabla \cdot \mathbf{u}^{n+1} = 0$.

This algorithm can be found for example in [1, 4]. In the pressure update step (9) a so-called explicit under-relaxation is often used: $p^{n+1} = p^n + \alpha_p p'$ with $\alpha_p \in \langle 0, 1 \rangle$, which allows to get a convergent scheme.

Pressure-Correction Method 2

In each iteration we define the pressure increment as

$$\psi^{n+1} = p^{n+1} - p^n + \nu \nabla \cdot \mathbf{u}^{n+1}. \tag{10}$$

Then we perform the following steps:

1. We compute the new velocity field $\mathbf{u}^{n+1}$ from

$$\frac{\mathbf{u}^{n+1} - \mathbf{u}^n}{\Delta t} + (\mathbf{u}^n \cdot \nabla)\mathbf{u}^{n+1} + \nabla(p^n + \psi^n) = \mathbf{f}. \tag{11}$$

2. We compute new pressure increment ψ^{n+1} from

$$\Delta \psi^{n+1} = \frac{1}{\Delta t}\nabla \cdot \mathbf{u}^{n+1}. \tag{12}$$

3. We update the pressure with the computed correction

$$p^{n+1} = p^n + \psi^{n+1} - \nu \nabla \cdot \mathbf{u}^{n+1}. \tag{13}$$

This version of the pressure-correction method is used for example in [6]. To be concise we do not describe the weak formulations of both decoupled methods. For more detailed information on this kind of methods, see [2].

The velocity under-relaxation is also included in both decoupled methods above. However, it is the implicit relaxation (see [4]), i.e., the relaxation parameter α_u is related to Δt by the formula $\Delta t = \frac{\alpha_u}{1-\alpha_u}$.

3.3 *Discretization*

To discretize problem (5) by means of the Galerkin method we define finite dimensional spaces $V^h \subset V$, $V_0^h \subset V_0$ and $W^h \subset L_2(\Omega)$ together with their bases. We are looking for a discrete solution $\mathbf{u}_h \in V^h$ and $p_h \in W^h$ such that the solution $\mathbf{u}_h$ is written as a linear combination of basis functions $R_i^u \in V^h$ and the solution p_h as a linear combination of basis functions $R_i^p \in W^h$, where R_i^u and R_i^p are B-spline/NURBS basis functions obtained from the B-spline/NURBS description of the computational domain. The spatial discretization of the coupled case is described in [3] in more details, the derivation of the discrete problem for the decoupled case is analogous.

4 Computational Examples

In this section we present some numerical results obtained by the isogeometric Navier–Stokes solver which we implemented. First, we compare the coupled and decoupled approach on a 3D backward facing step example. Then we compare the performance of the methods for flow in the geometry of Kaplan turbine which involves periodic boundary conditions and rotation of the domain.

4.1 *Backward Facing Step*

The first test case carried out is a flow in the 3D backward facing step. It consists of three patches with conforming interfaces. The discrete problem has 22,734 degrees of freedom, it means that in the coupled case we solve a system with matrix of size 22,734 × 22,734. The kinematic viscosity $\nu = 0.005$. A parabolic velocity profile with maximum magnitude equals to 1 is set as the inlet boundary condition.

In Figs. 1 and 2, we compare the number of iterations and solution time for both decoupled methods described above, where the relaxation parameters α_u and α_p are chosen from the interval (0.3, 1.0). More precisely, the calculations were obtained for the relaxation factors which vary independently from 0.3 to 1.0 by step 0.1, then the graphs represent interpolate functions. Remember that velocity and pressure under-relaxation is considered for the pressure-correction method 1, but only the velocity under-relaxation is considered for the pressure correction method 2.

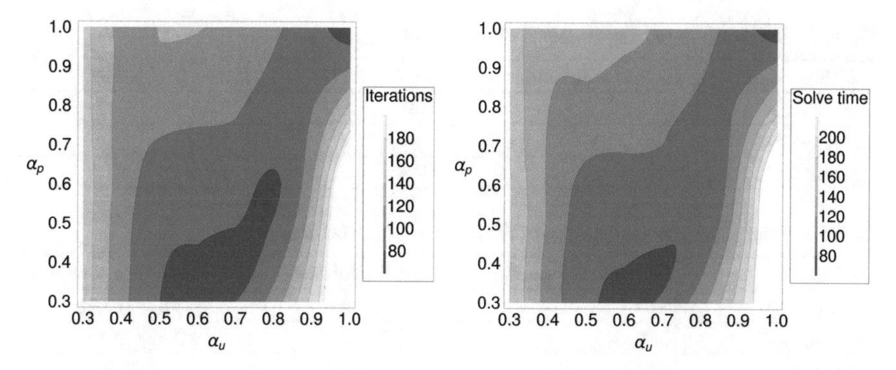

Fig. 1 Number of iterations (left) and solution time [s] (right) as a functions of relaxation parameters α_u and α_p solved by pressure-correction method 1 for backward facing step benchmark

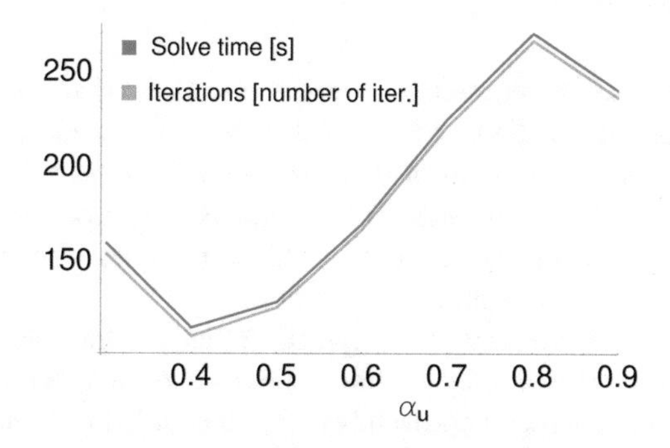

Fig. 2 Number of iterations (blue) and solution time (orange) as a functions of relaxation parameter α_u solved by the pressure-correction method 2 for the backward facing step

The calculations were terminated when the norm of the solution change was smaller than $\epsilon = 10^{-4}$ and all linear systems are solved with sparse LU decomposition method without reordering. Note that the pressure system was also solved using sparse LU decomposition, where Cholesky decomposition might also be used. From the results shown in Figs. 1 and 2 it can be seen that the fastest convergence is achieved for $\alpha_u \approx 0.6$ and $\alpha_p \approx 0.4$ for the pressure-correction method 1 and $\alpha_u = 0.4$ for the pressure-correction method 2.

Let us choose relaxation parameter $\alpha_u = 0.6$ and $\alpha_p = 0.4$ for the first projection method and $\alpha_u = 0.4$ for the second projection method and compare both decoupled approaches with the coupled method in Table 1. It is obvious that the coupled approach required the least number of iterations but the longest solution time.

Table 1 Backward facing step: comparison of the solution methods

Solver	Number of iterations	Total time (s)	Average time/iter.
Coupled	13	981	75.46
Decoupled1	77	81	1.05
Decoupled2	110	114	1.04

Table 2 Kaplan turbine: comparison of the solution methods for $\omega = 0/56$

Solver	Number of iterations	Total time (s)	Average time/iter.
Coupled	4/9	57/124	14.25/13.78
Decoupled1	219/517	774/2540	3.53/4.91
Decoupled2	44/202	160/967	3.64/4.79

4.2 Kaplan Turbine

The second test case is geometrically more complicated as periodic boundary conditions are set and rotation of the runner blades is considered for the flow in a Kaplan water turbine. It consists of three patches with conforming interfaces. The discrete problem has 15,683 degrees of freedom. The solution was computed for a fluid with $\nu = 0.1$ and constant velocity profile at the inlet. The axis of rotation is considered as x-axis with the angular velocity ω.

The numerical experiments were computed similarly to the backward facing step case according to the relaxation parameters. Let us choose $\alpha_u = 0.7$ and $\alpha_p = 0.6$ and compare both decoupled approaches with the coupled method for the angular velocity $\omega = 0$ and $\omega = 56$ in the Table 2. Although the solve time per iteration of the coupled method is much longer, the total time of the coupled method is more favourable as the number of the decoupled iterations is high.

It should be noted that nonzero off-diagonal blocks appear in the matrix formulation as the periodic conditions or rotation is considered (the velocity components are interconnected). In this case, one approach is to solve the equations for the interconnected components of velocity in a coupled way (the results are in the Table 2), but we lose the advantage of the decoupled methods. Another approach is to move the off-diagonal blocks to the right-hand side of the system by multiplying them with the corresponding components of the velocity solution from the previous step. Then we perform several inner iterations in each step of the method. In this case, the pressure-correction method 1 converges after $702s$ and the pressure-correction method 2 converges after $117s$ solving the fluid flow problem without rotation (note that as the inner iterations converge, the number of decoupled iterations is the same as for the first approach written in the Table 2). However, none of the decoupled methods converges if rotating flow is considered.

5 Conclusion

Two pressure-correction methods have been described and compared with the coupled method for the backward facing step benchmark and flow in a Kaplan water turbine. It has been shown that the benefit of the decoupled method is provided only if no periodic conditions or rotation is considered. Moreover, according to the results, the decoupled methods become less stable with rotation. Therefore, the coupled method still provides advantage for our computing purposes in the water turbines.

Acknowledgement This work has been supported by the European Union's Horizon 2020 research and innovation programme under grant agreement No. 678727.

References

1. Armfield, S., Street, R.: The fractional-step method for the Navier–Stokes equations on staggered grids: the accuracy of three variations. J. Comp. Physiol. **153**, 660–665 (1999)
2. Badia, S., Codina, R.: Algebraic pressure segregation methods for the incompressible Navier–Stokes equations. Arch. Comput. Meth. Eng. **15**, 1–52 (2007)
3. Bastl, B., Brandner, M., Egermaier, J., Michálková, K., Turnerová, E.: IgA-based solver for turbulence modelling on multipatch geometries. Adv. Eng. Softw. **113**, 7–18 (2017)
4. Klein, B., Kummer, F., Oberlack, M.: A SIMPLE based discontinuous Galerkin solver for steady incompressible flows. J. Comput. Phys. **237**, 235–250 (2013)
5. Piegl, L., Tiller, W.: The NURBS Book. Monographs in Visual Communication, ISBN 3-540-61545-8. Springer, Berlin (1997)
6. Šístek, J., Cirak, F.: Parallel iterative solution of the incompressible Navier–Stokes equations with application to rotating wings. Comput. Fluids **122**, 165–183 (2015)

High-Order Isogeometric Methods for Compressible Flows

I: Scalar Conservation Laws

Andrzej Jaeschke and Matthias Möller

Abstract Isogeometric analysis was applied very successfully to many problem classes like linear elasticity, heat transfer and incompressible flow problems but its application to compressible flows is very rare. However, its ability to accurately represent complex geometries used in industrial applications makes IGA a suitable tool for the analysis of compressible flow problems that require the accurate resolution of boundary layers. The convection-diffusion solver presented in this chapter, is an indispensable step on the way to developing a compressible solver for complex viscous industrial flows. It is well known that the standard Galerkin finite element method and its isogeometric counterpart suffer from spurious oscillatory behaviour in the presence of shocks and steep solution gradients. As a remedy, the algebraic flux correction paradigm is generalized to B-Spline basis functions to suppress the creation of oscillations and occurrence of non-physical values in the solution. This work provides early results for scalar conservation laws and lays the foundation for extending this approach to the compressible Euler equations in the next chapter.

Keywords Isogeometric analysis · Compressible flows · Algebraic flux correction

A. Jaeschke (✉)
Institute of Turbomachinery, Łódź University of Technology, Łódź, Poland
e-mail: andrzej.jaeschke@p.lodz.pl

M. Möller
Faculty of Electrical Engineering, Mathematics and Computer Science, Delft Institute of Applied Mathematics, Delft University of Technology, Delft, The Netherlands
e-mail: m.moller@tudelft.nl

H. van Brummelen et al. (eds.), *Numerical Methods for Flows*,
Lecture Notes in Computational Science and Engineering 132,
https://doi.org/10.1007/978-3-030-30705-9_3

1 Introduction

Isogeometric analysis (IGA) was proposed by Hughes et al. in [1]. Since its birth it was successfully applied in a variety of use case scenarios ranging from linear elasticity and incompressible flows to fluid-structure interaction problems [2]. There were, however, not many approaches to apply this method to compressible flow problems [3, 4]. Although this application did not gain the attention of many researches yet, it seems to be a promising field. Flow problems are usually defined on domains with complex but smooth shapes, whereby the exact representation of the boundary is indispensable due to the crucial influence of boundary layers on the flow behaviour. This is where IGA has the potential to demonstrate its strengths.

It is a well known fact that standard Galerkin finite element schemes (FEM) suffer from infamous instabilities when applied to convection-dominated problems, such as compressible flows. The same unwanted behaviour occurs for IGA-based standard Galerkin schemes [2] making it necessary to develop high-resolution high-order isogeometric schemes that overcome these limitations. From the many available approaches including the most commonly used ones, i.e., the streamline upwind Petrov–Galerkin (SUPG) method introduced by Brooks and Hughes in [5], we have chosen for the algebraic flux correction (AFC) methodology, which was introduced by Kuzmin and Turek in [6] and refined in a series of publications [7–13]. The family of AFC schemes is designed with the overall goal to prevent the creation of spurious oscillations by modifying the system matrix stemming from a standard Galerkin method in mass-conservative fashion. This algebraic design principle makes them particularly attractive for use in high-order isogeometric methods.

2 High-Resolution Isogeometric Analysis

This section briefly describes the basic construction principles of high-resolution isogeometric schemes for convection-dominated problems based on an extension of the AFC paradigm to B-Spline based discretizations of higher order.

2.1 Model Problem

Consider the stationary convection-diffusion problem [2]

$$-d\Delta u(\mathbf{x})) + \nabla \cdot (\mathbf{v}u(\mathbf{x})) = 0 \qquad \text{in } \Omega \tag{1}$$

$$u(\mathbf{x}) = \beta(\mathbf{x}) \qquad \text{on } \Gamma \tag{2}$$

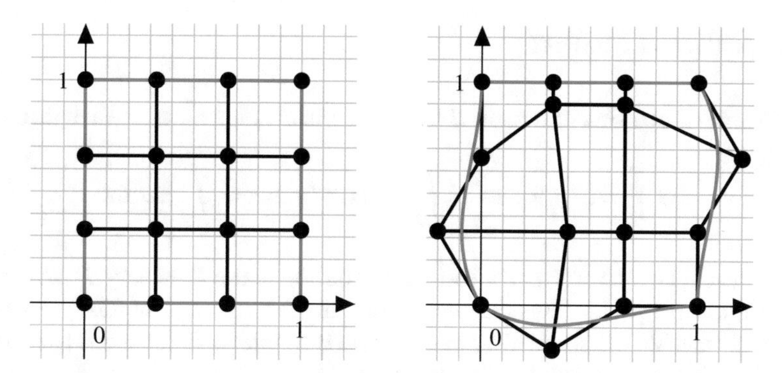

Fig. 1 Unit square (left) and deformed domain (right) modeled by tensor-product quadratic B-Spline basis functions defined on the open knot vector $\Xi = [0, 0, 0, 0.5, 1, 1, 1]$

with diffusion coefficient $d = 0.0001$ and constant velocity vector $\mathbf{v} = [\sqrt{2}, \sqrt{2}]^\top$. The problem is solved on the two domains depicted in Fig. 1.

Starting from the open knot vector $\Xi = [0, 0, 0, 0.5, 1, 1, 1]$, quadratic B-Spline basis functions $N_{a,2}(\xi)$ are generated by the Cox-de-Boor recursion formula [14]:

$$p = 0: \quad N_{a,0}(\xi) = \begin{cases} 1 \text{ if } \xi_a \leq \xi < \xi_{a+1}, \\ 0 \text{ otherwise,} \end{cases} \tag{3}$$

$$p > 0: \quad N_{a,p}(\xi) = \frac{\xi - \xi_a}{\xi_{a+p} - \xi_a} N_{a,p-1}(\xi) + \frac{\xi_{a+p+1} - \xi}{\xi_{a+p+1} - \xi_{a+1}} N_{a+1,p-1}(\xi), \tag{4}$$

where ξ_a are the entries in the knot vector Ξ. Their tensor product construction yields the bivariate B-Spline basis functions $\hat{\varphi}_j(\xi, \eta) = N_a(\xi)N_b(\eta)$ (with index map $j \mapsto (a, b)$), which are used to define the computational geometry model

$$\mathbf{x}(\xi, \eta) = \sum_j \mathbf{c}_j \hat{\varphi}_j(\xi, \eta), \quad (\xi, \eta) \in \hat{\Omega} = [0, 1]^2 \tag{5}$$

with control points $\mathbf{c}_j \in \mathbb{R}^2$ indicated by dots in Fig. 1. The mapping $\phi : \hat{\Omega} \to \Omega$ converts parametric values $\boldsymbol{\xi} = (\xi, \eta)$ into physical coordinates $\mathbf{x} = (x, y)$. The mapping should be bijective in order to possess a valid 'pull-back' operator $\phi^{-1} : \Omega \to \hat{\Omega}$.

For simplicity the boundary conditions are prescribed in the parametric domain:

$$\beta(\mathbf{x} = \phi(\boldsymbol{\xi})) = \begin{cases} 1 & \text{if } \eta \leq \frac{1}{5} - \frac{1}{5}\xi \\ 0 & \text{otherwise.} \end{cases} \tag{6}$$

2.2 Galerkin Method

Application of the Galerkin method to (1)–(6) yields: Find $u^h \in S^h$ such that

$$d\int_{\Omega} \nabla u^h \cdot \nabla v^h \mathrm{d}\mathbf{x} + \int_{\Omega} \nabla \cdot (\mathbf{v}u)^h v^h)\mathrm{d}\mathbf{x} = \int_{\Omega} R v^h \mathrm{d}\mathbf{x} \tag{7}$$

for all test functions $v^h \in V^h$ that vanish on the entire boundary Γ due to the prescription of Dirichlet boundary conditions. In the framework of IGA the discrete spaces S^h and V^h are spanned by multivariate B-Spline basis functions $\{\varphi_j(\mathbf{x})\}$.

Using Fletcher's group formulation [15], the approximate solution u^h and the convective flux $(\mathbf{v}u)^h$ can be represented as follows [12]:

$$u^h(\mathbf{x}) = \sum_j u_j \varphi_j(\mathbf{x}), \quad (\mathbf{v}u)^h(\mathbf{x}) = \sum_j (\mathbf{v}_j u_j)\varphi_j(\mathbf{x}). \tag{8}$$

Substitution into (7) and replacing v^h by all basis functions yields the matrix form

$$(S - K)\mathbf{u} = \mathbf{r}, \tag{9}$$

where $\mathbf{u}$ is the vector of coefficients u_i used in the expansion of the solution (8) and the entries of the discrete diffusion ($S = \{s_{ij}\}$) and convection ($K = \{k_{ij}\}$) operators and the discretized right-hand side vector ($\mathbf{r} = \{r_i\}$) are given by

$$k_{ij} = -\mathbf{v}_j \cdot \mathbf{c}_{ij}, \qquad \mathbf{c}_{ij} = \int_{\Omega} \nabla\varphi_j \varphi_i \mathrm{d}\mathbf{x}, \tag{10}$$

$$s_{ij} = d\int_{\Omega} \nabla\varphi_j \cdot \nabla\varphi_i \mathrm{d}\mathbf{x}, \qquad r_i = \int_{\Omega} R\varphi_i \mathrm{d}\mathbf{x}. \tag{11}$$

The above integrals are assembled by resorting to numerical quadrature over the unit square $\hat{\Omega} = [0, 1]^2$ using the 'pull-back' operator $\phi^{-1} : \Omega \to \hat{\Omega}$. For the entries of the physical diffusion matrix the final expression reads as follows [4]:

$$s_{ij} = d\int_{\hat{\Omega}} \nabla_{\boldsymbol{\xi}}\hat{\varphi}_j(\boldsymbol{\xi}) \cdot G(\boldsymbol{\xi})\nabla_{\boldsymbol{\xi}}\hat{\varphi}_j(\boldsymbol{\xi})\mathrm{d}\boldsymbol{\xi}, \tag{12}$$

where the geometric factor $G(\boldsymbol{\xi})$ is given in terms of the Jacobian $J = D\phi$:

$$G(\boldsymbol{\xi}) = |\det J(\boldsymbol{\xi})| J^{-1}(\boldsymbol{\xi}) J^{-\top}(\boldsymbol{\xi}). \tag{13}$$

It should be noted that expression (12) can be interpreted as the discrete counterpart of an anisotropic diffusion problem with symmetric diffusion tensor $dG(\boldsymbol{\xi})$ that is solved on the unit square $\hat{\Omega}$ using tensor-product B-Splines on a perpendicular grid.

2.3 Algebraic Flux Correction

The isogeometric Galerkin method (9) is turned into a stabilized high-resolution scheme by applying the principles of algebraic flux correction (AFC) of TVD-type, which were developed for lowest-order Lagrange finite elements in [7, 10].

In essence, the discrete convection operator K is modified in two steps:

1. Eliminate negative off-diagonal entries from K by adding a *discrete diffusion operator* D to obtain the modified discrete convection operator $L = K + D$.
2. Remove excess artificial diffusion in regions where this is possible without generating spurious wiggles by applying *non-linear anti-diffusion*: $K^*(\mathbf{u}) = L + \bar{F}(\mathbf{u})$.

Discrete Diffusion Operator The optimal entries of $D = \{d_{ij}\}$ are given by [6]:

$$d_{ij} = d_{ji} = \max\{0, -k_{ij}, -k_{ji}\}, \quad d_{ii} = -\sum_{j \neq i} d_{ij}, \tag{14}$$

yielding a symmetric operator with zero column and row sums. The latter enables the decomposition of the diffusive contribution to the ith degree of freedom

$$(D\mathbf{u})_i = \sum_{j \neq i} f_{ij}, \quad f_{ij} = d_{ij}(u_j - u_i), \tag{15}$$

whereby the diffusive fluxes $f_{ij} = -f_{ji}$ are skew-symmetric by design [6].

In a practical implementation, operator D is not constructed explicitly, but the entries of $L := K$ are modified in a loop over all pairs of degrees of freedoms (i, j) for which $j \neq i$ and the basis functions have overlapping support $\mathrm{supp}\hat{\varphi}_i \cap \mathrm{supp}\hat{\varphi}_j \neq \emptyset$. For univariate B-Spline basis functions of order p, we have $\mathrm{supp}\hat{\varphi}_i = (\xi_i, \xi_{i+p+1})$, where ξ_i denotes the ith entry of the knot vector Ξ. Hence, the loops in (14) and (15) extend over all $j \neq i$ with $|j - i| \leq p$ in one spatial dimension, which can be easily generalized to tensor-product B-Splines in multiple dimensions.

The modified convection operator $L = K + D$ yields the stabilized linear scheme

$$(S - L)\mathbf{u} = \mathbf{r}. \tag{16}$$

Anisotropic Physical Diffusion The discrete diffusion matrix S might also cause spurious oscillations in the solution since it is only 'harmless' for lowest order finite elements under the additional constraint that triangles are nonobtuse (all angles smaller than or equal to $\pi/2$) and quadrilaterals are nonnarrow (aspect ratios smaller than or equal to $\sqrt{2}$ [16]), respectively. Kuzmin et al. [12, 17] propose stabilization techniques for anisotropic diffusion problems, which can be applied to (12) directly. It should be noted, however, that we did not observe any spurious wiggles in all our numerical tests even without any special treatment of the diffusion matrix S.

Nonlinear Anti-diffusion According to Godunov's theorem [18], the linear scheme (16) is limited to first-order accuracy. Therefore a nonlinear scheme must be constructed by adaptively blending between schemes (9) and (16), namely [7, 10]:

$$(S - K^*(\mathbf{u}))\mathbf{u} = \mathbf{r}. \tag{17}$$

Here, the nonlinear discrete convection operator reads

$$K^*(\mathbf{u}) = L + \bar{F}(\mathbf{u}) = K + D + \bar{F}(\mathbf{u}), \tag{18}$$

which amounts to applying a modulated anti-diffusion operator $\bar{F}(\mathbf{u})$ to avoid the loss of accuracy in smooth regions due to excessive artificial diffusion. The raw anti-diffusion, $-D$, features all properties of a discrete diffusion operator, and hence, its contribution to a single degree of freedom can be decomposed as follows [7, 10]:

$$f_i(\mathbf{u}) := (\bar{F}(\mathbf{u})\mathbf{u})_i = \sum_{j \neq i} \alpha_{ij}(\mathbf{u}) d_{ij}(u_i - u_j), \tag{19}$$

where $\alpha_{ij}(\mathbf{u}) = \alpha_{ji}(\mathbf{u})$ is an adaptive flux limiter. Clearly, for $\alpha_{ij} \equiv 1$ the anti-diffusive fluxes will restore the original Galerkin scheme (9) and $\alpha_{ij} \equiv 0$ will lead to the linear scheme (16). Kuzmin et al. [7, 10] proposed a TVD-type multi-dimensional limiting strategy for lowest-order Lagrange finite elements, which ensures that the resulting scheme (17) yields accurate solutions that are free of spurious oscillations. The flux limiter was extended to non-nodal basis functions in [4] and utilized for computing the numerical results presented in Sect. 3.

Like with the diffusion operator D, we do not construct $K^*(\mathbf{u})$ explicitly but include the anti-diffusive correction $\bar{\mathbf{f}}(\mathbf{u}) = \{\bar{f}_i(\mathbf{u})\}$ into the right-hand side [7, 10]

$$(S - L)\mathbf{u} = \mathbf{r} + \bar{\mathbf{f}}(\mathbf{u}). \tag{20}$$

The nonlinear scheme can be solved by iterative defect correction [7] possibly combined with Anderson acceleration [19] or by an inexact Newton method [20].

Non-nodal degrees of freedom make it necessary to first project the prescribed boundary values (6) onto the solution space S^h so that the coefficients of the degrees of freedoms that are located at the Dirichlet boundary part can be overwritten accordingly. Since the standard L_2 projection can lead to non-physical under- and overshoots near discontinuities and steep gradients, the constrained data projection approach proposed in [11] for lowest order nodal finite elements is used.

3 Numerical Results

This section presents the numerical results for the model problem (1)–(6), which were computed using the open-source isogeometric analysis library G+Smo [21].

The tensor-product B-Spline basis (4×4 basis functions of degree $p = 2$) that was used for the geometry models depicted in Fig. 1 was refined by means of knot insertion [2] to generate 18×18 quadratic B-Spline basis functions for approximating the solution. It should be noted that this type of refinement, which is an integral part of the Isogeometric Analysis framework, preserves the shape of the geometry exactly. Consequently, the numerical solution does not suffer from an additional error stemming from an approximated computational domain as it is the case for, say, higher-order Lagrange finite elements defined on simplex or quadrilateral meshes.

For the diffusion coefficient $d = 0.0001$ and the considered basis the element Péclet number is equal to $Pe_h \approx 555$, which states that the problem is highly convection-dominated. For the deformed geometry, the actual value varies slightly from one 'element' to the other but stays in the same order of magnitude.

The numerical solution that was computed on the unit square is depicted in Fig. 2 (left), whereas the approximate solution for the deformed geometry is shown on the right. In both cases the minimum and maximum bounds of the exact solution, that is, $u_{min} = 0$ and $u_{max} = 1$ are preserved by the numerical counterpart, which results from the successful application of the AFC stabilization of TVD-type.

It should be noted that the internal layer is smeared stronger than the boundary layer, which is due to the constrained L_2 projection of the Dirichlet boundary data into the space S^h. The discontinuous profile (6) along the left boundary cannot be represented exactly by quadratic B-Splines, and hence, it is smeared across multiple 'elements'. A possible remedy is to locally reduce the approximation order to $p = 1$ by inserting a knot at the boundary location $\eta_b = 1/5$ and increasing its multiplicity to $m_b = 2$, which will reduce the continuity to $C^{p-m_b} = C^0$ locally. The varying thickness of the boundary layer on the deformed geometry stems from the fact the distance of the rightmost vertical internal 'grid line' to the boundary also varies.

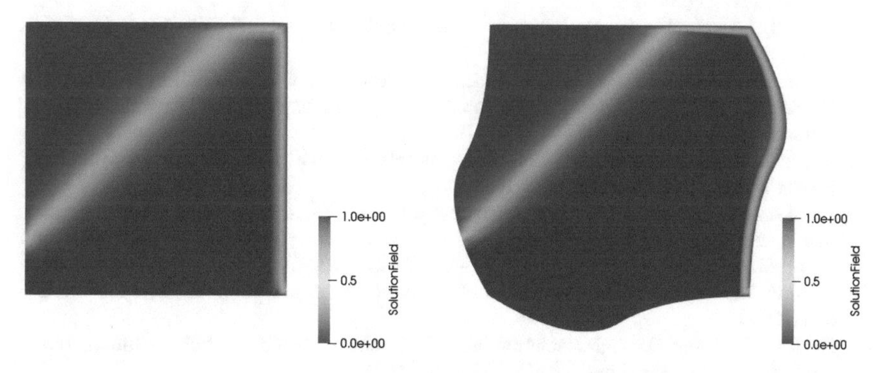

Fig. 2 Numerical solutions computed on the unit square (left) and deformed domain (right)

4 Conclusions

The high-resolution isogeometric scheme presented in this work for the stationary convection-diffusion equation is a first step to establish isogeometric methods for convection-dominated problems and, in particular, compressible flows, which are addressed in more detail in chapter "High-Order Isogeometric Methods for Compressible Flows. II: Compressible Euler Equations". This chapter extends the family of algebraic flux correction schemes to quadratic B-Spline discretizations thereby demonstrating that the algebraic design principles that were originally derived for low-order nodal Lagrange finite elements carry over to non-nodal Spline basis functions.

Ongoing research focuses on the extension of this approach to truncated hierarchical B-Splines [22] possibly combined with the local increase of the knot multiplicities, which seems to be a viable approach for refining the spline spaces S^h and V^h adaptively in the vicinity of shocks and steep gradients to compensate for the local reduction of the approximation order by algebraic flux correction (h-refinement) and to prevent excessive spreading of these localized features (continuity reduction).

Acknowledgement This work has been supported by the European Unions Horizon 2020 research and innovation programme under grant agreement No. 678727.

References

1. Hughes, T.J.R., Cottrell, J.A., Bazilevs, Y.: Isogeometric analysis: CAD, finite elements, NURBS, exact geometry and mesh refinement. Comput. Methods Appl. Mech. Eng. **194**(39–41), 4135–4195 (2005)
2. Cottrell, J.A., Hughes, T.J.R., Bazilevs, T.J.R.: Isogeometric Analysis: Toward Integration of CAD and FEA. Wiley, Hoboken (2009)
3. Trontin, P.: Isogeometric analysis of Euler compressible flow. Application to aerodynamics. Conference: 50th AIAA Aerospace Sciences Meeting Including the New Horizons Forum and Aerospace Exposition (2012)
4. Jaeschke, A.: Isogeometric analysis for compressible flows with application in turbomachinery. Master's Thesis, Delft University of Technology (2015)
5. Brooks, A.N., Hughes, T.J.R.: Streamline upwind/Petrov-Galerkin formulations for convection dominated flows with particular emphasis on the incompressible Navier-Stokes equations. Comput. Methods Appl. Mech. Eng. **32**(1–3), 199–259 (1982)
6. Kuzmin, D., Turek, S.: Flux correction tools for finite elements. J. Comput. Phys. **175**(2), 525–558 (2002)
7. Kuzmin, D., Turek, S.: High-resolution FEM-TVD schemes based on a fully multidimensional flux limiter. J. Comput. Phys. **198**(1), 131–158 (2004)
8. Kuzmin, D., Möller, M.: Flux-corrected transport, chapter algebraic flux correction I. In: Scalar Conservation Laws. Springer, Berlin (2005)
9. Kuzmin, D.: On the design of general-purpose flux limiters for finite element schemes. I. Scalar Convection. J. Comput. Phys. **219**(2), 513–531 (2006)

10. Kuzmin, D.: Algebraic flux correction for finite element discretizations of coupled systems. In: Proceedings of the International Conference on Computational Methods for Coupled Problems in Science and Engineering II, CIMNE, Barcelona, pp. 653–656 (2007)
11. Kuzmin, D., Möller, M., Shadid, J.N., Shashkov, M.: Failsafe flux limiting and constrained data projections for equations of gas dynamics. J. Comput. phys. **229**(23), 8766–8779, 11 (2010)
12. Kuzmin, D.: Flux-corrected transport, chapter Algebraic flux correction I. In: Scalar Conservation Laws. Springer, Berlin (2012)
13. Kuzmin, D., Möller, M., Gurris, M.: Flux-corrected transport, chapter Algebraic flux correction II. Compressible Flow Problems. Springer, Berlin (2012)
14. de Boor, C.: Subroutine package for calculating with B-splines. Technical Report LA-4728-MS. Los Alamos National Laboratory, New Mexico (1971)
15. Fletcher, C.A.J.: The group finite element formulation. Comput. Methods Appl. Mech. Eng. **37**, 225–243 (1983)
16. Farago, I., Horvath, R., Korotov, S.: Discrete maximum principle for linear parabolic problems solved on hybrid meshes. Appl. Numer. Math. **53**(2–4), 249–264, 05 (2004)
17. Kuzmin, D., Shashkov, M.J., Svyatski, D.: A constrained finite element method satisfying the discrete maximum principle for anisotropic diffusion problems on arbitrary meshes. J. Comput. Phys. **228**, 3448–3463 (2009)
18. Godunov, S.K.: Finite difference method for numerical computation of discontinuous solutions of the equations of fluid dynamics. Mat. Sbornik **47**, 271–306 (1959)
19. Walker, H.F., Ni, P.: Anderson acceleration for fixed-point iterations. SIAM J. Numer. Anal. **49**(4),1715–1735, 08 (2011)
20. Möller, M.: Efficient solution techniques for implicit finite element schemes with flux limiters. Int. J. Numer. Methods Fluids **55**(7), 611–635, 11 (2007)
21. Jüttler, B., Langer, U., Mantzaflaris, A., Moore, A., Zulehner, W.: Geometry + simulation modules: implementing isogeometric analysis. Proc. Appl. Math. Mech. **14**(1), 961–962 (2014); Special Issue: 85th Annual Meeting of the Int. Assoc. of Appl. Math. and Mech. (GAMM), Erlangen (2014)
22. Giannelli, C., Jüttler, B., Speleers, H.: Thb-splines: the truncated basis for hierarchical splines. Comput. Aided Geom. Des. **29**(7), 485–498, 10 (2012)

High-Order Isogeometric Methods for Compressible Flows

II: Compressible Euler Equations

Matthias Möller and Andrzej Jaeschke

Abstract This work extends the high-resolution isogeometric analysis approach established in chapter "High-Order Isogeometric Methods for Compressible Flows. I: Scalar Conservation Laws" (Jaeschke and Möller: High-order isogeometric methods for compressible flows. I. Scalar conservation Laws. In: Proceedings of the 19th International Conference on Finite Elements in Flow Problems (FEF 2017)) to the equations of gas dynamics. The group finite element formulation is adopted to obtain an efficient assembly procedure for the standard Galerkin approximation, which is stabilized by adding artificial viscosities proportional to the spectral radius of the Roe-averaged flux-Jacobian matrix. Excess stabilization is removed in regions with smooth flow profiles with the aid of algebraic flux correction (Kuzmin et al., Flux-corrected transport, chapter Algebraic flux correction II. Compressible Flow Problems. Springer, Berlin, 2012). The underlying principles are reviewed and it is shown that linearized FCT-type flux limiting (Kuzmin, J Comput Phys 228(7):2517–2534, 2009) originally derived for nodal low-order finite elements ensures positivity-preservation for high-order B-Spline discretizations.

Keywords Isogeometric analysis · High-order methods · High-resolution methods · Compressible flows

M. Möller (✉)
Faculty of Electrical Engineering, Mathematics and Computer Science, Delft Institute of Applied Mathematics, Delft University of Technology, Delft, The Netherlands
e-mail: m.moller@tudelft.nl

A. Jaeschke
Institute of Turbomachinery, Łódź University of Technology, Łódź, Poland
e-mail: andrzej.jaeschke@p.lodz.pl

H. van Brummelen et al. (eds.), *Numerical Methods for Flows*,
Lecture Notes in Computational Science and Engineering 132,
https://doi.org/10.1007/978-3-030-30705-9_4

1 Introduction

Compressible fluid flow problems have traditionally been solved by low-order finite element and finite volume schemes, which were equipped with stabilization techniques like, e.g., SUPG, FCT/TVD and nonlinear shock-capturing, in order to resolve flow patterns like shock waves and contact discontinuities without producing nonphysical undershoots and overshoots. A recent trend in industrial CFD applications especially involving turbulent flows is the use of high-order methods, which enable a more accurate representation of curved geometries and thus a better resolution of boundary layers and, in general, provide better accuracy per degree of freedom (DOF). Moreover, high-order methods have a favorable compute-to-data ratio, which makes them particularly attractive for use on modern high-performance computing platforms, where memory transfers are typically the main bottleneck.

Despite the huge success of isogeometric analysis (IGA) [1] in structural mechanics, incompressible fluid mechanics and fluid-structure interaction, publications on its successful application to *compressible* flows are rare [2–4]. This may be attributed to the challenges encountered in developing shock-capturing techniques for *continuous* high-order finite element methods. In this work we present an isogeometric approach for solving the compressible Euler equations within an IGA framework, thereby adopting the concept of algebraic flux correction (AFC) as stabilization technique [5, 6]. In particular, we show that the linearized FCT-type limiting strategy introduced in [7, 8] carries over to B-Spline discretizations.

2 High-Resolution Isogeometric Analysis

This section briefly describes the design principles of our IGA-AFC scheme and highlights the novelties and main differences to its nodal finite element counterpart. For a comprehensive description of FEM-AFC the reader is referred to [5, 6].

2.1 Governing Equations

Consider the d-dimensional compressible Euler equations in divergence form

$$\frac{\partial U}{\partial t} + \nabla \cdot \mathbf{F}(U) = 0. \tag{1}$$

Here, $U : \mathbb{R}^d \to \mathbb{R}^{d+2}$ denotes the state vector of conservative variables

$$U = \left[U_1, \ldots, U_{d+2}\right]^\top = \left[\rho, \rho\mathbf{v}, \rho E\right]^\top, \tag{2}$$

and $\mathbf{F} : \mathbb{R}^{d+2} \to \mathbb{R}^{(d+2)\times d}$ stands for the tensor of inviscid fluxes

$$\mathbf{F} = \begin{bmatrix} F_1^1 & \dots & F_1^d \\ \vdots & \ddots & \vdots \\ F_{d+2}^1 & \dots & F_{d+2}^d \end{bmatrix} = \begin{bmatrix} \rho\mathbf{v} \\ \rho\mathbf{v}\otimes\mathbf{v} + pI \\ \rho E\mathbf{v} + p\mathbf{v} \end{bmatrix} \tag{3}$$

with density ρ, velocity $\mathbf{v}$, total energy E, and I denoting the d-dimensional identity tensor. For an ideal polytropic gas, the pressure p is given by the equation of state

$$p = (\gamma - 1)\left(\rho E - 0.5\rho|\mathbf{v}|^2\right), \tag{4}$$

where γ denotes the heat capacity ratio, which equals $\gamma = 1.4$ for dry air. The governing equations are equipped with initial conditions prescribed at time $t = 0$

$$U(\mathbf{x}, 0) = U_0(\mathbf{x}) \quad \text{in } \Omega, \tag{5}$$

and boundary conditions of Dirichlet and Neumann type, respectively

$$U = G(U, U_\infty) \quad \text{on } \Gamma_D, \qquad \mathbf{n}\cdot\mathbf{F} = F_n(U, U_\infty) \quad \text{on } \Gamma_N. \tag{6}$$

Here, $\mathbf{n}$ is the outward unit normal vector and U_∞ denotes the vector of 'free stream' solution values, which are calculated and imposed as outlined in [6].

2.2 *Spatial Discretization by Isogeometric Analysis*

Application of the Galerkin method to the variational form of the first-order conservation law system (1) yields the following system of semi-discrete equations [6]

$$\sum_j \left(\int_\Omega \varphi_i\varphi_j \mathrm{d}\mathbf{x}\right)\frac{dU_j}{dt} - \sum_j \left(\int_\Omega \nabla\varphi_i\varphi_j \mathrm{d}\mathbf{x}\right)\cdot\mathbf{F}_j + \int_{\Gamma_n} \varphi_i F_n \,\mathrm{d}s = 0, \tag{7}$$

where $\mathbf{F}_j(t) := \mathbf{F}(U_j(t))$ denotes the value of (3) evaluated at the j-th solution coefficient $U_j(t)$ at time t. This approach is known as Fletcher's group formulation [9], which amounts to expanding $U^h \approx U$ and $\mathbf{F}^h \approx \mathbf{F}$ into the same basis $\{\varphi_j\}$, that is

$$U^h(\mathbf{x}, t) = \sum_j U_j(t)\varphi_j(\mathbf{x}), \quad \mathbf{F}^h(\mathbf{x}, t) = \sum_j \mathbf{F}_j(t)\varphi_j(\mathbf{x}). \tag{8}$$

To further simplify the notation, let us define the consistent mass matrix $M_C := \{m_{ij}\}$ and the discretized divergence operator $\mathbf{C} := \{\mathbf{c}_{ij}\}$ as follows

$$m_{ij} = \int_\Omega \varphi_i \varphi_j \,\mathrm{d}\mathbf{x}, \qquad \mathbf{c}_{ij} = \int_\Omega \varphi_i \nabla \varphi_j \,\mathrm{d}\mathbf{x}. \tag{9}$$

Then the semi-discrete system (7) can be written in compact matrix form as

$$M_C \frac{dU_k}{dt} - \sum_{l=1}^{d} \left[C^l \right]^\top F_k^l + S_k(U) = 0, \tag{10}$$

where superscript $l = 1, \ldots, d$ refers to the l-th spatial component of the discrete divergence operator $\mathbf{C}$ and the tensor of inviscid fluxes $\mathbf{F}$, respectively, and subscript $k = 1, \ldots, d + 2$ stands for the component that corresponds to the k-th variable. Here, $S_k(U)$ accounts for the contribution of boundary fluxes; see [6] for more details.

As explained in Sections 2.1 and 2.2 of chapter "High-Order Isogeometric Methods for Compressible Flows. I: Scalar Conservation Laws" [10], we adopt tensor-product B-Spline basis functions for approximating the numerical solution (8) and modeling the domain Ω. As a consequence, the integrals in (9) are evaluated by introducing the 'pull-back' operator $\phi^{-1} : \Omega \to \hat{\Omega}$ and applying numerical quadrature on the reference domain $\hat{\Omega} = [0, 1]^d$ as it is common practice in the IGA community.

2.3 *Temporal Discretization by Explicit Runge–Kutta Methods*

The semi-discrete system (10) is discretized in time by an explicit strong stability preserving (SSP) Runge–Kutta time integration schemes of order three [11]

$$MU^{(1)} = MU^n + \Delta t R(U^n) \tag{11}$$

$$MU^{(2)} = \frac{3}{4} MU^n + \frac{1}{4} \left(MU^{(1)} + \Delta t R(U^{(1)}) \right) \tag{12}$$

$$MU^{n+1} = \frac{1}{3} MU^n + \frac{2}{3} \left(MU^{(2)} + \Delta t R(U^{(2)}) \right), \tag{13}$$

where $M = I \otimes M_C$ is the block counterpart of the mass matrix and $R(U)$ represents all remaining terms involving inviscid and boundary fluxes. If M_C is replaced by its row-sum lumped counterpart (see below) then the above Runge–Kutta schemes reduce to scaling the right-hand sides by the inverse of a diagonal matrix.

2.4 Algebraic Flux Correction

The Galerkin method (7) is turned into a high-resolution scheme by applying algebraic flux correction of linearized FCT-type [8], thereby adopting the primitive variable limiter introduced in [7]. The description is intentionally kept short and addresses mainly the extensions of the core components of FEM-AFC to IGA-AFC.

Row-Sum Mass Lumping A key ingredient in all flux correction algorithms for time-dependent problems is the decoupling of the unknowns in the transient term of (7) by performing row-sum mass lumping, which turned out to be one of the main problems in generalizing FEM-AFC to higher order nodal Lagrange finite elements since the presence of negative off-diagonal entries leads to singular matrices. The following theorem shows that IGA-AFC is free of this problem by design.

Theorem 1 *All diagonal entries and the non-zero off-diagonal entries of the consistent mass matrix M_C are strictly positive if tensor-product B-Spline basis functions φ_j are adopted. Hence, the row-sum lumped mass matrix $M_L := diag(m_i)$ is unconditionally invertible with strictly positive diagonal entries*

$$m_i := \sum_j \int_\Omega \varphi_i(\mathbf{x})\varphi_j(\mathbf{x})\,\mathrm{d}\mathbf{x} = \sum_j \int_{\hat{\Omega}} \hat{\varphi}_i(\boldsymbol{\xi})\hat{\varphi}_j(\boldsymbol{\xi})\,|\det J(\boldsymbol{\xi})|\,\mathrm{d}\boldsymbol{\xi} > 0. \tag{14}$$

Proof *Since the forward mapping $\phi : \hat{\Omega} \to \Omega$ must be bijective for its inverse $\phi^{-1} : \Omega \to \hat{\Omega}$ to exist, the determinant of $J = D\Phi$ is unconditional non-zero. The strict positivity of B-Spline basis function over their support completes the proof.*

□

Galerkin Flux Decomposition [6] The contribution of the residual $R(U)$ to a single DOF, say, i can be decomposed into a sum of solution differences between U_i and U_j, where j extends over all DOFs for which the mass matrix satisfies $m_{ij} \neq 0$:

$$R_i = \sum_j \mathbf{c}_{ji} \cdot \mathbf{F}_j = \sum_{j \neq i} \mathbf{e}_{ij} \cdot (\mathbf{F}_j - \mathbf{F}_i) = \sum_{j \neq i} \mathbf{e}_{ij} \cdot \mathbf{A}_{ij}(U_j - U_i). \tag{15}$$

Here, $\mathbf{e}_{ij} = 0.5(\mathbf{c}_{ji} - \mathbf{c}_{ij})$ and $\mathbf{A}_{ij} = A(U_i, U_j)$ denotes the flux-Jacobian matrix $\mathbf{A}(U) = \partial\mathbf{F}(U)/\partial U$ evaluated for the density averaged Roe mean values [12].

The derivation procedure utilizes the partition of unity property of basis $\{\varphi_j\}$, which remains valid for tensor-product as well as THB-Splines [13] thus enabling adaptive refinement. The underlying tensor-product construction makes it possible to fully unroll the '$j \neq i$ loop' by exploiting the fact that the support of univariate B-Spline functions of order p extends over a knot span of size $p + 1$.

Artificial Viscosities [6] Expression (15) is augmented by artificial viscosities

$$\tilde{R}_i := R_i + \sum_{j \neq i} D_{ij}(U_j - U_i), \quad D_{ij} := \|\mathbf{e}_{ij}\| \, R_{ij} \, |\Lambda_{ij}| \, R_{ij}^{-1} \tag{16}$$

where Λ_{ij} and R_{ij} are matrices of eigenvalues and eigenvectors of $\tilde{\mathbf{A}}_{ij}$, respectively.

Linearized FCT Algorithm [8] Row-sum mass lumping and application of stabilizing artificial viscosities lead to the positivity-preserving predictor scheme

$$M_L \frac{d\tilde{U}}{dt} = \tilde{R}(\tilde{U}), \tag{17}$$

which is advanced in time by the SSP-RK procedure (11)–(13) starting from the end-of-step solution U^n of the previous time step. The solution $\tilde{U}^{n+1}$ and the approximation $\dot{U}^{n+1} \approx M_L^{-1} \tilde{R}(\tilde{U}^{n+1})$ are used to linearize the antidiffusive fluxes

$$F_{ij} = m_{ij} \left(\dot{U}_i^{n+1} - \dot{U}_j^{n+1} \right) + \tilde{D}_{ij}^{n+1} \left(\tilde{U}_i^{n+1} - \tilde{U}_j^{n+1} \right), \quad F_{ji} = -F_{ij}, \tag{18}$$

where $\tilde{D}_{ij}^{n+1} = D_{ij}(\tilde{U}_i^{n+1}, \tilde{U}_j^{n+1})$ stands for the evaluation of the viscosity operator at the density-averaged Roe mean values based on the predicted solution.

The philosophy of FCT schemes is to multiply F_{ij} by a symmetric correction factor $0 \leq \alpha_{ij} \leq 1$ to obtain the amount of constrained antidiffusive correction

$$\bar{F}_i = \sum_{j \neq i} \alpha_{ij} F_{ij}, \tag{19}$$

which can be safely added to the coefficients of the positivity-preserving predictor

$$U_i^{n+1} = \tilde{U}_i^{n+1} + \frac{\Delta t}{m_i} \bar{F}_i \tag{20}$$

without generating spurious oscillations in the updated end-of-step solution. The flux limiting procedure developed in [7] calculates individual correction factors α_{ij}^u for user-definable scalar control variables $u(U)$, e.g., density and pressure since $\rho > 0$ and $p > 0$ implies $\rho E > 0$ following directly from the equation of state (4). A safe choice for the final correction factor in (19) is to set $\alpha_{ij} = \min\{\alpha_{ij}^\rho, \alpha_{ij}^p\}$.

It should be noted that the original limiting procedure has been designed for low-order *nodal* finite elements, where it ensures that the nodal values of the end-of-step solution are bounded from below and above by the local minimal/maximal nodal values of the positivity-preserving predictor. That is, the following holds for all i:

$$\tilde{u}_i^{\min} := \min_{j \neq i} \tilde{u}_j \leq u_i^{n+1} \leq \max_{j \neq i} \tilde{u}_j =: \tilde{u}_i^{\max}. \tag{21}$$

The piece-wise linearity of P_1 finite elements ensures, that the corrected solution does not exceed the imposed bounds inside the elements. This is not the case for higher-order Lagrange FEM, but can be easily shown for B-Spline based FEM:

Theorem 2 *If the flux limiter [7] is applied to the weights of the B-Spline expansion*

$$\tilde{u}(\mathbf{x}) = \sum_j \tilde{u}_j \varphi_j(\mathbf{x}) \tag{22}$$

then the flux corrected end-of-step solution (of the control variable u)

$$u^{n+1}(\mathbf{x}) = \sum_j u_j^{n+1} \varphi_j(\mathbf{x}) \tag{23}$$

remains bounded from below and above by the bounding functions

$$\tilde{u}^{\max}_{\min}(\mathbf{x}) = \sum_j \tilde{u}_j^{\max}_{\min} \varphi_j(\mathbf{x}). \tag{24}$$

Proof *Assume that for some $\mathbf{x}^* \in \Omega$ we have that $u^{n+1}(\mathbf{x}^*) > \tilde{u}^{\max}(\mathbf{x}^*)$. Then*

$$0 > \tilde{u}^{\max}(\mathbf{x}^*) - u^{n+1}(\mathbf{x}^*) = \sum_j \underbrace{[\tilde{u}_j^{\max} - u_j^{n+1}]}_{\geq 0} \underbrace{\varphi(\mathbf{x}^*)}_{\geq 0} \geq 0, \tag{25}$$

which is a contradiction. A similar argument holds for the lower bound. □

3 Numerical Results

The proposed IGA-AFC approach has been applied to Sod's two dimensional shock tube problem [14], which has been solved on the unit square domain and the VKI U-bend test case geometry proposed in [15]. In both cases, tensor-product bi-quadratic B-Splines combined with the third-order SSP-Runge Kutta time stepping scheme have been employed. Density and pressure have been adopted as control variables and the final correction factors have been computed as their minimum.

Benchmark I Figure 1 (left) shows the numerical solution sampled along the line $y = 0.5$. The solution was computed on the unit square, which was discretized by 66×66 equidistantly distributed B-Spline basis functions and marched forward in time with time step size $\Delta t = 0.0005$ until the final time $T = 0.231$ was reached. This small time-step size was needed to prevent the highly oscillatory Galerkin scheme from breaking down completely. The density and pressure profiles that were computed by IGA-AFC stay within the physical bounds and show a crisp resolution

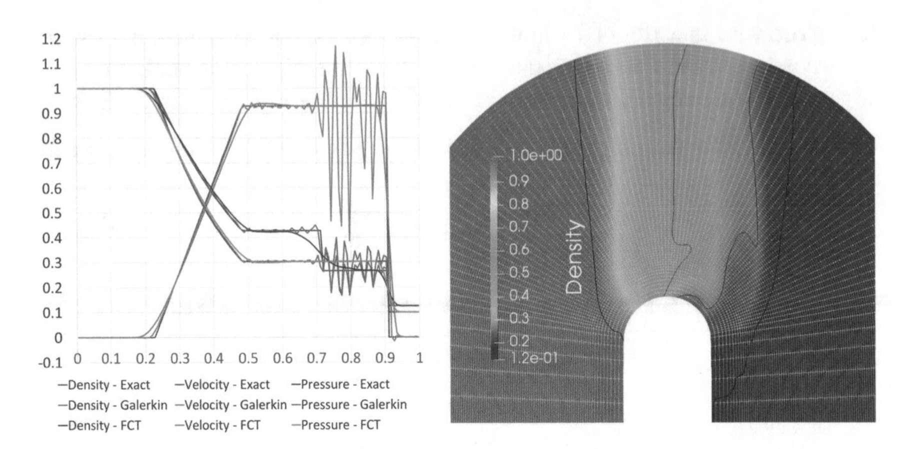

Fig. 1 Numerical solution to Sod's shock tube problem at $T = 0.231$ computed on the unit square (left) and the VKI U-bend test case geometry (right) with bi-quadratic B-Spline basis functions

of the shock wave and the expansion fan. However, the contact discontinuity is smeared over several layers, which needs to be improved in forthcoming research.

Benchmark II Figure 1 (right) shows part of the density profile that is obtained from simulating Sod's shock tube problem on the VKI U-bend geometry [15] with all other settings remaining unchanged except for the time step size $\Delta t = 0.001$. The black contour lines indicate the location of the three characteristic wave types, i.e., the rarefaction wave, the contact discontinuity and the shock wave from left to right. As in the rectangular case, the IGA-AFC scheme succeeds in preserving the positivity of the control variables (ρ and p, and consequently ρE), thereby demonstrating its practical applicability on non-equidistant curved 'meshes'. For illustration purposes the computational mesh has been approximated and illustrated by white lines to give an impression of the mesh width, which varies locally.

It should be noted that the C^1 continuous parameterization of the curved boundary by a quadratic B-Spline function yields a unique definition of the outward unit normal vector $\mathbf{n}(\mathbf{x}) = \mathbf{n}(\phi(\boldsymbol{\xi}))$ in every point on the boundary Γ. Moreover, $\mathbf{n}$ is a continuous function of the boundary parameter values $\boldsymbol{\xi}$. As a consequence, the approximate solution is free of 'numerical artifacts', which often occur for polygonal boundary representations in P_1/Q_1 finite elements, where the C^0 'kinks' between two consecutive boundary segments serve is microscopic compression or expansion corners and, moreover, give rise to undetermined normal vectors in the nodes. In practice, this problem is often overcome by averaging the local normal vectors of the adjacent boundary segments but, still, the so-defined normal vector is not a continuous function of the boundary parameterization, which is the case for IGA.

4 Conclusions

In this work, we have extended the high-resolution isogeometric scheme presented in chapter "High-Order Isogeometric Methods for Compressible Flows. I: Scalar Conservation Laws" [10] to systems of conservation laws, namely, to the compressible Euler equations. The main contribution is the positivity proof of the linearized FCT algorithm for B-Spline based discretizations, which provides the theoretical justification of our IGA-AFC approach. Future work will focus on reducing the diffusivity of the flux limiter and using THB-Splines [13] to perform adaptive refinement

Acknowledgement This work has been supported by the European Unions Horizon 2020 research and innovation programme under grant agreement No. 678727.

References

1. Cottrell, J.A., Hughes, T.J.R., Bazilevs, Y.: Isogeometric Analysis: Toward Integration of CAD and FEA. Wiley, Hoboken (2009)
2. Jaeschke, A.: Isogeometric analysis for compressible flows with application in turbomachinery. Master's Thesis, Delft University of Technology (2015)
3. Duvigneau, R.: Isogeometric analysis for compressible flows using a discontinuous Galerkin method. Comput. Methods Appl. Mech. Eng. **333**, 443– 461 (2018)
4. Trontin, P.: Isogeometric analysis of Euler compressible flow. Application to aerodynamics. In: Conference: 50th AIAA Aerospace Sciences Meeting including the New Horizons Forum and Aerospace Exposition (2012)
5. Kuzmin, D.: Flux-corrected transport, chapter Algebraic flux correction I. In: Scalar Conservation Laws. Springer, Berlin (2012)
6. Kuzmin, D., Möller, M., Gurris, M.: Flux-corrected transport, chapter Algebraic flux correction II. In: Compressible Flow Problems. Springer, Berlin (2012)
7. Kuzmin, D., Möller, M., Shadid, J.N., Shashkov, M.: Failsafe flux limiting and constrained data projections for equations of gas dynamics. J. Comput. Phys.**229**(23), 8766–8779, 11 (2010)
8. Kuzmin, D.: Explicit and implicit FEM-FCT algorithms with flux linearization. J. Comput. Phys. **228**(7), 2517–2534 (2009)
9. Fletcher, C.A.J.: The group finite element formulation. Comput. Methods Appl. Mech. Eng. **37**, 225–243 (1983)
10. Jaeschke, A., Möller, M.: High-order isogeometric methods for compressible flows. I. Scalar conservation Laws. In: Proceedings of the 19th International Conference on Finite Elements in Flow Problems (FEF 2017)
11. Gottlieb, S., Shu, C., Tadmor, E.: Strong stability-preserving high-order time discretization methods. SIAM Rev. **43**(1), 89–112 (2001)
12. Roe, P.L.: Approximate Riemann solvers, parameter vectors, and difference schemes. J. Comput. Phys. **43**(2), 357–372 (1981)
13. Giannelli, C., Jüttler, B., Speleers, H.: Thb-splines: the truncated basis for hierarchical splines. Comput. Aided Geom. Des. **29**(7), 485–498, (2012). Geometric Modeling and Processing (2012)
14. Sod, G.A.: A survey of several finite difference methods for systems of nonlinear hyperbolic conservation laws. J. Comput. Phys. **27**(1), 1–31 (1978)
15. Verstraete, T.: The VKI U-bend optimization test case. In: Technical Report. Von Karman Institute for Fluid Dynamics, Sint-Genesius-Rode (2016)

Simulations of Non-hydrostatic Flows by an Efficient and Accurate p-Adaptive DG Method

G. Tumolo and L. Bonaventura

Abstract We review recent results in the development of a class of accurate, efficient, high order, dynamically p-adaptive Discontinuous Galerkin methods for geophysical flows. The proposed methods are able to capture phenomena at very different spatial scales, while minimizing the computational cost by means of a dynamical degree adaptation procedure and of a novel, fully second order, semi-implicit semi-Lagrangian time discretization. We then present novel results of the application of this technique to high resolution simulations of idealized non-hydrostatic flows.

Keywords Semi-Lagrangian · Semi-implicit time discretization · Discontinuous Galerkin · p-Adaptivity · Atmospheric flows

1 Introduction

Discontinuous Galerkin methods have been increasingly applied to the simulation of geophysical flows over the last two decades. High order finite element methods imply however stringent stability restrictions on explicit time discretization methods, thus increasing sensibly the computational cost of their effective application to large scale geophysical problems. Implicit methods have been successfully applied to reduce this cost [3]. In recent works by the authors [7–9], a comprehensive strategy for the reduction of the computational cost of DG methods for geophysical applications has been proposed. This strategy successfully reduces the computational cost by a combination of two techniques. On one hand, a semi-implicit,

G. Tumolo
ECMWF, Reading, UK
e-mail: giovanni.tumolo@ecmwf.int

L. Bonaventura (✉)
MOX - Dipartimento di Matematica, Politecnico di Milano, Milano, Italy
e-mail: luca.bonaventura@polimi.it

H. van Brummelen et al. (eds.), *Numerical Methods for Flows*,
Lecture Notes in Computational Science and Engineering 132,
https://doi.org/10.1007/978-3-030-30705-9_5

semi-Lagrangian time discretization is employed, that allows the use of much longer time steps than explicit schemes, while retaining full second order accuracy in time. This is achieved by a semi-Lagrangian extension of the TR-BDF2 method for ODE problems, whose accuracy and stability properties have been analysed theoretically in [4] for the more standard ODE version and confirmed empirically for the semi-Lagrangian extension in [8]. This approach is complemented by a dynamically adaptive choice of the polynomial degree used in each element. Results in [8, 9], have shown that, on Cartesian meshes, this adaptive strategy reduces the number of degrees of freedom and the associated computational cost by a factor of up to 50% in two-dimensional tests, while retaining the same level of accuracy as the non adaptive discretization. A mass conservative variant was then introduced in [7], where this approach was also extended to logically Cartesian meshes with non uniform element size, retaining high order accuracy in spite of the mesh non uniformity. In this paper, we will present results of the application of this adaptive technique to idealized high resolution non-hydrostatic flows, that constitute typical benchmarks for modern dynamical cores of NWP and climate models. These represent, to the best of our knowledge, the first dynamically adaptive simulations of lee wave phenomena presented in the literature. The results show that the method is able to capture phenomena at very different spatial scales and at the same time to reduce significantly the computational cost with respect to standard high order discretizations.

2 Outline of the Numerical Method

We consider here a version of the fully compressible, non-hydrostatic Euler equations of motion on a vertical (x, z) plane. We refer to [2] for a complete derivation. Notice that the same equation set has been employed in a number of Met Office meteorological models [1]. Here, p_0 denotes a reference pressure value, $\Theta = T\left(\frac{p}{p_0}\right)^{-R/c_p}$ is the potential temperature, p is the pressure, $\Pi = \left(\frac{p}{p_0}\right)^{R/c_p}$ is the Exner pressure, while c_p, c_v, R are the constant pressure and constant volume specific heats and the gas constant of dry air, respectively. Hydrostatic reference profiles have been introduced so that $\Pi(x, y, z, t) = \pi^*(z)+\pi(x, y, z, t)$, $\Theta(x, y, z, t) = \theta^*(z) + \theta(x, y, z, t)$ and $c_p\theta^*\frac{d\pi^*}{dz} = -g$, where g denotes the acceleration of gravity. The Coriolis force is omitted for simplicity. As a result, the model equations can be written as

$$\frac{D\Pi}{Dt} = -(c_p/c_v - 1)\Pi\nabla\left(\frac{\partial u}{\partial x} + \frac{\partial w}{\partial z}\right), \tag{1}$$

$$\frac{Du}{Dt} = -c_p\Theta\frac{\partial \pi}{\partial x}, \tag{2}$$

$$\frac{Dw}{Dt} = -c_p \Theta \frac{\partial \pi}{\partial z} + g \frac{\theta}{\theta^*}, \tag{3}$$

$$\frac{D\theta}{Dt} = -\frac{d\theta^*}{dz} w. \tag{4}$$

Notice that u, w denote the horizontal and vertical velocity components, respectively, while the Lagrangian derivative is defined as

$$\frac{D}{Dt} = \frac{\partial}{\partial t} + u\frac{\partial}{\partial x} + w\frac{\partial}{\partial z}.$$

The semi-implicit, semi-Lagrangian discretization of the model equations is based on the application of the so called TR-BDF2 method. We refer to [4] for the complete references on the history and properties of the method and to [8] for the details of its semi-Lagrangian extension. It is sufficient to recall here that the method is second order accurate and unconditionally absolutely stable (A-stable) for any value of the stability parameter $\gamma \in [0, 1]$. A stronger stability property that implies complete asymptotic dissipation of high frequency modes (L-stability) is only guaranteed for the optimal value $\gamma = 1 - \sqrt{2}/2$. We denote by $E\left(t^n, \Delta\tau\right)$ the numerical evolution operator representing the approximation of the flow trajectory that is the basis of the semi-Lagrangian method. We also denote by γ the averaging parameter employed by the time discretization method. Firstly, a trapezoidal rule stage is computed, so as to obtain:

$$\begin{aligned} &\pi^{n+2\gamma} + \gamma \Delta t \,\left(c_p/c_v - 1\right) \Pi^n \nabla \cdot \mathbf{u}^{n+2\gamma} \\ &= -\pi^* + E\left(t^n, 2\gamma\Delta t\right)\left[\Pi - \gamma\Delta t\left(c_p/c_v - 1\right)\Pi \;\nabla\cdot\mathbf{u}\right], \end{aligned} \tag{5}$$

$$u^{n+2\gamma} + \gamma\Delta t \; c_p \Theta^n \frac{\partial \pi}{\partial x}^{n+2\gamma} = E(t^n, 2\gamma\Delta t)\left[u - \gamma\Delta t \; c_p\Theta\frac{\partial\pi}{\partial x}\right], \tag{6}$$

$$\begin{aligned} &w^{n+2\gamma} + \gamma\Delta t\left(c_p\Theta^n\frac{\partial\pi}{\partial z}^{n+2\gamma} - g\frac{\theta^{n+2\gamma}}{\theta^*}\right) \\ &= E(t^n, 2\gamma\Delta t)\left[w - \gamma\Delta t\left(c_p\Theta\frac{\partial\pi}{\partial z} - g\frac{\theta}{\theta^*}\right)\right], \end{aligned} \tag{7}$$

$$\theta^{n+2\gamma} + \gamma\Delta t\frac{d\theta^*}{dz}w^{n+2\gamma} = E(t^n, 2\gamma\Delta t)\left[\theta - \gamma\Delta t\frac{d\theta^*}{dz}w\right]. \tag{8}$$

Notice that this first stage could also be performed by the off-centered trapezoidal rule without decreasing the overall second order accuracy of the method [4]. The second stage consists in the semi-Lagrangian counterpart of the BDF2 method applied to (1)–(4), so as to obtain:

$$\pi^{n+1} + \gamma_2 \Delta t \left(c_p/c_v - 1\right) \Pi^{n+2\gamma} \nabla \cdot \mathbf{u}^{n+1}$$
$$= -\pi^* + (1-\gamma_3)[E\left(t^n, \Delta t\right) \Pi] + \gamma_3[E\left(t^n + 2\gamma\Delta t, (1-2\gamma)\Delta t\right) \Pi], \quad (9)$$

$$u^{n+1} + \gamma_2 \Delta t \; c_p \Theta^{n+2\gamma} \frac{\partial \pi}{\partial x}^{n+1}$$
$$= (1-\gamma_3)[E\left(t^n, \Delta t\right) u] + \gamma_3[E\left(t^n + 2\gamma\Delta t, (1-2\gamma)\Delta t\right) u], \quad (10)$$

$$w^{n+1} + \gamma_2 \Delta t \left(c_p \Theta^{n+2\gamma} \frac{\partial \pi}{\partial z}^{n+1} - g \frac{\theta^{n+1}}{\theta^*} \right)$$
$$= (1-\gamma_3)[E\left(t^n, \Delta t\right) w] + \gamma_3[E\left(t^n + 2\gamma\Delta t, (1-2\gamma)\Delta t\right) w], \quad (11)$$

$$\theta^{n+1} + \gamma_2 \Delta t \frac{d\theta^*}{dz} w^{n+1} \quad (12)$$
$$= (1-\gamma_3)[E\left(t^n, \Delta t\right) \theta] \gamma_3[E\left(t^n + 2\gamma\Delta t, (1-2\gamma)\Delta t\right) \theta].$$

For each of the two stages, a spatial discretization by the DG approach outlined in [8] is carried out first. Notice that this approach yields a discretization that is not mass conservative. A mass conservative variant is outlined instead in [7]. The discretization can employ any unstructured mesh of quadrilaterals and could be easily extended to triangular meshes as well. Subsequently, substitution of the potential temperature and velocity degrees of freedom into the Exner pressure equation along the lines of the strategy derived from [2] yields a single linear system for the Exner pressure values for each stage. As remarked in [8], the total computational effort required by the solution of these two systems is only marginally larger than that required the solution of a single system as resulting from an off-centered Crank Nicolson method. Considering the superior accuracy of the TR-BDF2 method with respect to the off-centered Crank Nicolson method, the proposed approach is significantly more efficient. Some estimates of the computational gains will be given in Sect. 3. These estimates are to be considered approximate and prudential, since we have been working so far with a preliminary, proof-of-concept and non optimized implementation. For example, in the present implementation the polynomial degree can be different for each prognostic variable, but is still bound to be the same in all spatial directions. A more advanced implementation is currently being developed, which will allow to choose different degrees independently in each direction, thus further reducing the computational costs, especially in the case of the highly anisotropic meshes usually employed in atmospheric modelling.

3 Numerical Results

A number of classical benchmarks have been considered in order to assess the accuracy and efficiency of the proposed method. Several results have been reported in [8]. In this paper, we will exclusively focus on dynamically adaptive runs. Notice that, in all tests, the purely inviscid equations were considered. Thanks to the intrinsic numerical diffusion introduced by the L-stable time discretization, no explicit diffusion or filtering was required to run any of the reported tests. Intrinsic numerical diffusion is also introduced by the interpolation step of the semi-Lagrangian method, but this is much less significant due to the high order of the polynomial approximations employed. In the description of all the tests, a key parameter to measure the efficiency of the semi-implicit approach are the values of the acoustic Courant number, defined as the maximum of $c_s p\Delta t/h$, where c_s denotes the speed of sound, h the element size and p the polynomial degree. As a first more quantitative assessment of the accuracy of the proposed method, the normalized vertical momentum flux was computed. This non dimensional quantity is a standard diagnostic for non-hydrostatic model performance, see e.g. [6] for a precise definition and reference analytic values, as well as [2] and the literature references therein to similar assessments. We first consider a linear lee wave benchmark, with isothermal reference profiles

$$\theta^*(z) = T_0 \exp\left(\frac{N^2}{g} z\right) \qquad \pi^*(z) = 1 - \frac{g^2}{c_p T_0 N^2}\left[1 - \exp\left(-\frac{N^2}{g} z\right)\right]$$

with $T_0 = 300\,\mathrm{K}$, and a Brunt-Väisäla frequency of $N = 0.0179\,\mathrm{s}^{-1}$. The mountain has aspect ratio 10^{-4}. Profiles of the horizontally averaged vertical momentum flux, normalized with its theoretical value, are displayed in Fig. 1, showing that convergence is achieved as the steady state is approached. In this test case, the minimum size of the elements is 2 km in each direction and the maximum polynomial degree is $p = 4$ for all prognostic variables. A time step of $\Delta t = 7\,\mathrm{s}$ yields in this case maximum acoustic Courant number 7 in the horizontal direction and 9 in the vertical direction, respectively.

A more challenging test was run on a domain 40 km wide and 20 km high, setting $T_0 = 273\,\mathrm{K}$, and a Brunt-Väisäla frequency of $N = 0.02\,\mathrm{s}^{-1}$. The mountain aspect ratio was approximately 0.5, so that non-hydrostatic and nonlinear effects were not negligible. An absorbing layer starting at height 9 km has been used to avoid spurious reflections from the top boundary. Profiles of the horizontally averaged vertical momentum flux, normalized with its theoretical value, are displayed in Fig. 2. In this test, convergence to the theoretical value 0.65 is only starting to occur over the simulated time span. Plots of the horizontal velocity, vertical velocity and local polynomial degree at time 5 h are displayed in Figs. 3, 4, and 5. In this test case, the minimum size of the elements is 200 m in each direction and the maximum polynomial degree is $p = 4$ for all prognostic variables, which entails a maximum acoustic Courant number 13 in the horizontal direction and 25 in the

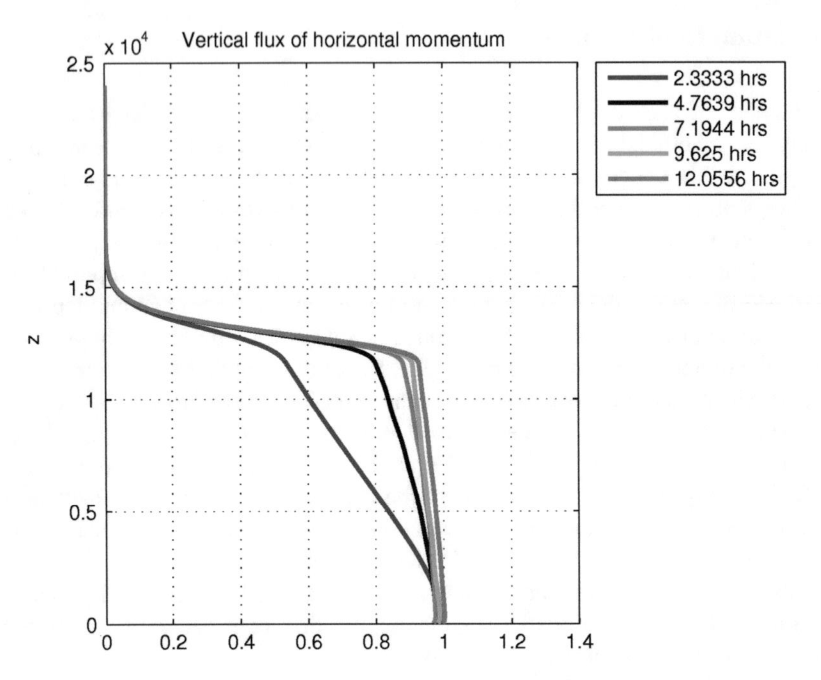

Fig. 1 Convergence of vertical momentum flux in linear hydrostatic lee wave test, z in meters

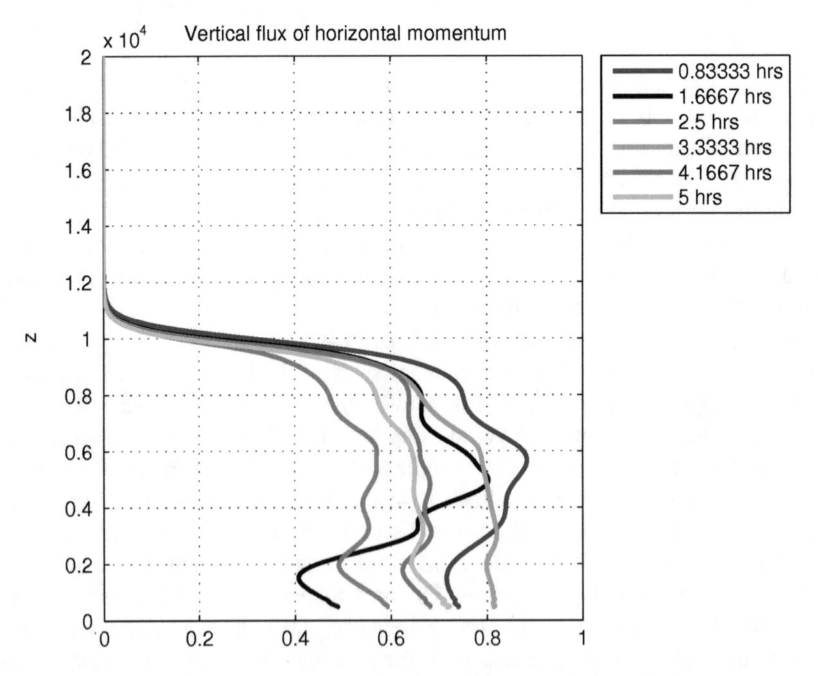

Fig. 2 Convergence of vertical momentum flux in nonlinear, nonhydrostatic lee wave test, z in meters

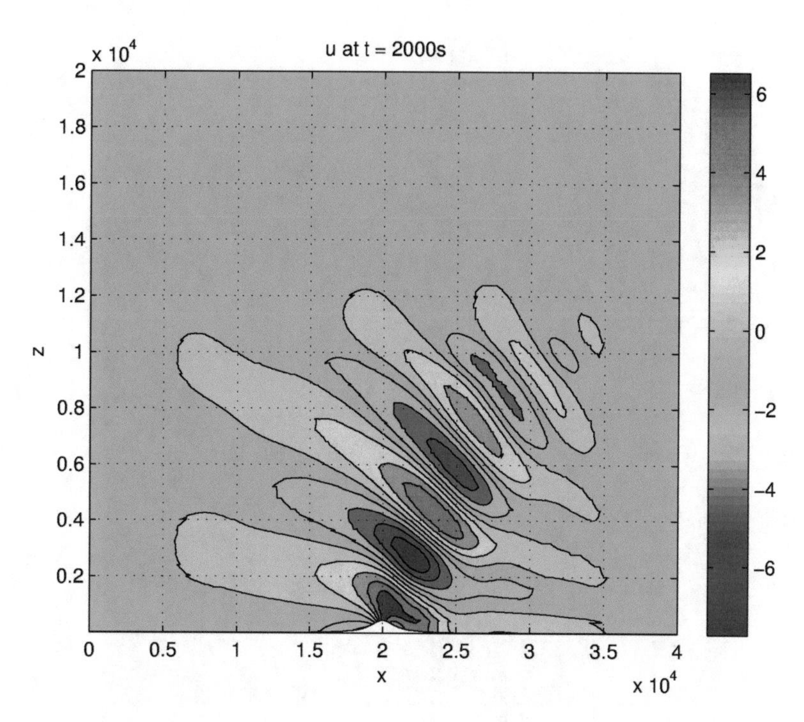

Fig. 3 Nonlinear nonhydrostatic lee wave test: horizontal velocity, x, z in meters

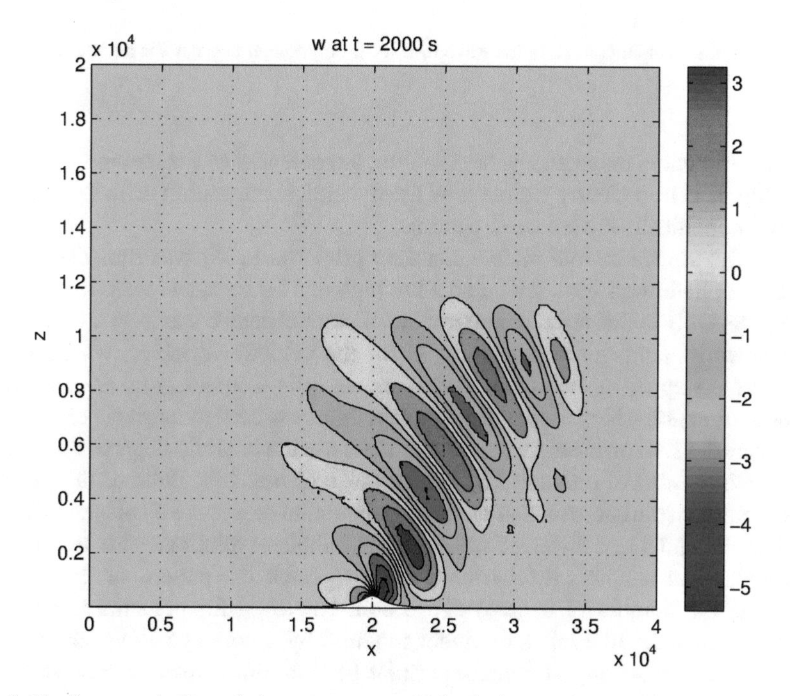

Fig. 4 Nonlinear nonhydrostatic lee wave test: vertical velocity, x, z in meters

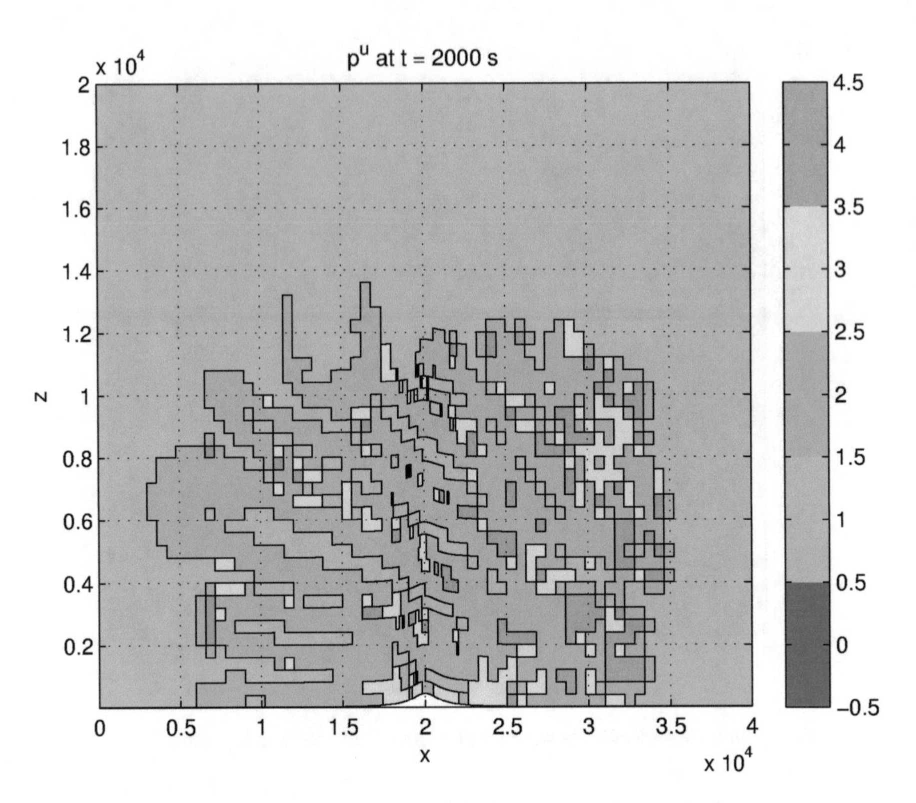

Fig. 5 Nonlinear nonhydrostatic lee wave test: local polynomial degrees for horizontal velocity, x, z in meters

vertical direction, respectively, highlighting the potential of the method in terms of stability. It is also clearly visible how the dynamical adaptation criterion is able to track automatically the lee wave pattern.

Finally, the warm/cold bubble test case proposed in [5] was run. The computational domain was 1 km wide and 1 km high and 50 elements were used in each direction. Only in this test, the maximum polynomial degree was $p = 4$ for pressure and potential temperature and $p = 5$ for the velocity variables, which entails, for a time step of $\Delta t = 1$ s, a maximum acoustic Courant number equal to 87 in both directions. Notice that degree adaptation was carried out independently on the degrees used for pressure and potential temperature, while degrees for velocity were chosen always one unit larger than those of pressure. Plots of the potential temperature perturbation and of the local polynomial degree are displayed in Figs. 6, 7, 8, 9, 10 and 11 at different stages of the bubbles evolution. Also in this case, the dynamical adaptation criterion is able to track the pattern of the bubbles, reducing the number of degrees of freedom employed for potential temperature to approximately 30% of the number required by a corresponding non adaptive simulation when setting a tolerance of order 10^{-3} for the adaptation criterion on the

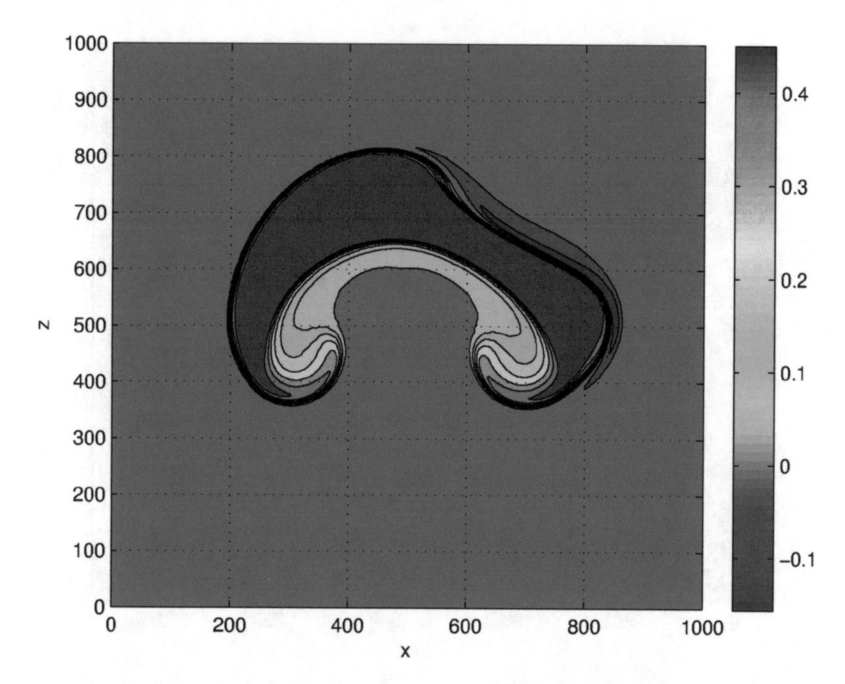

Fig. 6 Double bubble test case: potential temperature perturbation, initial stage, x, z in meters

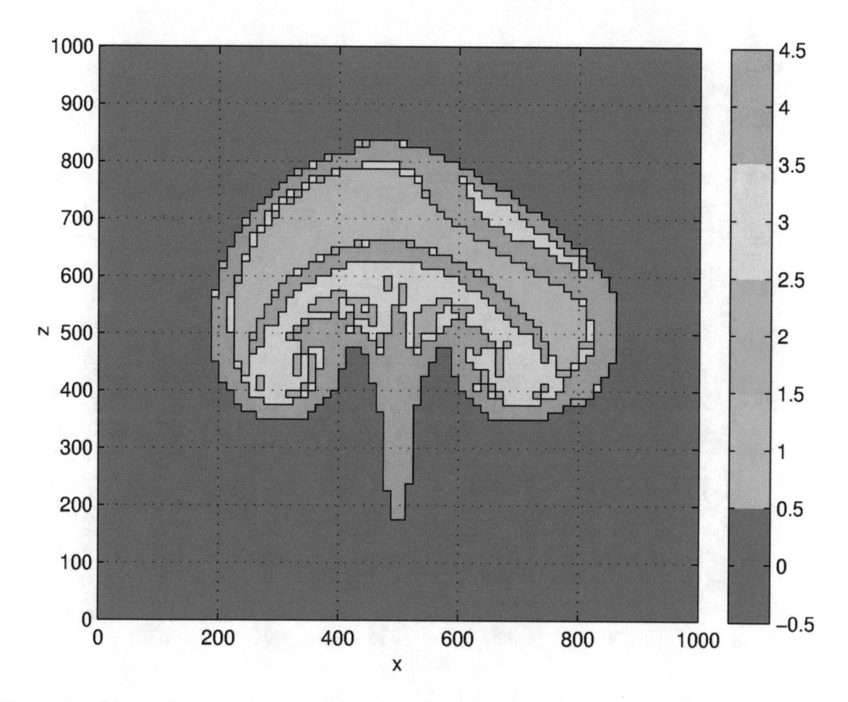

Fig. 7 Double bubble test case: local polynomial degrees for potential temperature, initial stage, x, z in meters

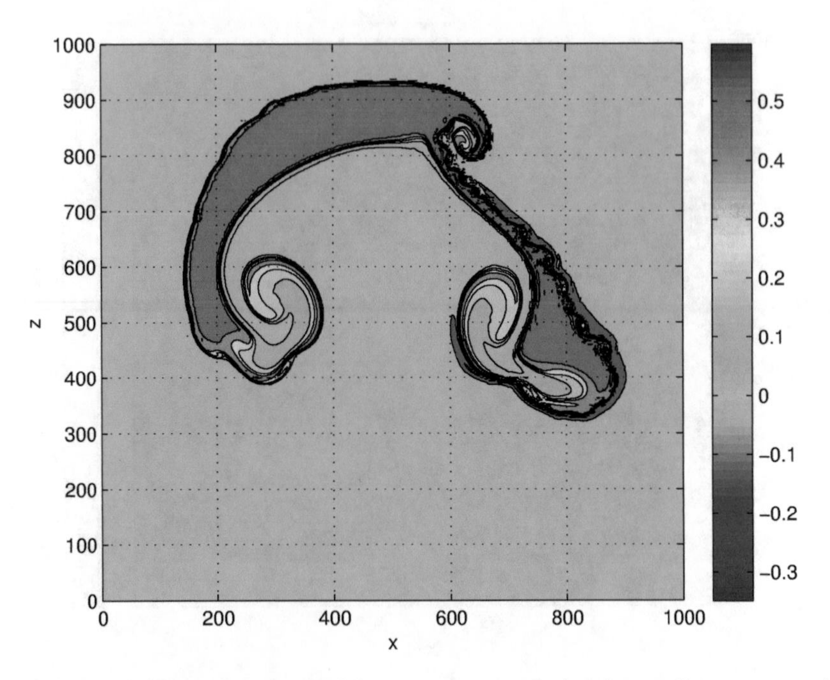

Fig. 8 Double bubble test case: potential temperature perturbation, intermediate stage, x, z in meters

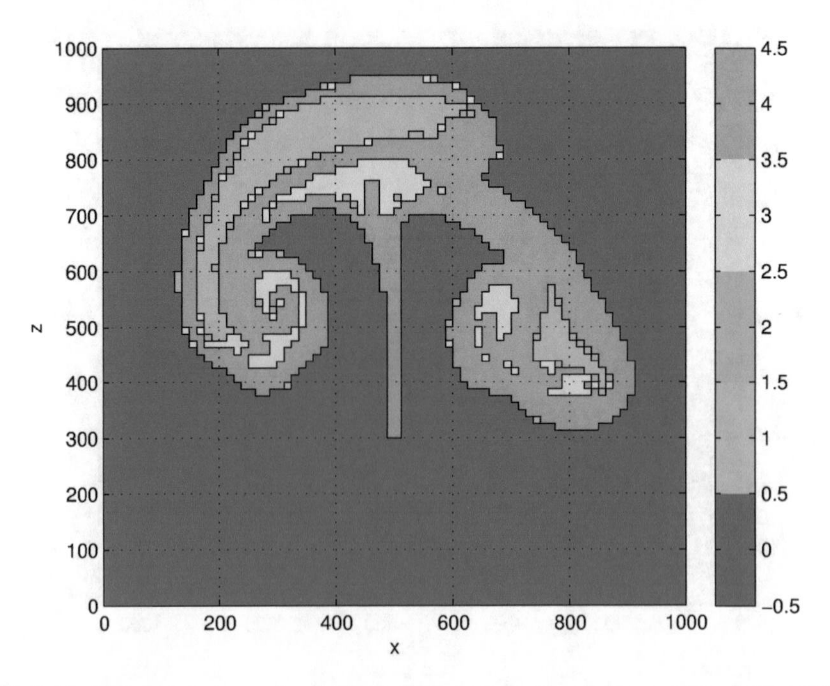

Fig. 9 Double bubble test case: local polynomial degrees for potential temperature, intermediate stage

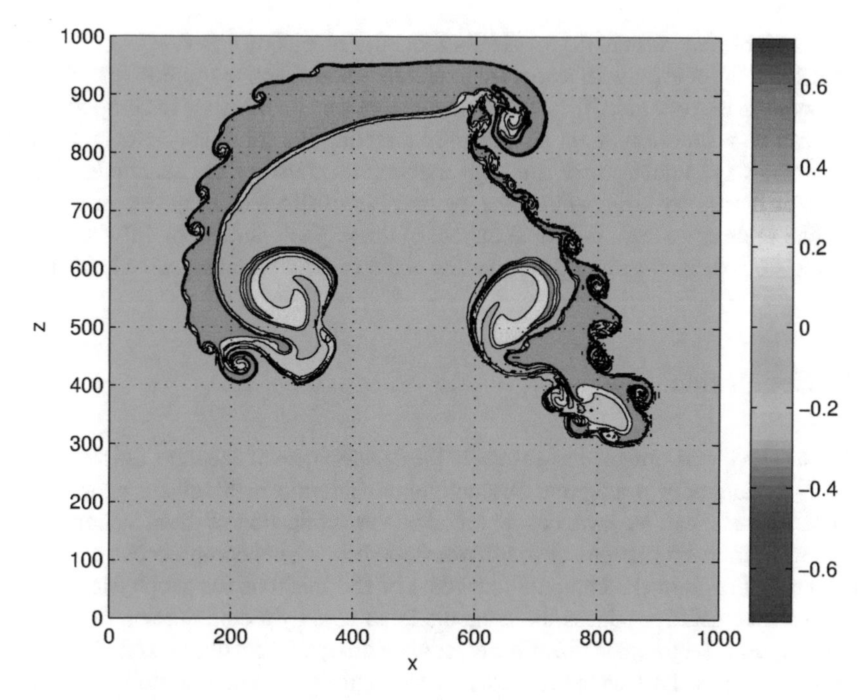

Fig. 10 Double bubble test case: potential temperature perturbation, final stage

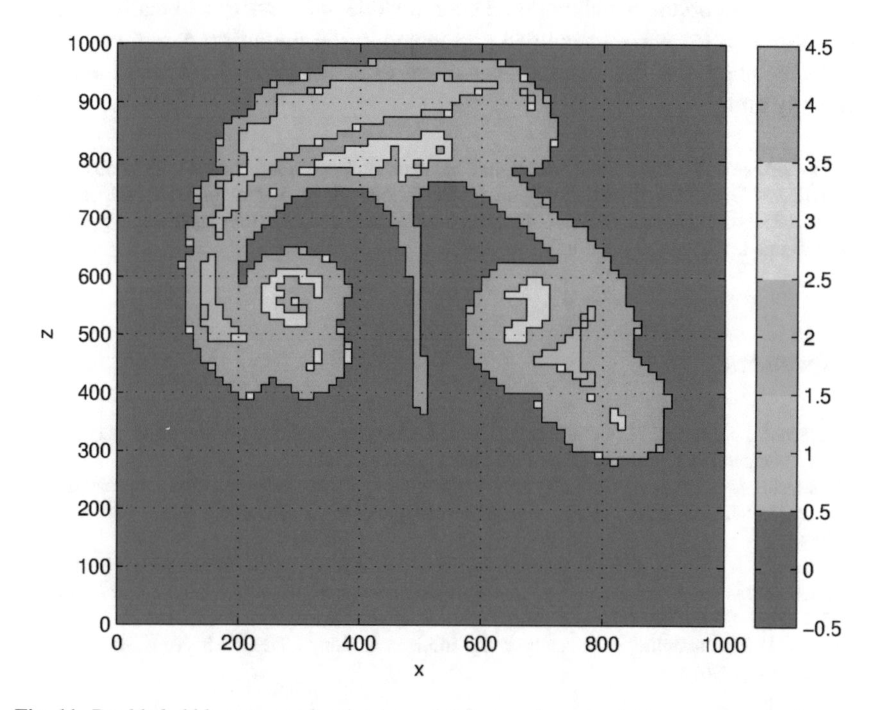

Fig. 11 Double bubble test case: local polynomial degrees for potential temperature, final stage

same variable (we refer to [9] for details about the adaptation criterion). On the other hand, the relative error with respect to a reference non adaptive solution for potential temperature was of order 10^{-2}, thus showing that a major decrease in computational cost can be achieved without significantly affecting the solution accuracy. Further numerical experiments were run with maximum acoustic Courant number up to 200. For these very large values, the method is still stable and reproduces most of the fine scale structures, but some impact of the implicit numerical diffusion starts to be visible and some among the smallest scale vortices are not resolved any more.

4 Conclusion

We have reviewed some recent results in the development of accurate, efficient, high order, dynamically p-adaptive Discontinuous Galerkin methods for geophysical flows. In particular, we have concentrated on the application of these techniques to high resolution simulations of non-hydrostatic flows. This comprehensive strategy for the reduction of the computational cost of DG methods for geophysical applications successfully reduces the computational cost by a combination of a semi-implicit, semi-Lagrangian time discretization with dynamical degree adaptivity. The methods proposed by the authors are able to capture phenomena at different spatial scales, while minimizing the computational cost, as shown by numerical results in several important benchmarks. These include the first dynamically adaptive simulations of lee wave phenomena presented in the literature. A more advanced implementation and the development of an effective parallelization strategy are presently under way.

Acknowledgements The authors would like to thank Filippo Giorgi for his continuous support and INDAM-GNCS for financial support in the framework of several projects and individual grants. Useful discussions with F.X. Giraldo on the topics addressed in this paper are also gratefully acknowledged.

References

1. Benacchio, T., Wood, N.: Semi-implicit semi-Lagrangian modelling of the atmosphere: a met office perspective. Commun. Appl. Ind. Math. **7**, 4–25 (2016)
2. Bonaventura, L.: A semi-implicit, semi-Lagrangian scheme using the height coordinate for a nonhydrostatic and fully elastic model of atmospheric flows. J. Comput. Phys. **158**, 186–213 (2000)
3. Giraldo, F.X., Kelly, J.F., Constantinescu, E.M.: Implicit-explicit formulations of a three-dimensional nonhydrostatic unified model of the atmosphere (NUMA). SIAM J. Sci. Comput. **35**, 1162–1194 (2013)
4. Hosea, M.E., Shampine, L.F.: Analysis and implementation of TR-BDF2. Appl. Numer. Math. **20**, 21–37 (1996)

5. Robert, A.: Bubble convection experiments with a semi-implicit formulation of the Euler equations. J. Atmos. Sci. **50**, 1865–1873 (1993)
6. Smith, R.B.: The influence of mountains on the atmosphere. Adv. Geophys. **21**, 87–230 (1979)
7. Tumolo, G.: A mass conservative TR-BDF2 semi-implicit semi-Lagrangian DG discretization of the shallow water equations on general structured meshes of quadrilaterals. Commun. Appl. Ind. Math. 7:165–190, 2016.
8. Tumolo, G., Bonaventura, L.: A semi-implicit, semi-Lagrangian, DG framework for adaptive numerical weather prediction. Q. J. Roy. Meteorol. Soc. **141**, 2582–2601 (2015)
9. Tumolo, G., Bonaventura, L., Restelli, M.: A semi-implicit, semi-Lagrangian, p-adaptive discontinuous Galerkin method for the shallow water equations. J. Comput. Phys. **232**, 46–67 (2013)

A Fully Semi-Lagrangian Method for the Navier–Stokes Equations in Primitive Variables

Luca Bonaventura, Elisa Calzola, Elisabetta Carlini, and Roberto Ferretti

Abstract We propose a semi-Lagrangian method for the numerical solution of the incompressible Navier–Stokes equations. The method is based on the Chorin–Temam fractional step projection method, combined with a fully semi-Lagrangian scheme to approximate both advective and diffusive terms in the momentum equation. A standard finite element method is used instead to solve the Poisson equation for the pressure. The proposed method allows to employ large time steps, while avoiding the solution of large linear systems to compute the velocity components, which would be required by a semi-implicit approach. We report numerical results obtained in two dimensions using triangular meshes on classical benchmarks, showing good agreement with reference solutions in spite of the very large time step employed.

Keywords Semi-Lagrangian · Navier–Stokes equations · Incompressible flows · Projection methods · Finite element methods

L. Bonaventura (✉)
MOX - Politecnico di Milano, Milano, Italy
e-mail: luca.bonaventura@polimi.it

E. Calzola · E. Carlini
Dipartimento di Matematica, Università Sapienza, Roma, Italy
e-mail: carlini@mat.uniroma1.it

R. Ferretti
Dipartimento di Matematica e Fisica, Università Roma Tre, Roma, Italy
e-mail: ferretti@mat.uniroma3.it

H. van Brummelen et al. (eds.), *Numerical Methods for Flows*,
Lecture Notes in Computational Science and Engineering 132,
https://doi.org/10.1007/978-3-030-30705-9_6

1 Introduction

In this paper, we propose a fully semi-Lagrangian (SL) scheme for the approximation of the incompressible Navier–Stokes equations (NSE) in two space dimensions

$$\begin{cases} \frac{\partial \boldsymbol{u}}{\partial t} + (\boldsymbol{u} \cdot \nabla)\, \boldsymbol{u} - \nu \Delta \boldsymbol{u} + \nabla p = 0 & (x,t) \in \Omega \times (0,T)\,, \\ \nabla \cdot \boldsymbol{u} = 0 & (x,t) \in \Omega \times (0,T)\,, \\ \boldsymbol{u}\,(x,t) = \boldsymbol{b}\,(x,t) & (x,t) \in \partial\Omega \times (0,T)\,, \\ \boldsymbol{u}\,(x,0) = \boldsymbol{u}_0\,(x) & x \in \Omega. \end{cases} \tag{1}$$

Here the unknowns are the velocity field $\boldsymbol{u} : \Omega \times [0,T] \to \mathbb{R}^2$ and the pressure $p : \Omega \times [0,T] \to \mathbb{R}$, T is the final solution time, $\Omega \subset \mathbb{R}^2$ is an open bounded domain, $\boldsymbol{u}_0 : \Omega \times [0,T] \to \mathbb{R}^2$ an initial velocity field, $\boldsymbol{b} : \partial\Omega \times [0,T] \to \mathbb{R}^2$ a boundary condition such that $\int_{\partial\Omega} \boldsymbol{n} \cdot \boldsymbol{b} d\gamma = 0$ (with $\boldsymbol{n}$ external normal of $\partial\Omega$).

In the proposed approach, the classical Chorin–Teman projection method [7, 12] is combined with a semi-Lagrangian approach for the discretization of both advection and diffusion terms, thus extending to the primitive variable formulation of the NSE the method presented in [5] in the case of the vorticity–streamfunction formulation. For the approximation of momentum advection, a classical semi-Lagrangian approach is employed, see e.g. the review in [8]. However, a novel implicit approach for the fully nonlinear approximation of the characteristic lines is proposed, which allows to simplify this part of the computation and to avoid linearization of the momentum advection terms. For the approximation of the diffusion terms, we follow the approach introduced in [3]. A conservative variant of the same approach has also been presented in [4]. The proposed technique allows the use of large time steps, while avoiding the solution of large linear systems for the velocity components that is required by implicit or semi-implicit approaches. Furthermore, it is easily extensible to full three-dimensional problems. A review of more conventional Lagrange-Galerkin approaches for the NSE equation is reported in [2].

2 The Fully Semi-Lagrangian Scheme

We define a time grid $t_n = n\Delta t$, $n = 0, \ldots, N_T$, where N_T denotes the number of time steps and $\Delta t = T/N_T$ is the time step. In its time-discrete version, the classical Chorin–Temam projection method consists of two steps. First, an intermediate velocity field $\boldsymbol{u}^{n+1/2}$ is computed by solving the momentum equation without the pressure term, which, using an explicit Euler time stepping, reads

$$\begin{aligned} \frac{\boldsymbol{u}^{n+1/2} - \boldsymbol{u}^n}{\Delta t} &= -\left(\boldsymbol{u}^n \cdot \nabla\right) \boldsymbol{u}^n + \nu \Delta \boldsymbol{u}^n, \\ \boldsymbol{u}^{n+1/2}|_{\partial\Omega} &= \boldsymbol{b}^{n+1}. \end{aligned} \tag{2}$$

Next, in order to project the intermediate velocity $\boldsymbol{u}^{n+1/2}$ on the space of solenoidal vector fields, a suitable pressure term P^{n+1} is computed as

$$
\begin{aligned}
-\Delta P^{n+1} &= -(\Delta t)^{-1} \nabla \cdot \boldsymbol{u}^{n+1/2}, \\
\boldsymbol{n} \cdot \nabla P^{n+1}|_{\Gamma} &= 0,
\end{aligned} \tag{3}
$$

and the velocity field is then updated as

$$
\boldsymbol{u}^{n+1} = \boldsymbol{u}^{n+1/2} - \Delta t \nabla P^{n+1}.
$$

Higher order extensions and improvements of the original approach are reviewed e.g. in [10]. In our novel, fully semi-Lagrangian approach, Eq. (2) is replaced by

$$
\boldsymbol{u}_i^{n+1/2} = \frac{1}{4} \sum_{k=1}^{4} I\left[\boldsymbol{u}^n\right]\left(\boldsymbol{x}_i - \boldsymbol{u}_i^{n+1/2}\Delta t + \boldsymbol{\delta}_k\right). \tag{4}
$$

Here, $\boldsymbol{u}_i^{n+1/2}$ denotes the fully discrete approximation of $\boldsymbol{u}^{n+1/2}$ at the mesh node $\boldsymbol{x}_i$, $I\left[\boldsymbol{u}^n\right]$ denotes the interpolation operator applied to the fully discrete field $\boldsymbol{u}_i^n$ and the displacements $\boldsymbol{\delta}_k$, associated to the diffusive term, are defined as

$$
\boldsymbol{\delta}_k = \begin{pmatrix} \pm\delta \\ 0 \end{pmatrix} \text{ for } k = 1, 2, \quad \boldsymbol{\delta}_k = \begin{pmatrix} 0 \\ \pm\delta \end{pmatrix} \text{ for } k = 3, 4 \tag{5}
$$

with $\delta = \sqrt{4\nu\Delta t}$ (see a complete description and justification of this approach in [3, 5]). Notice that, while the characteristic line associated to the advection process is approximated by a straight line, the displacement value resulting from this approach is fully coupled to the value of the velocity field at time level $n + \frac{1}{2}$. This requires some fixed point iterations to determine each $\boldsymbol{u}_i^{n+1/2}$, which are however completely decoupled across the computational mesh and can be computed independently for each mesh node. Notice also that, as already remarked in [3, 5], the resulting method is only first order in time. However, higher order accuracy extensions can be achieved by multi-stage approaches like those presented in [1, 13].

The handling of boundary conditions also follows the procedure outlined in [5]. Denoting $z_i = \boldsymbol{x}_i - \boldsymbol{u}_i^{n+1/2}\Delta t$, whenever this position vector lies outside of the computational domain, it is redefined to coincide to the closest boundary point, according to what is usually done in most semi-Lagrangian approaches. Concerning the parabolic displacements, if for some k the vector $z_i + \boldsymbol{\delta}_k$ lies outside the computational domain, the corresponding $\boldsymbol{\delta}_k$ is redefined so that it lies at the intersection of the boundary with the straight line connecting $z_i + \boldsymbol{\delta}_k$ and z_i, as

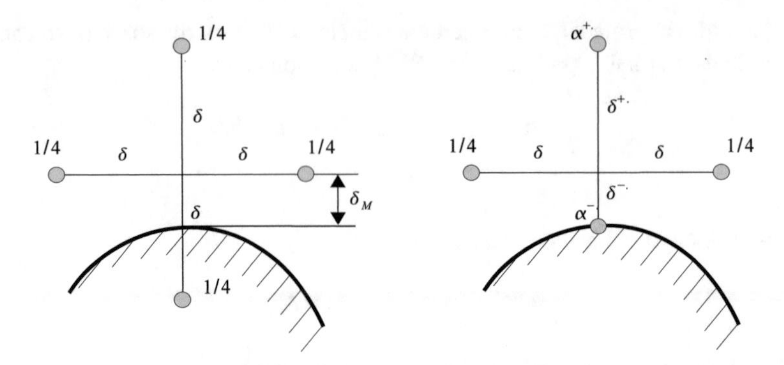

Fig. 1 Redefinition of displacements and weights at the boundary

sketched in Fig. 1. In these cases, formula (4) is replaced by

$$\boldsymbol{u}_i^{n+1/2} = \sum_{k=1}^{4} \alpha_k I\left[\boldsymbol{u}^n\right]\left(\boldsymbol{x}_i - \boldsymbol{u}_i^{n+1/2}\Delta t + \boldsymbol{\delta}_k\right), \tag{6}$$

in which, keeping the notation of Fig. 1, the weights α_k and the coordinates of the displacements are defined as

$$\delta^- = \delta_M, \quad \delta^+ = \frac{4\Delta t \nu}{\delta_M}, \quad \alpha^- = \frac{1}{2}\frac{\delta^+}{\delta^+ + \delta^-}, \quad \alpha^+ = \frac{1}{2} - \alpha^-. \tag{7}$$

3 Numerical Tests

The proposed scheme has been implemented in Matlab. Triangular meshes have been generated using the function `initmesh` of the Matlab PDE toolbox. A triangle-based cubic interpolation, implemented in the `cubic` option of the Matlab function `griddata`, has then been used to reconstruct the solution at the foot of the characteristic lines. The Poisson equation (3) has been approximated by the linear finite element method and solved using the corresponding Matlab PDE toolbox solver. Two numerical simulations have been carried out corresponding to two classical benchmarks for the NSE, specifically, the lid-driven cavity flow and the flow around a cylinder.

The lid-driven cavity flow benchmark was considered in the same configuration discussed in [5, 6]. In this test, the phenomena of interest are taking place especially at the corners of the domain. Therefore, the mesh was refined in these regions, thus leading to relatively high values for the Courant number $C_{\Delta x} = \Delta t\|u\|_\infty/\Delta x$ and the parabolic stability parameter $P_{\Delta x}^{Re} = \Delta t/(Re\Delta x^2)$, see Table 1. The stream

Table 1 Mesh time step data, Courant numbers and parabolic stability parameter in the case $Re = 100$ and $Re = 1000$ for the lid-driven cavity test

Vertices	$\Delta x_{\max}$	$\Delta x_{\min}$	Δt	$C_{\Delta x_{\max}}$	$C_{\Delta x_{\min}}$	$P^{100}_{\Delta x_{\max}}$	$P^{100}_{\Delta x_{\min}}$	$P^{1000}_{\Delta x_{\max}}$	$P^{1000}_{\Delta x_{\min}}$
6469	2.0e−2	4.1e−3	5.0e−2	2.5	12	1.2	29	1.2e−2	2.9
21,221	1.0e−2	2.4e−3	2.5e−2	2.5	10	2.5	43	2.5e−1	4.3

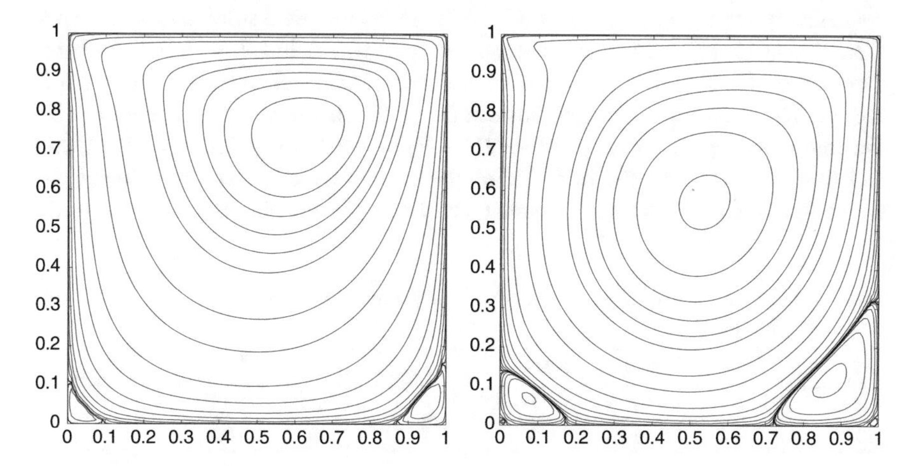

Fig. 2 Stream function contours at time $T = 50$ in the $Re = 100$ case (left) and in the $Re = 1000$ case (right)

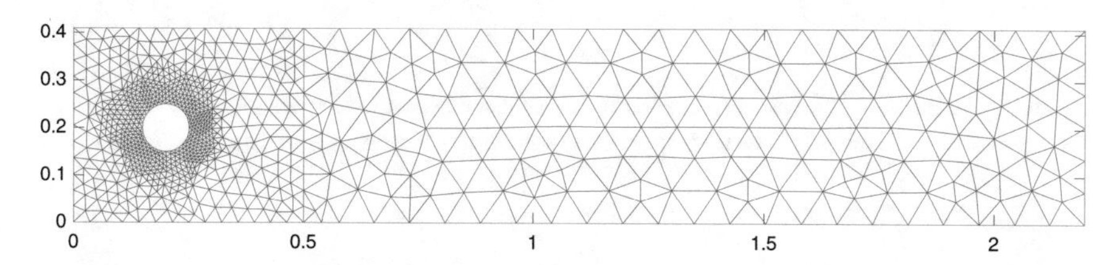

Fig. 3 Mesh on the state domain Ω

function contours in the cases $Re = 100$ and $Re = 1000$ are reported in Fig. 2, showing good agreement with the reference solutions presented in [6].

For the flow around the cylinder, the configuration presented in [11] was considered. More specifically, the domain is a rectangular channel Ω of height $H = 0.41$ and length $L = 2.2$ in non-dimensional units, with a circular hole of diameter $D = 0.1$ placed slightly off the cylinder axis. A parabolic normal velocity profile $u_{in} = 4U_m x_2(H - x_2)/H^2$ is imposed at the inflow boundary, while a homogeneous Dirichlet condition is imposed on the upper and lower part of the boundary and on the boundary of the obstacle, whereas on the rightmost outflow part of the boundary a parabolic normal velocity profile, chosen as on the inflow boundary, has been imposed. The kinematic viscosity was taken to be $\nu = 10^{-3}\,\mathrm{m}^2/\mathrm{s}$. The computational mesh employed is shown in Fig. 3. The Reynolds number is defined for this problem as $Re = 2U_m D/3\nu$. Two different tests were carried out. In the first, a value $U_m = 0.3\,\text{m/s}$ was employed, which yields $Re = 20$, for which a steady flow is observed. In the other test, $U_m = 1.5\,\text{m/s}$ was considered, which yields $Re = 100$, for which an unsteady flow is observed. In Table 2, we show the mesh parameters and Reynolds numbers chosen in our simulation, together with the corresponding Courant numbers and parabolic stability constants. In Figs. 4 and 5, respectively, the time evolution of the norm of the numerical solution **u** is shown at time $T = 0.75, 1.5, 2.25, 3$ for the case $Re = 20$ and at time $T = 1, 3, 5, 8$ for the case $Re = 100$. Figure 4 shows the stationary regime reached by the velocity **u** in the case $Re = 20$. Figure 5 shows instead the vortex shedding in the case $Re = 100$. We compute the pressure drop $\Delta P = P_A - P_B$ between two points $A = (0.15, 0.2)$ and $B = (0.25, 0.2)$ on the front and on the back of the cylinder, respectively, and we compare it with reference values taken from [9]. For the case $Re = 20$, the

Table 2 Mesh time step data, Courant numbers and parabolic stability parameter for the test of a flow around a cylinder

Re	$\Delta x_{\max}$	$\Delta x_{\min}$	Δt	$C_{\Delta x_{\max}}$	$C_{\Delta x_{\min}}$	$P^{Re}_{\Delta x_{\max}}$	$P^{Re}_{\Delta x_{\min}}$
20	2.0e−2	4.1e−3	3.0e−2	0.45	2.20	3.70	89
100	2.0e−2	4.1e−3	1.0e−2	0.75	3.60	0.25	5.9

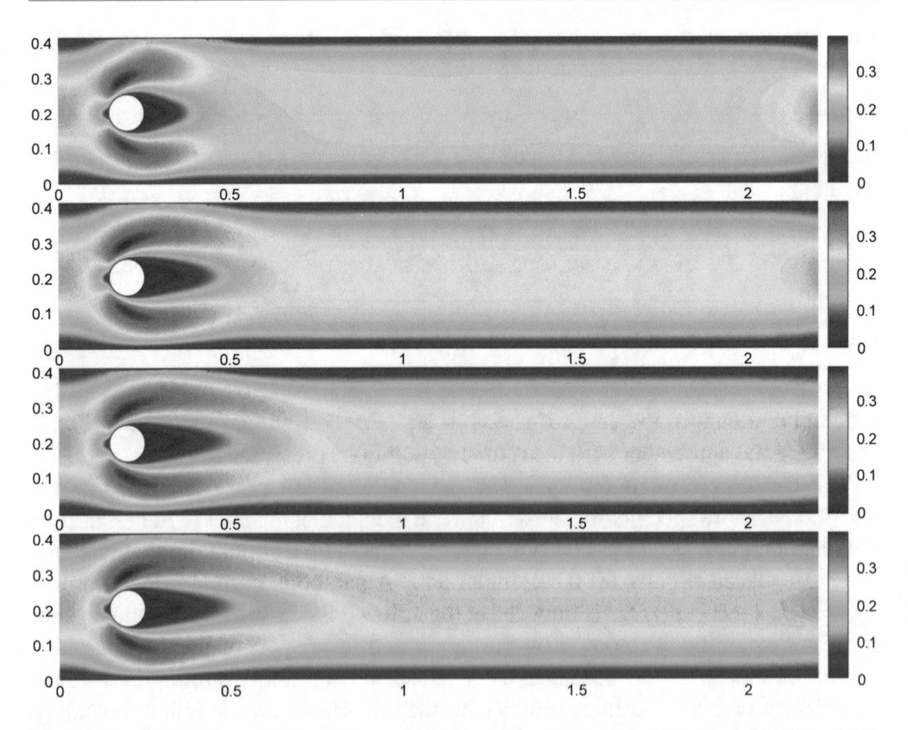

Fig. 4 Norm of the numerical solution **u** with $Re = 20$ at time $T = 0.75$, $T = 1.5$, $T = 2.25$, $T = 3$

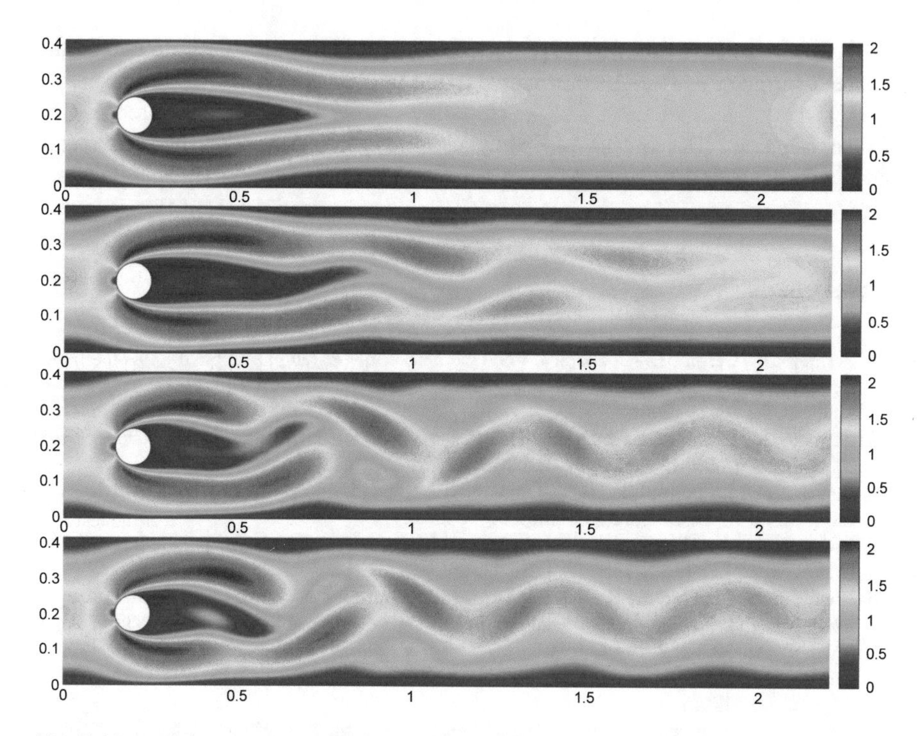

Fig. 5 Norm of the numerical solution **u** with $Re = 100$ at time $T = 1, 3, 5, 8$

reference pressure drop is $\Delta P_{ref} = 0.1175$ and the pressure drop resulting from our computation is $\Delta P_{sim} = 0.1122$. For the case $Re = 100$, the reference and simulated pressure drops are $\Delta P_{ref} = 2.4800$ and $\Delta P_{sim} = 2.4823$, respectively.

4 Conclusions and Future Developments

The classical Chorin–Teman projection method has been combined with a semi-Lagrangian approach for the discretization of both advection and diffusion terms, along the lines of [3], thus extending to the primitive variable formulation of the NSE the method presented in [5] in the case of the vorticity–streamfunction formulation. Furthermore, a novel implicit approach for the fully nonlinear approximation of the characteristic lines has been employed, which allows to simplify this part of the computation and to avoid linearization of the momentum advection terms. Numerical results show that the proposed technique allows to obtain accurate results even when using very large time steps, at the same time avoiding the solution of large linear systems for the velocity components, as required by implicit or semi-implicit approaches. While only first order in time, the proposed approach can be extended to higher order accuracy by multi-stage approaches like those presented in [1, 13]. Also extensions to three-dimensional problems and discontinuous finite element formulations will be considered in the future.

Acknowledgement This work has been partly supported by INDAM-GNCS in the framework of the GNCS 2017 project *Metodi numerici per equazioni iperboliche e cinetiche e applicazioni.*

References

1. Bermejo, R., Carpio, J.: An adaptive finite element semi-Lagrangian implicit-explicit Runge-Kutta-Chebyshev method for convection dominated reaction-diffusion problems. Appl. Num. Mathem. **58**, 16–39 (2008)
2. Bermejo, R., Saavedra, L.: Lagrange-Galerkin methods for the incompressible Navier-Stokes equations: a review. Comm. Appl. Ind. Math. **7**, 26–55 (2016)
3. Bonaventura, L., Ferretti, R.: Semi-Lagrangian methods for parabolic problems in divergence form. SIAM J. Sci. Comp. **36**, A2458–A2477 (2014)
4. Bonaventura, L., Ferretti, R.: Flux form semi-Lagrangian methods for parabolic problems. Comm. Appl. Ind. Math. **7**, 53–70 (2016)
5. Bonaventura, L., Ferretti, R., Rocchi, L.: A fully semi-Lagrangian discretization for the 2D incompressible Navier–Stokes equations in the vorticity–streamfunction formulation. Appl. Math. Comp. **323**, 132–144 (2018)
6. Botella, O., Peyret, R.: Benchmark spectral results on the lid-driven cavity flow. Comp. Fluids **27**, 421–433 (1998)
7. Chorin, A.: Numerical solution of the Navier-Stokes equations. Math. Comp. **22**, 745–762 (1968)
8. Falcone, M., Ferretti R., Semi-Lagrangian Approximation Schemes for Linear and Hamilton–Jacobi Equations. SIAM, Philadelphia (2014)
9. Feng, X., Köster, M., Zhang, L.: Cylinder Flow Benchmark with Commercial Software Packages: A Comparative Study, Lecture "Mathematische Softwarewerkzeuge". University of Dortmund, Germany (2005)
10. Guermond, J.L., Minev, P., Shen, J.: An overview of projection methods for incompressible flows. Comp. Meth. Appl. Mech. Eng. **195**, 6011–6045 (2006)
11. Schäfer, M., Turek, S., Durst, F., Krause, E., Rannacher, R.: Benchmark computations of laminar flow around a cylinder. In: Flow Simulation with High-Performance Computers II, pp. 547–566. Vieweg Teubner Verlag, Wiesbaden (1996)
12. Temam, R.: Sur l'approximation de la solution des equations de Navier-Stokes par la methode des pas fractionnaires (II). Arch. Ration. Mech. Anal. **33**, 377–385 (1969)
13. Tumolo, G., Bonaventura, L.: A semi-implicit, semi-Lagrangian, DG framework for adaptive numerical weather prediction, Q. J. Roy. Met. Soc. **141** 2582–2601 (2015)

Mesh Adaptation for k-Exact CFD Approximations

Alain Dervieux, Eléonore Gauci, Loic Frazza, Anca Belme, Alexandre Carabias, Adrien Loseille, and Frédéric Alauzet

Abstract This paper illustrates the application of error estimates based on k-exactness of approximation schemes for building mesh adaptive approaches able to produce better numerical convergence to continuous solution. The cases of $k = 1$ and $k = 2$, i.e. second-order and third-order accurate approximations with steady and unsteady flows are considered.

Keywords Computational fluid dynamics · Mesh adaptation · Adjoint state

A. Dervieux (✉)
Université Côte d'Azur, Inria, Sophia-Antipolis, France

Société Lemma, Biot, France
e-mail: alain.dervieux@inria.fr

E. Gauci
Université Côte d'Azur, Inria, Sophia-Antipolis, France
e-mail: Eleonore.Gauci@inria.fr

L. Frazza · A. Loseille
INRIA Saclay Ile-de-France 1, Palaiseau, France
e-mail: loic.frazza@inria.fr; Adrien.Loseille@inria.fr

A. Belme
Sorbonne Université, Université Paris 06, CNRS, UMR 719, Institut Jean le Rond d'Alembert, Paris, France
e-mail: belme@dalembert.upmc.fr

A. Carabias
Société Lemma, Paris, France
e-mail: Alexandre.Carabias@lemma-ing.com

F. Alauzet
INRIA Saclay Ile-de-France 1, Palaiseau, France

Société Lemma, Biot, France
e-mail: Frederic.Alauzet@inria.fr

H. van Brummelen et al. (eds.), *Numerical Methods for Flows*, Lecture Notes in Computational Science and Engineering 132, https://doi.org/10.1007/978-3-030-30705-9_7

1 Introduction

The purpose of mesh adaptation research is, thanks to an improved accuracy, to be able to compute new phenomena and also to master the numerical uncertainties which have been up to now unsufficiently controlled. An important strategy is to minimize the approximation error with respect to the mesh. A central question is then to find a good representation of the approximation error. A family of approximations plays a particular role in Computational Fluid Dynamics. The k-exact approximations provide a zero error when the exact solution is a polynomial of degree k. They involve finite elements like continuous and discontinuous Galerkin and ENO finite-volume approximations. Typically, k-exact approximations have a truncation error of order $k + 1$.

The purpose of this paper is to adapt mesh using error estimates for a few k-exact approximations in CFD.

We focus on methods which prescribe an anisotropic mesh under the form of a parametrization of it by a Riemannian metric. A Riemannian metric is a continuous symmetric matrix field defined on the computational domain Ω, for example in two dimensions:

$$\mathcal{M} : \Omega \subset \mathbb{R}^2 \rightarrow \mathbb{R}^{2^2} \quad \mathbf{x} \mapsto \mathcal{M}(\mathbf{x}) = \mathcal{R}(\mathbf{x})^t \begin{pmatrix} \frac{1}{\Delta\xi(\mathbf{x})^2} & 0 \\ 0 & \frac{1}{\Delta\eta(\mathbf{x})^2} \end{pmatrix} \mathcal{R}(\mathbf{x}).$$

Notation $\mathcal{R}$ holds for a rotation for prescribing mesh stretching directions and $\Delta\xi, \Delta\eta$ for prescribing mesh size in these directions. A mesh obeying these prescriptions is called a *unit mesh* for $\mathcal{M}$. We observe that the very complex and *discrete* thing which is a mesh is replaced by a *continuous* function to be found as the minimum of a numerical error. Then we have to organise a process

$$\text{Metric} \rightarrow \text{Mesh and discrete solution} \rightarrow \text{Error} \rightarrow \text{New metric}$$

which can be thought of as either a pure *discrete* process, or the discretization of a *continuous* process. Let us recall why metrics are particularly adapted to 1-exact approximations. These approximations involve most second-order methods based on continuous P_1 finite-element approximation, namely Galerkin, SUPG, Residual distribution and vertex-centered MUSCL approximations. First, the P_1-interpolation error plays a central role in error estimates. Second, this interpolation error can be converted in terms of the mesh metric. We recall, following [20, 21], the main features of the *continuous metric-based analysis* initiated in several papers like [2, 10, 12]. For a function u defined on the computational domain, we use the continuous interpolation error $u - \pi_{\mathcal{M}} u$ instead of the discrete interpolation error $u - \Pi_{\mathcal{M}} u$:

$$u - \pi_{\mathcal{M}} u = |tr(\mathcal{M}^{-\frac{1}{2}} |H_u| \mathcal{M}^{-\frac{1}{2}})| \quad ; \quad |u - \pi_{\mathcal{M}} u| \approx const. |u - \Pi_{\mathcal{M}} u|, \tag{1}$$

where H_u is the Hessian of u and $\mathcal{M}$ also denotes a unit mesh for metric $\mathcal{M}$. We consider minimizing:

$$j(\mathcal{M}) = \|u - \pi_{\mathcal{M}} u\|_{\mathbf{L}^1(\Omega_h)}, \tag{2}$$

and we define as optimal metric the one which minimizes the right-hand side under the constraint of a total number of vertices equal to a parameter N. After solving analytically this optimization problem, we get—solely using the fact that H is a positive symmetric matrix—the unique optimal $(\mathcal{M}_{\mathbf{L}^1}(\mathbf{x}))_{\mathbf{x}\in\Omega}$ as:

$$\mathcal{M}_{\mathbf{L}^1} = N \left(\int_{\Omega} (\det(H_u))^{\frac{1}{4}} \right)^{-1} (\det(H_u))^{\frac{-1}{4}} H_u. \tag{3}$$

Knowing the continuous function u, we can derive the continuous optimal metric. In practice, u is solution of a PDE and the whole process of computing u and then deriving the optimal metric is operated by a fixed point iteration involving the generation of a mesh according to the metric, the solution of the PDE on the mesh and the building of a discrete metric.

In this paper, we discuss three types of functional j and the extension to higher-order approximations. For most cases, we propose to evaluate the method by measuring the mesh-adaptive convergence order α, defined by:

$$error = O(N^{-\frac{\alpha}{dim}}), \tag{4}$$

where N is the total number of nodes and dim the dimension of computational domain, and $\alpha = k + 1$ for a k-exact approximation.

2 Features-, Goal-, Norm-Oriented Formulations

These formulations are presented for the continuous case and applied to the second-order accurate particular case of P^1 approximations on triangles and tetrahedra.

2.1 Feature-Based (FB) Adaptation

The continuous feature-based anisotropic method (2)–(3) is generally defined by replacing the local interpolation error by the application of the recovered Hessian of the solution times a local mesh size defined by the continuous metric, see [3, 10, 12, 16, 18, 22, 24].

A typical example is the prediction of the sonic boom signature of a supersonic aircraft (see [22] for specific features). Let us consider the C25D geometry of the workshop [1]. We use the Mach number M as sensor, i.e. $j(\mathcal{M}) = \|M -$

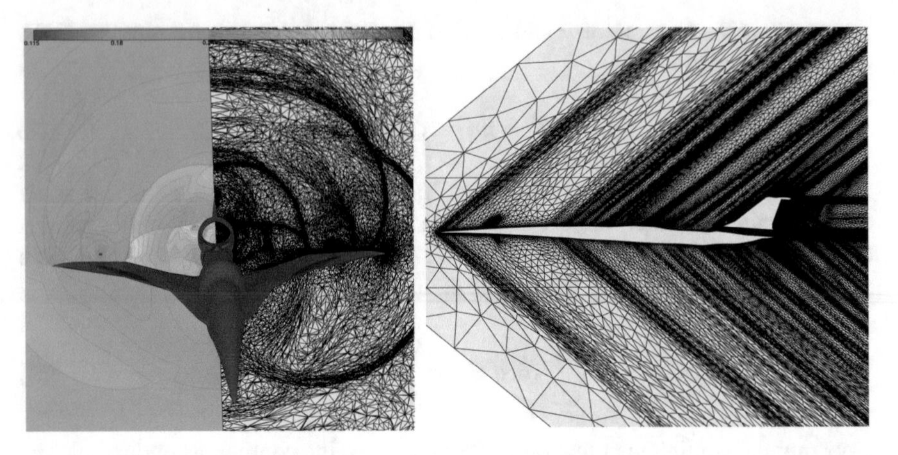

Fig. 1 Lowboom C25 computation with a feature-based mesh adaptation: cut plane $x = 30$ (left) and on symmetry plane (right)

$\pi_{\mathcal{M}} M\|_{\mathbf{L}^1(\Omega_h)}$. Cuts of mesh and solution are depicted in Fig. 1. The FB approach is particularly attractive due to its simplicity and its ability in taking into account physical aspects through the choice of the sensors. However, for systems, the choice of sensors is extremely sensitive.

2.2 *The Goal-Oriented (GO) Formulation*

The GO mesh adaptation focuses on deriving the optimal mesh for computing a prescribed scalar quantity of interest (QoI). Many papers deals with *a posteriori* goal-based error formulation to drive adaptivity, using adjoint formulations or gradients, e.g. [13, 25]. We investigate *a priori* based GO formulations for steady and unsteady problems. Loseille et al. [23] derived the goal-based error estimate in a steady context for Euler flows, showing that the QoI error estimate is expressed as a weighted interpolation error on solution flow fields. This leads to an optimal metric computed as a sum of Hessians of Euler fluxes weighted by gradient components of the adjoint state and permits to focus on the capture of important features with respect to the chosen functional, such as sonic boom print at ground, see Fig. 2.

2.3 *Numerical Corrector and Norm-Oriented (NO) Formulation*

Given a discrete problem, a mesh (of metric) $\mathcal{M}_h$ and the discrete solution W_h computed with the mesh, we call "numerical corrector" a discrete field W_h' such

Fig. 2 A typical comparison of feature-based (left) and goal-oriented (right) mesh adaptation for the computation of sonic boom. On left the whole flow is captured, on right, focus is put on the shock structures influencing the boom path at bottom

that the sum $W_h + W_h'$ is a significantly more accurate approximation of the exact solution than W_h. In other words, W_h' is a good approximation of the error. Clearly, W_h' is useful for estimating the error, for correcting it, and for building a norm-oriented mesh adaptation algorithm.

A trivial way to compute W_h' could be to first compute an extremely accurate and extremely costly $W_{h/2^k}$ (k large) solution computed on a mesh $\mathcal{M}_{h/2^k}$ obtained by dividing k times the elements of $\mathcal{M}_h$, and finally to put $W_h' = W_{h/2^k} - W_h$. But the interesting feature of a numerical corrector should be that its computational cost is *not much higher* than the computational cost of W_h. We describe now a corrector evaluation of low computational cost relying on the application of a Defect Correction principle and working on the initial mesh $\mathcal{M}_h$:

$$\Psi_h(W_h + \overline{W}'_{h,DC}) \approx -R_{h/2\to h}\Psi_{h/2}(R_{h\to h/2}W_h) \quad ; \quad W'_{h,DC} = \overline{W}'_{h,DC} - (\pi_h W_h - W_h),$$

where $\pi_h W_h - W_h$ is a recovery-based evaluation of the interpolation error (see [23] for details). The notation $R_{h/2\to h}$ holds for the transfer (extension by linear interpolation) operator from the twice finer mesh $\mathcal{M}_{h/2}$ to the initial mesh $\mathcal{M}_h$, while $R_{h\to h/2}$ holds for the transfer operator from the initial mesh $\mathcal{M}_h$ to the twice finer mesh $\mathcal{M}_{h/2}$. The finer-mesh residual $\Psi_{h/2}(W_h)$ can be assembled by defining the sub-elements of $\mathcal{M}_{h/2}$ only locally around any vertex of $\mathcal{M}_h$. Applications of this method to the Navier–Stokes model can be found in [15]. We present an application with the Euler model used for sonic boom prediction. We consider again the C25 geometry. The important input is the pressure signature at one-body length below the aircraft. Figure 3 depicts the pressure signal and the local error bar, from the non-linear corrector, for a tailored mesh (mesh aligned with the Mach cone) and for adapted meshes. The tailored mesh calculation may seem converged but the

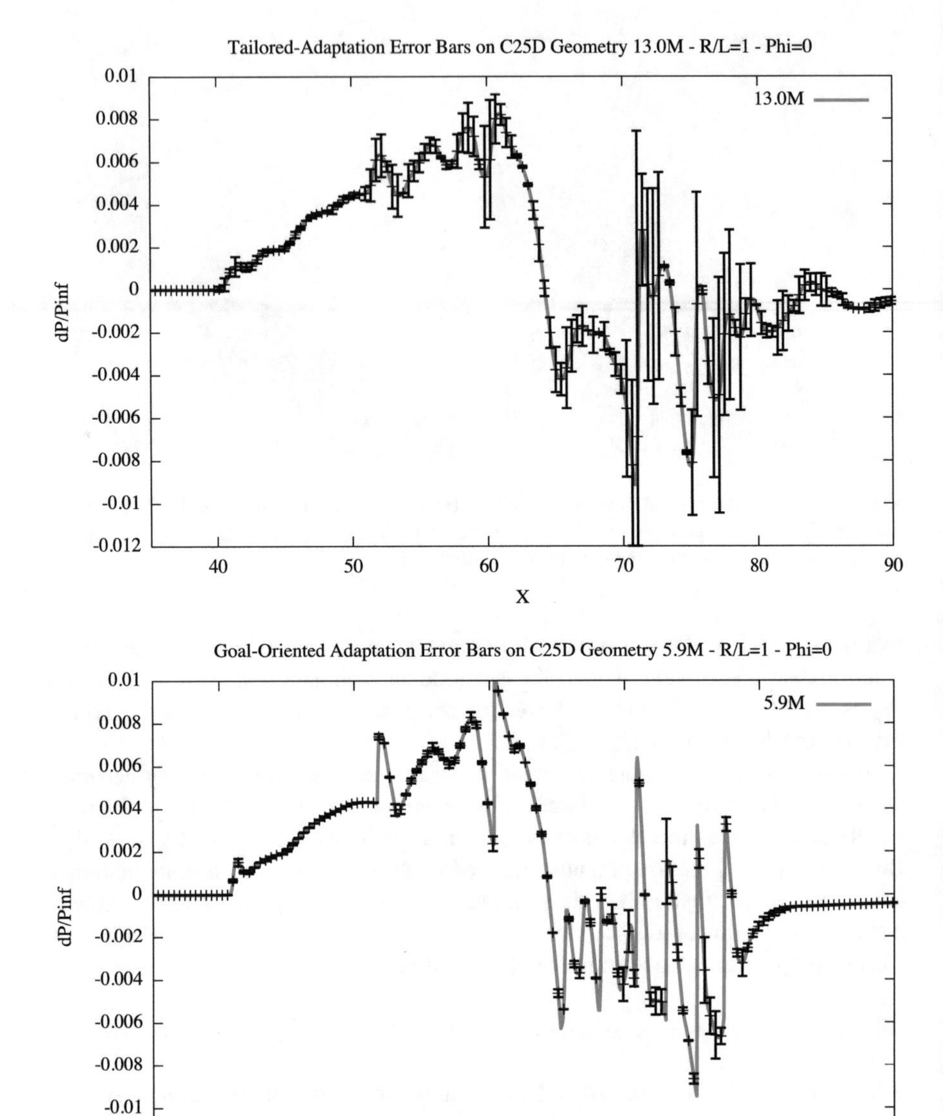

Fig. 3 Flow around a lowboom C25 geometry: Pressure levels (red) and non-linear corrector (black) error intervals for the pressure ($z = 0$, $y = -C$). Top: on an adhoc tailored mesh. Bottom: on a self-adaptive mesh (right). On right figure, the good convergence is indicated by much smaller intervals

corrector remains large. The right-hand side shows a more coherent convergence, with a much smaller corrector.

We can now introduce the NO formulation. We base it on the L^2-norm of approximation error. It consists in the minimization of the following expression with respect to the mesh $\mathcal{M}$:

$$j(\mathcal{M}) = ||W - W_{\mathcal{M}}||^2_{L^2(\Omega)} \quad \text{with} \quad \Psi_{\mathcal{M}}(W_{\mathcal{M}}) = 0. \tag{5}$$

Introducing $g = W - W_{\mathcal{M}}$, we get a formulation similar to a GO formulation $j(\mathcal{M}) = (g, W - W_{\mathcal{M}})$. The central idea of NO is to replace g by the numerical corrector W'_h as defined in this section. The rest of the NO process follows the GO algorithm with $g = W'_h$. The whole NO adaptation algorithm finally writes:

Step 1: solve state equation for W
Step 2: solve corrector equation for W'_h
Step 3: solve adjoint equation for W^*
Step 4: evaluate optimal metric as a function of W and W^*
Step 5: generate unit mesh for $\mathcal{M}_{opt,norm}$ and go to *Step 1*.

In order to give an idea of how this NO works, we consider as *benchmark* a test case from [14] featuring a 2D boundary layer (Fig. 4). The Laplace equation is solved with a RHS inducing the boundary layer depicted in the figure. FB and NO mesh-adaptive methods are compared by displaying the convergence curve related to Criterion (4). In abscissae the number of nodes used for computing the discrete solution u_h is shown, and in ordinates the L^2-norm of the approximation error $u - u_h$ which is measured from the known analytic solution. When the FB method is applied, a tremendous improvement of the error is obtained with 128 vertices, then a uniform element division and further FB adaptation are applied in alternation. While the element division is applied, the error is as expected divided by 4. In contrast, for 512, 1024, 2048, ... vertices (abscissae in the figure), the effect of FB adaptation

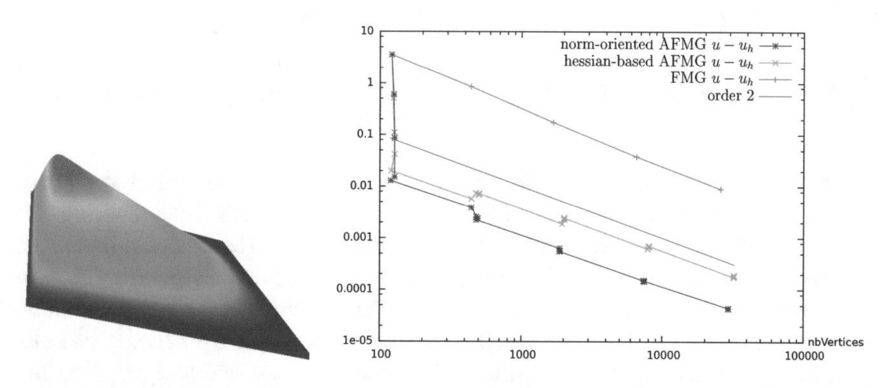

Fig. 4 Elliptic test case of a 2D boundary layer. A comparison between uniform refinement ("FMG"), feature/Hessian-based, and norm-oriented mesh adaptation methods: error $|u - u_h|_{L^2}$ in terms of number of vertices

is to *increase* the error, and the second-order convergence is lost. On the contrary, with this test case, each NO mesh-adaptation phase improves (even slightly) the error norm, producing an asymptotic numerical convergence of order two.

3 Estimates for k-Exact Approximations

Due to their error size and characteristics (dispersion, dissipation), second-order accurate approximations are unable to compute many phenomena. For smooth contexts, high-order methods bring crucial improvements.

3.1 Higher-Order (HO) Estimates

Main existing HO schemes satisfy the so-called k-exactness property expressing the fact that if the exact solution is a polynomial of order k then the approximation scheme will give the exact solution as answer. The assembly of these schemes involves a step of polynomial reconstruction (e.g., ENO schemes), or of polynomial interpolation (e.g., Continuous/discontinuous Galerkin). The main part of the error can then be expressed in terms of a $(k+1)$-th term of a Taylor series where the spatial increment is related with local mesh size. We want to stress that this is the key of an easy extension of metric-based adaptation to HO schemes. We illustrate this with the computation of 2D Euler flows. Considering a triangulation of the computational domain and its dual cells C_i built with triangle medians, the exact solution W of Euler equations verifies (omitting initial conditions):

$$B(W, V_0) = 0, \ \ \forall \ V_0 \in \mathcal{V}_0 = \{V_0 \ \text{ constant by cell}\}, \ \ \text{with}$$

$$\begin{aligned} B(W, V_0) &= \int_0^T \sum_i E_i(W, V_0)dt + \int_0^T \int_{\partial\Gamma} \mathcal{F}_\Gamma(W) \cdot \mathbf{n} V_0 d\Gamma dt \\ E_i(W, V_0) &= \int_{C_i} V_0 \frac{\partial \pi_0 W}{\partial t} d\Omega + \frac{1}{2} \sum_j \int_{\partial C_i \cap \partial C_j} V_0(\mathcal{F}(W)|_{\partial C_i} + \mathcal{F}(W)|_{\partial C_j}) \cdot \mathbf{n} d\sigma. \end{aligned}$$

Here we denote by π_0 the operator replacing a function by its mean on each cell, $\mathcal{F}$ the Euler fluxes, and the second sum is taken over the cells j neighboring cell i. Let us define a *quadratic Central-ENO scheme* [4, 19]. The computational cost of this scheme is rather large but acceptable for 2D calculations (its extension to 3D is even more computationally expensive). This scheme is based on a quadratic reconstruction on any integration cell C_i using the means of the variable on cells around C_i. Let us denote by R_2^0 the global reconstruction operator mapping the constant-by-cell discrete field into its quadratic-by-cell reconstruction. The CENO scheme writes in short:

$$\text{Find } W_0 \text{ constant by cell s.t. } B(R_2^0 W_0, V_0) = 0, \ \forall \ V_0 \text{ constant by cell.}$$

A representative functional of goal-oriented error is:

$$\delta j = (g, R_2^0 \pi_0 W - R_2^0 W_0).$$

Lemma 1 (Carabias et al. [9]) *Introducing the adjoint state $W_0^* \in \mathcal{V}_0$, solution of:*

$$\frac{\partial B}{\partial W}(R_2^0 W_0)(R_2^0 V_0, W_0^*) = (g, R_2^0 V_0), \ \forall \ V_0 \in \mathcal{V}_0, \tag{6}$$

we have the following equivalence:

$$(g, R_2^0 \pi_0 W - R_2^0 W_0) \approx \frac{\partial B}{\partial W}(W)(R_2^0 \pi_0 W - W, W_0^*). \quad \square \tag{7}$$

This estimate is typical of a k-exact variational scheme and permits to express the error as a Taylor term of rank $k+1$, rank 3 in our case, with respect to directional mesh size $\delta \mathbf{x}$, which we replace by the power $3/2$ of a quadratic term:

$$\delta j \preceq \sup_{\delta \mathbf{x}} \mathbb{T}(\delta \mathbf{x})^3 \approx \left(\sup_{\delta \mathbf{x}} |\tilde{H}|(\delta \mathbf{x})^2 \right)^{\frac{3}{2}} \quad \forall \delta \mathbf{x} \in \mathbb{R}^2. \tag{8}$$

In [9] we fit the second-order tensor $\tilde{H}$ to the third-order tensor $\mathbb{T}$ by least-squares, and the optimal metric is computed in a similar way to the second-order accurate case. An *a priori* better option for accounting higher-order interpolation error, not tested here, is to apply the strategy of [11].

Remark The *a priori* estimate of Lemma 1 is inspired by the *a posteriori* estimates of Barth and Larson [5] in which the authors explain that the analysis extends to many k-exact approximations of high-order. This is also true for the present *a priori* analysis. In particular, the analogous estimate for a k-exact *discontinuous P_k-Galerkin* approximation writes:

$$(g, \Pi_k W - W_k) \approx \frac{\partial B}{\partial W}(W)(\Pi_k W - W_k, W_k^*), \tag{9}$$

where W_k and W_k^* are the DG$_k$ discrete state and adjoint and Π_k the elementwise interpolation of degree k. $\square$

3.2 High-Order Accurate Unsteady Mesh Adaptation

We illustrate the use of the above estimate (6)–(8) with an application to an unsteady flow. For many propagation phenomena, the discretisation grid (space and time) necessary for a complete representation is very heavy. We consider here an acoustic wave propagation based on the Euler equations.

In order to apply an unsteady mesh adaptation, we adopt the so-called Global-Fixed-Point algorithm [6]. In short, the time interval is divided in n_{adap} sub-intervals. After a computation of state and adjoint on the whole time interval, an optimal space-time metric is evaluated as a set of spatial metrics for each of the time sub-intervals.

In Fig. 5, the propagation of a noise from a road (bottom left) to a balcony (near top, right) around an anti-noise wall (middle of bottom) is computed. The functional is the pressure integral on an interval of the balcony. Since a few wavelengths are emitted at the noise source, the mesh adaptation process will concentrate on the part of the wave train which will hit the balcony. This dramatically reduces the region of the computational domain which needs to be refined. With 30,980 vertices (mean of the 20 meshes used over the time interval) the resolution is about 10 points per half wave and would require five millions vertices if the mesh were a uniform mesh of same maximal fineness. As for Criterion (4), we have measured for this case $\alpha = 2.45$, which is not satisfactory with respect to the theoretical order of approximation, which is 3, but already carries an important improvement with respect to analogous adaptation based on a second-order finite-volume approximation, see [8].

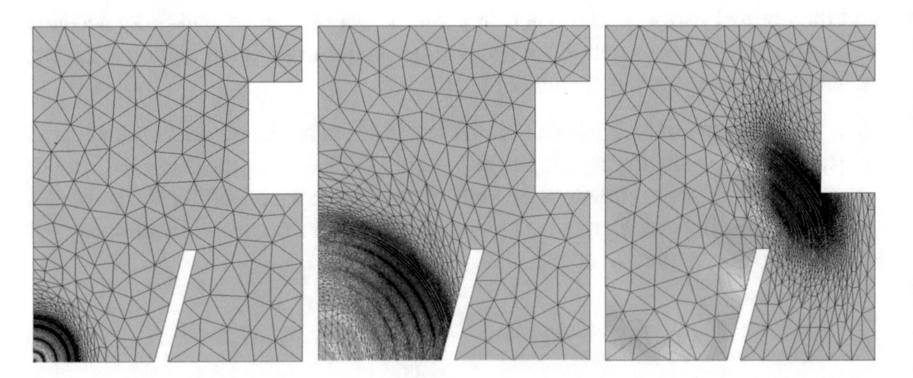

Fig. 5 Goal Oriented unsteady calculation of nonlinear acoustics propagation with third-order goal-oriented adaptation. Pressure at three different time levels and the corresponding meshes

4 Conclusions

The k-exact analysis described in this note allows us to express errors in terms of interpolation errors. This holds for various k-exact approximations like FVM, FEM, DG, ENO. This also holds for three types of adaptation strategies, namely the feature-based, the goal-oriented, and the norm-oriented. Applications with P_1-Galerkin and P_2-CENO approximations are demonstrated. This method can be complemented with a special treatment of singularities, [17], and combined with a FMG process, [7].

Acknowledgements This work has been supported by French National Research Agency (ANR) through project MAIDESC no ANR-13-MONU-0010. This work was granted access to the HPC resources of CINES under the allocations 2017-A0022A05067 and 2017-A0022A06386 made by GENCI (Grand Equipement National de Calcul Intensif).

References

1. 2nd AIAA Sonic Boom Prediction Workshop. Grapevine, Texas (2017)
2. Agouzal, A., Lipnikov, K., Vasilevskii, Y.: Adaptive generation of quasi-optimal tetrahedral meshes. East-West J. **7**(4), 223–244 (1999)
3. Alauzet, F., Loseille, A.: High order sonic boom modeling by adaptive methods. J. Comp. Phys. **229**, 561–593 (2010)
4. Barth, T.J., Frederickson, P.O.: Higher order solution of the Euler equations on unstructured grids using quadratic reconstruction. AIAA Paper 90-13 (1990)
5. Barth, T.J., Larson, M.G.: A-posteriori error estimation for higher order Godunov finite volume methods on unstructured meshes. In: Herbin, R., Kröner, D. (eds.) Finite Volumes for Complex Applications III, pp. 41–63. HERMES Science, London (2002)
6. Belme, A., Dervieux, A., Alauzet, F.: Time accurate anisotropic goal-oriented mesh adaptation for unsteady flows. J. Comp. Phys. **231**(19), 6323–6348 (2012)
7. Brèthes, G., Allain, O., Dervieux, A.: A mesh-adaptative metric-based full multigrid for the Poisson problem. Int. J. Num. Meth. in Fluids **79**(1), 30–53 (2015)
8. Carabias, A.: Analyse et adaptation de maillage pour des schémas non-oscillatoires d'ordre élevé (in French). PhD, Université de Nice-Sophia-Antipolis, France (2013)
9. Carabias, A., Belme, A., Loseille, A., Dervieux, A.: Anisotropic goal-oriented error analysis for a third-order accurate CENO Euler discretization. Int. J. Num. Meth. in Fluids **86**(6), 392–413 (2018)
10. Castro-Díaz, M.J., Hecht, F., Mohammadi, B., Pironneau, O.: Anisotropic unstructured mesh adaptation for flow simulations. Int. J. Num. Meth. in Fluids **25**, 475–491 (1997)
11. Coulaud, O., Loseille, A.: Very high order anisotropic metric-based mesh adaptation in 3D. 25th International Meshing Roundtable. Procedia Eng. **163**, 353–365 (2016)
12. Dompierre, J., Vallet, M.G., Fortin, M., Bourgault, Y., Habashi, W.G.: Anisotropic mesh adaptation: towards a solver and user independent CFD. In: AIAA 35th Aerospace Sciences Meeting and Exhibit, AIAA-1997-0861, Reno (1997)
13. Ferro, N., Micheletti, S., Perotto, S.: Anisotropic mesh adaptation for crack propagation induced by a thermal shock in 2D. Comput. Methods Appl. Mech. Engrg. **331**, 138–158 (2018)
14. Formaggia, L., Perotto, S.: Anisotropic a priori error estimates for elliptic problems. Numer. Math. **94**, 67–92 (2003)

15. Frazza, L., Loseille, A., Alauzet, F. Dervieux, A.: Nonlinear corrector for RANS equations. AIAA 2018-3242 (2018)
16. Frey, P.J., Alauzet, F.: Anisotropic mesh adaptation for CFD computations. Comp. Meth. Applied Mech. Eng. **194**(48–49), 5068–5082 (2005)
17. Gauci, E., Belme, A., Carabias, A., Loseille, A., Alauzet, F., Dervieux, A.: A priori error-based mesh adaptation in CFD (SCPDE17, Hong Kong). J. Methods Appl. Anal. Special Issue (2019, in progress)
18. Huang, W.: Metric tensors for anisotropic mesh generation. J. Comp. Phys. **204**, 633–665 (2005)
19. Ivan, L., Groth, C.P.T.: High-order solution-adaptive central essentially non-oscillatory (CENO) method for viscous flows. J. Comp. Phys. **257**, 830–862 (2014)
20. Loseille, A., Alauzet, F.: Continuous mesh framework. Part I: well-posed continuous interpolation error. SIAM J. Numer. Anal. **49**(1), 38–60 (2011)
21. Loseille, A., Alauzet, F.: Continuous mesh framework. Part II: validations and applications. SIAM Num. Anal. **49**(1):61–86 (2011)
22. Loseille, A., Löhner, R.: Adaptive anisotropic simulations in aerodynamics. In: 48th AIAA Aerospace Sciences Meeting and Exhibit, AIAA-2010-169, Orlando (2010)
23. Loseille, A., Dervieux, A., Alauzet, F.: Anisotropic norm-oriented mesh adaptation for compressible flows. In: 53rd AIAA Aerospace Sciences Meeting. AIAA-2015-2037, Kissimmee (2015)
24. Vasilevski, Y.V., Lipnikov, K.N.: Error bounds for controllable adaptive algorithms based on a Hessian recovery. Computational Mathematics and Mathematical Physics, **45**(8), 1374–1384, (2005)
25. Yano, M., Darmofal, D.: An optimization framework for anisotropic simplex mesh adaptation: application to aerodynamics flows. AIAA Paper, 2012-0079 (2012)

Entropy Stable Discontinuous Galerkin Finite Element Moment Methods for Compressible Fluid Dynamics

M. R. A. Abdelmalik and Harald van Brummelen

Abstract In this work we propose numerical approximations of the Boltzmann equation that are consistent with the Euler and Navier–Stoke–Fourier solutions. We conceive of the Euler and the Navier–Stokes–Fourier equations as moment approximations of the Boltzmann equation in renormalized form. Such renormalizations arise from the so-called Chapman-Enskog analysis of the one-particle marginal in the Boltzmann equation. We present a numerical approximation of the Boltzmann equation that is based on the discontinuous Galerkin method in position dependence and on the renormalized-moment method in velocity dependence. We show that the resulting discontinuous Galerkin finite element moment method is entropy stable. Numerical results are presented for turbulent flow in the lid-driven cavity benchmark.

Keywords Entropy stability · Continuum fluid dynamics · Moment closure · Moment systems · Discontinuous Galerkin finite elements

---

M. R. A. Abdelmalik (✉)
Department of Mechanical Engineering, Eindhoven University of Technology, Eindhoven, Netherlands

Institute of Computational Sciences and Engineering, University of Texas at Austin, Austin, TX, USA
e-mail: m.abdel.malik@tue.nl

H. van Brummelen
Department of Mechanical Engineering, Eindhoven University of Technology, Eindhoven, Netherlands
e-mail: E.H.v.Brummelen@tue.nl

H. van Brummelen et al. (eds.), *Numerical Methods for Flows*,
Lecture Notes in Computational Science and Engineering 132,
https://doi.org/10.1007/978-3-030-30705-9_8

1 Introduction

The Boltzmann equation provides a description of the molecular dynamics of fluid flows based on their one-particle phase-space distribution. However, the Boltzmann equation also encapsulates all conventional continuum flow models in the sense that its hydrodynamic-limit solutions correspond to solutions of the compressible Euler and Navier–Stokes equations [3, 11], the incompressible Euler and Navier–Stokes equations [14, 18], the incompressible Stokes equations [19] and the incompressible Navier–Stokes–Fourier system [17]; see [24] for an overview.

Numerical approximation of the Boltzmann equation poses a formidable challenge, on account of its high-dimensional phase-space setting since the one-particle marginal depends on time, position and microscopic velocity [7]. Therefore, in D spatial dimensions, the one-particle phase-space is $2D + 1$ dimensional. However, when interest is restricted to the continuum flow regime, lower-dimensional continuum flow models may be used. Continuum flow models, such as Euler and Navier–Stokes, may be understood as weighted velocity averages of the Boltzmann equation [7]. Such an averaging engenders a system of evolution equations for the moments of the one-particle distribution.

The relationship between the Boltzmann equation and the continuum flow models may be used to design numerical schemes for the latter. Such schemes have been developed for the compressible Euler and Navier–Stokes–Fourier equations; see for example [8, 9] for the so-called kinetic flux-vector splitting methods following the Chapman-Enskog expansion, [20, 22] for the so-called gas-kinetic schemes that follow a Hilbert expansion. In [4] the Boltzmann equation was used to design a discontinuous Galerkin finite element method for the compressible Euler equations and the so-called 10-moment systems.

In this work we construct entropy stable discontinuous Galerkin finite element (DGFE) moment methods to approximate solutions of the compressible Euler and Navier–Stokes–Fourier equations. We show that such compressible equations can be conceived of as renormalized solutions of the Boltzmann equation, where the renormalization map is derived from the Chapman-Enskog analysis. In comparison with previous work in [8, 9] we show that the discontinuous Galerkin approximation only requires the upwinded distribution to result in an entropy stable numerical method. In comparison to the work in [4] for the compressible Euler equations we do not require Gauss-quadrature methods to approximate the edge-wise integrals, instead all such integrals presented in this work are carried out analytically.

The remainder of this work is organized as follows. Section 2 introduces the Boltzmann equation and its properties that facilitate the Chapman-Enskog analysis to derived the compressible Euler and Navier–Stokes–Fourier equations. Section 3 introduces the reinterpretation of the compressible Euler and Navier–Stokes–Fourier equations as Galerkin approximations of a renormalized Boltzmann equation. Section 3 proceeds to construct DGFE methods for the compressible equations and shows that the resulting methods are entropy stable. In Sect. 4 we demonstrate the proposed method by numerically simulating the lid-driven cavity benchmark. Finally, Sect. 5 presents a concluding discussion.

2 The Boltzmann Equation

We study the evolution of the so-called one-particle marginal, denoted by f and governed by

$$\partial_t f + v_i \partial_{x_i} f = \mathcal{C}(f), \tag{1}$$

where the collision operator $f \mapsto \mathcal{C}(f)$ acts only on the velocity, $\boldsymbol{v} = (v_1, \ldots, v_D)$, dependence of f locally at each time t and position $\boldsymbol{x}$. We consider a class of collision operators $f \mapsto \mathcal{C}$ that possess certain conservation, dissipation and symmetry properties, viz., conservation of mass, momentum and energy, dissipation of appropriate entropy functionals and invariance under Galilean transformations. These fundamental properties are treated in further detail below. Our treatment of these properties is standard (see, for instance, [16]) and is presented merely for coherence and completeness.

2.1 *Properties of the Collision Operator*

To elaborate the conservation properties of the collision operator, let $\langle\cdot\rangle$ denote integration in the velocity dependence of any scalar, vector or matrix valued measurable function over D-dimensional Lebesgue measure. A function $\psi : \mathbb{R}^D \to \mathbb{R}$ is called a collision invariant of $\mathcal{C}$ if

$$\langle \psi\, \mathcal{C}(f) \rangle = 0 \quad \forall f \in \mathscr{D}(\mathcal{C}), \tag{2}$$

where $\mathscr{D}(\mathcal{C}) \subset L^1(\mathbb{R}^D, \mathbb{R}_{\geq 0})$ denotes the domain of $\mathcal{C}$. Equation (1) associates a scalar conservation law with each collision invariant:

$$\partial_t \langle \psi f \rangle + \partial_{x_i} \langle v_i \psi f \rangle = 0. \tag{3}$$

We require that $\{1, v_1, \ldots, v_D, |\boldsymbol{v}|^2\}$ are collision invariants of $\mathcal{C}$ and that the span of this set contains all collision invariants, i.e.

$$\langle \psi\, \mathcal{C}(f) \rangle = 0 \quad \forall f \in \mathscr{D}(\mathcal{C}) \quad \Leftrightarrow \quad \psi \in \operatorname{span}\{1, v_1, \ldots, v_D, |\boldsymbol{v}|^2\} =: \mathscr{I}. \tag{4}$$

To elucidate the conservation laws in (3), we introduce a parametrization of the moment densities and fluxes according to

$$\langle f\rangle = \rho, \tag{5a}$$

$$\langle v_i f\rangle = \rho u_i, \tag{5b}$$

$$\langle |\boldsymbol{v}|^2 f\rangle = \rho|\boldsymbol{u}|^2 + (D+2)\rho\theta, \tag{5c}$$

$$\langle v_i v_j f\rangle = \rho u_i u_j + \rho\theta\delta_{ij} + \sigma_{ij}, \tag{5d}$$

$$\langle |\boldsymbol{v}|^2 v_i f\rangle = \rho|\boldsymbol{u}|^2 u_i + (D+2)\rho\theta u_i + \sigma_{ij}u_j + q_i, \tag{5e}$$

where ρ, $\boldsymbol{u}$, θ, $\boldsymbol{\sigma}$ and $\boldsymbol{q}$ denote the macroscopic fluid density, velocity, temperature, stress and heat flux respectively, and δ_{ij} denotes the Kronecker delta. We infer that the moments $\langle f\rangle$, $\langle v_i f\rangle$ and $\langle |\boldsymbol{v}|^2 f\rangle$, correspond to the mass density, the (components of) momentum-density and energy-density, respectively. Accordingly, the conservation law (3) implies that solutions of (1) conserve mass, momentum and energy, i.e. solutions of (1) satisfy

$$\partial_t\rho + \partial_{x_i}(\rho u_i) = 0 \tag{6a}$$

$$\partial_t(\rho u_j) + \partial_{x_i}(\rho u_i u_j + \rho\theta + \sigma_{ij}) = 0 \tag{6b}$$

$$\partial_t(\rho|\boldsymbol{u}|^2 + D\rho\theta) + \partial_{x_i}(\rho|\boldsymbol{u}|^2 u_i + (D+2)\rho\theta u_i + \sigma_{ij}u_j + q_i) = 0. \tag{6c}$$

Note that the system of conservation laws (6) is not closed since there are $(D^2 + 3D)/2$ independent variables and only $D+2$ relations. The closure of (6) requires a constitutive modeling assumption that characterizes σ_{ij} and q_i in terms of ρ, u_i and θ.

The entropy dissipation property of $\mathcal{C}$ follows from the local dissipation relation assumption

$$\langle \ln(f)\,\mathcal{C}(f)\rangle \le 0 \quad \forall f \in \mathscr{D}(\mathcal{C}). \tag{7}$$

Note that equality in (7) holds if and only if $\ln(f) \in \mathscr{I}$ in accordance with (4). In addition, we assume that for every $f \in \mathscr{D}(\mathcal{C})$ the following equivalences hold:

$$\mathcal{C}(f) = 0 \quad \Leftrightarrow \quad \langle \ln(f)\,\mathcal{C}(f)\rangle = 0 \quad \Leftrightarrow \quad \ln(f) \in \mathscr{I} \tag{8}$$

Relation (7) leads to an abstraction of Boltzmann's H-theorem for (1): weighting the Boltzmann equation (1) with $\ln(f)$, which is the derivative of the entropy $f\ln(f) - f$, asserts that solutions of the Boltzmann equation (1) satisfy the local entropy-dissipation law:

$$\partial_t\langle f\ln(f) - f\rangle + \partial_{x_i}\langle v_i(f\ln(f) - f)\rangle = \langle \ln(f)\,\mathcal{C}(f)\rangle \le 0\,. \tag{9}$$

The first equivalence in (8) characterizes local equilibria of $\mathcal{C}$ by vanishing entropy dissipation. By virtue of (4), the second equivalence indicates that the form of such local equilibria is given by the so-called Maxwellian distributions

$$\mathcal{M}_{(\rho,\boldsymbol{u},\theta)}(\boldsymbol{v}) := \frac{\rho}{(2\pi\theta)^{\frac{D}{2}}} \exp\left(-\frac{|\boldsymbol{v}-\boldsymbol{u}|^2}{2\theta}\right) \tag{10}$$

for some $(\rho, \boldsymbol{u}, \theta) \in \mathbb{R}_{>0} \times \mathbb{R}^D \times \mathbb{R}_{>0}$, therefore it holds that $\log \mathcal{M} \in \mathscr{I}$.

The assumed symmetry properties of the collision operator pertain to commutation with translational and rotational transformations. In particular, for all vectors $\boldsymbol{u} \in \mathbb{R}^D$ and all orthogonal tensors $\boldsymbol{O} : \mathbb{R}^D \to \mathbb{R}^D$, we define the translation transformation $T_{\boldsymbol{u}} : \mathscr{D}(\mathcal{C}) \to \mathscr{D}(\mathcal{C})$ and the rotation transformation $T_{\boldsymbol{O}} : \mathscr{D}(\mathcal{C}) \to \mathscr{D}(\mathcal{C})$ by:

$$(T_{\boldsymbol{u}} f)(\boldsymbol{v}) = f(\boldsymbol{u} - \boldsymbol{v}) \quad \forall f \in \mathscr{D}(\mathcal{C}) \tag{11}$$

$$(T_{\boldsymbol{O}} f)(\boldsymbol{v}) = f(\boldsymbol{O}^*\boldsymbol{v}) \quad \forall f \in \mathscr{D}(\mathcal{C}) \tag{12}$$

with $\boldsymbol{O}^*$ the Euclidean adjoint of $\boldsymbol{O}$. Note that the above transformations act on the $\boldsymbol{v}$-dependence only. It is assumed that $\mathcal{C}$ possesses the following symmetries:

$$\mathcal{C}(\mathcal{T}_{\boldsymbol{u}} f) = \mathcal{T}_{\boldsymbol{u}} \mathcal{C}(f), \quad \mathcal{C}(\mathcal{T}_{\boldsymbol{O}} f) = \mathcal{T}_{\boldsymbol{O}} \mathcal{C}(f). \tag{13}$$

The symmetries (13) imply that (1) complies with Galilean invariance, i.e. if $f(t, \boldsymbol{x}, \boldsymbol{v})$ satisfies the Boltzmann equation (1), then so do $f(t, \boldsymbol{x} - \boldsymbol{u}t, \boldsymbol{v} - \boldsymbol{u})$ and $f(t, \boldsymbol{O}^*\boldsymbol{x}, \boldsymbol{O}^*\boldsymbol{v})$.

2.2 Properties of the Linearized Collision Operator

Deriving the Euler and Navier–Stokes–Fourier equations from (1) involves a perturbation analysis, and consequent linearization, of (1) about Maxwellian distributions (10). To facilitate such an analysis we elaborate the properties of the collision operator linearized about and modulated by $\mathcal{M}$ in the direction of g

$$\mathcal{L}_{\mathcal{M}}(g) = \frac{\partial_\delta \mathcal{C}(\mathcal{M} + \delta g)|_{\delta=0}}{\mathcal{M}}. \tag{14}$$

The conservation, dissipation and symmetry properties of $\mathcal{L}$ follow from those of $\mathcal{C}$ [16]. The fact that conservation laws (3) are retained in the linearized setting of (1) follows from the vanishing of the variation of $\mathcal{C}(\mathcal{M})$ over all Maxwellians since $\mathcal{C}(\mathcal{M}) = 0$, therefore $\mathscr{I} \subset \ker(\mathcal{L})$.

The entropy dissipation of (1) in the linearized setting follows from the local dissipation relation

$$\langle g\mathcal{M}\mathcal{L}_{\mathcal{M}}(g)\rangle = \frac{1}{2}\partial_{\delta^2}^2\langle \ln(\mathcal{M}(1+\delta g))\mathcal{C}(\mathcal{M}(1+\delta g))\rangle|_{\delta=0} \leq 0. \tag{15}$$

The first equality in (15) follows from the fact that the first and second variations of $\mathcal{C}(\mathcal{M})$ over all Maxwellians vanish. The last inequality follows from (7) and (8) which show that the relation in (7) attains a maximum when $f = \mathcal{M}$.

To elucidate the Galilean symmetry of $\mathcal{L}_{\mathcal{M}}$, consider the orthogonal transformation [16] $\mathcal{O}_{\boldsymbol{O}} \equiv \mathcal{T}_{\boldsymbol{u}}\mathcal{T}_{\boldsymbol{O}}\mathcal{T}_{\boldsymbol{u}}^{-1}$. It follows from linearizing (13) about $\mathcal{M}$ that

$$\mathcal{O}_{\boldsymbol{O}}\mathcal{L}_{\mathcal{M}}g = \mathcal{L}_{\mathcal{M}}\mathcal{O}_{\boldsymbol{O}}g \tag{16}$$

In addition, to guarantee invertibility of $\mathcal{L}_{\mathcal{M}}$, we assume that $\mathcal{L}_{\mathcal{M}}$ satisfies a Fredholm alternative under the following conditions

(i) $\ker(\mathcal{L}_{\mathcal{M}}) \equiv \mathscr{I}$,
(ii) $\mathcal{L}_{\mathcal{M}}$ is self adjoint,
(iii) $\ker(\mathcal{L}_{\mathcal{M}}) \oplus \mathrm{rng}(\mathcal{L}_{\mathcal{M}})$ is a Hilbert space with an inner product weighted with $\mathcal{M}$.

2.3 *Hydrodynamic Limits*

We conclude this section by deriving moment-closure approximations to the macroscopic conservation laws (3) that lead to the compressible Euler and Navier–Stokes equations. To that end we consider an asymptotic expansion of f that solves (1) in re-scaled form given by

$$f(t, \boldsymbol{x}, \boldsymbol{v}) = f_\epsilon(\hat{t}, \hat{\boldsymbol{x}}, \boldsymbol{v}), \text{ with } (\hat{t}, \hat{\boldsymbol{x}}) = (\epsilon t, \epsilon \boldsymbol{x}), \tag{17}$$

where we take ϵ to be the so-called Knudsen number which is a measure of rarefaction of the fluid flow. The re-scaled Boltzmann equation (1) reads

$$\partial_{\hat{t}} f_\epsilon + v_i \partial_{\hat{x}_i} f_\epsilon = \frac{1}{\epsilon}\mathcal{C}(f_\epsilon) \tag{18}$$

where we assume that

$$\epsilon \ll 1 \tag{19}$$

to derive a continuum description of fluid dynamics. To that end, we consider a Chapman-Enskog expansion of f_ϵ in powers of ϵ:

$$f_\epsilon(\hat{t}, \hat{\boldsymbol{x}}, \boldsymbol{v}) = \mathcal{M}(f_\epsilon) \sum_{k \geq 0} \epsilon^k g_k(t, \hat{\boldsymbol{x}}, \hat{\boldsymbol{v}}) \quad \text{for } i = 0, 1, \ldots, D, D+1, \tag{20}$$

where the fluctuations g_k are functions of $\boldsymbol{v}$ that also depend on $(\hat{t}, \hat{\boldsymbol{x}})$ through $\rho(\hat{t}, \hat{x})$, $\boldsymbol{u}(\hat{t}, \hat{\boldsymbol{x}})$, and $\theta(\hat{t}, \hat{\boldsymbol{x}})$, and their $\hat{x}$-derivatives evaluated at $(\hat{t}, \hat{\boldsymbol{x}})$. Furthermore, $\mathcal{M}(f_\epsilon)$ is the local Maxwellian conforming to (10) with the same invariant moments as f_ϵ:

$$\langle \psi_j \mathcal{M}(f_\epsilon) \rangle = \langle \psi_j f_\epsilon \rangle. \tag{21}$$

where $\psi_0 = 1$, $\psi_j = v_j$ $(j \in \{1, \ldots, D\})$, $\psi_{D+1} = |\boldsymbol{v}|^2$. We aim to derive fluid dynamic equations governing the evolution of the invariant moments from the leading order terms in (20).

Remark 1 We implicitly assume that the coefficients g_k in (20) are smooth and rapidly decaying for $|\boldsymbol{v}| \to \infty$.

Constraints for the expansion (20) follow from the conservation of moments $\langle \psi_i f_\epsilon \rangle$ in the sense that $\langle \psi_i f_\epsilon \rangle$ satisfy the formal system of conservation laws of the form (3):

$$\partial_{\hat{t}} \langle \psi_j f_\epsilon \rangle + \partial_{\hat{x}_i} \langle v_i \psi_j f_\epsilon \rangle = 0. \tag{22}$$

Closure of (22) would correspond to a system of macroscopic conservation laws that govern the evolution of the fluid density, momentum and energy. The closure of (22), corresponding to the compressible Euler and Navier–Stokes equations, is obtained by substituting (20) into (18) and comparing the leading order coefficients of ϵ.

2.4 Euler Equations

The leading order terms yield

$$\mathcal{C}(\mathcal{M} g_0) = 0 \implies g_0[\langle \psi_i f_\epsilon \rangle (\hat{t}, \hat{\boldsymbol{x}})](\boldsymbol{v}) = 1 \tag{23}$$

Therefore, the leading order conservation laws (22) formally satisfy

$$\partial_{\hat{t}} \langle \psi_j \mathcal{M}(f_\epsilon) \rangle + \partial_{\hat{x}_i} \langle v_i \psi_j \mathcal{M}(f_\epsilon) \rangle = 0 \quad \text{up to } O(\epsilon). \tag{24}$$

Neglecting higher order coefficients of ϵ, (24) reduces to the compressible Euler equations [16]

$$\partial_t \rho + \partial_{x_i}(\rho u_i) = 0 \tag{25a}$$

$$\partial_t(\rho u_j) + \partial_{x_i}(\rho u_i u_j + \rho\theta\delta_{ij}) = 0 \tag{25b}$$

$$\partial_t(\rho|\boldsymbol{u}|^2 + D\rho\theta) + \partial_{x_i}(\rho|\boldsymbol{u}|^2 u_i + (D+2)\rho\theta u_i) = 0. \tag{25c}$$

One may note that the closure relation (23) corresponds to an ideal gas.

2.4.1 Navier–Stokes–Fourier Equations

The first correction to (23) is governed by

$$(\partial_{\hat{t}} + v_i\partial_{\hat{x}_i})\mathcal{M}(f_\epsilon) = \epsilon^{-1}\mathcal{C}\left(\mathcal{M}(f_\epsilon)\,(1+\epsilon g_1) + O(\epsilon^2)\right) + O(\epsilon) \implies (\partial_{\hat{t}} + v_k\partial_{\hat{x}_k})\ln\mathcal{M}(f_\epsilon)$$
$$= \mathcal{L}_{\mathcal{M}(f_\epsilon)}(g_1) + O(\epsilon). \tag{26}$$

From the assumptions in Sect. 2.2 we note that the evolution of $\ln\mathcal{M}(f_\epsilon)$ is contained in $\mathscr{I}^\perp$. Therefore, introducing the orthogonal projection $\Pi_{\mathscr{I}}$ onto $\mathscr{I}$ [16],

$$\Pi_{\mathscr{I}}(g) = \frac{\langle\mathcal{M}g\rangle}{\rho} + \frac{(\boldsymbol{v}-\boldsymbol{u})\cdot\langle(\boldsymbol{v}-\boldsymbol{u})\mathcal{M}g\rangle}{\rho\theta} + \left(\frac{|\boldsymbol{v}-\boldsymbol{u}|^2}{2\theta} - \frac{D}{2}\right)\frac{2}{D}\left\langle\left(\frac{|\boldsymbol{v}-\boldsymbol{u}|^2}{2\theta} - \frac{D}{2}\right)\mathcal{M}g\right\rangle, \tag{27}$$

we can simplify relation (26) by noting that

$$\mathcal{L}_{\mathcal{M}}(g_1) = (\mathrm{Id} - \Pi_{\mathscr{I}})\frac{v_k\partial_{\hat{x}_k}\mathcal{M}}{\mathcal{M}} + O(\epsilon) \tag{28a}$$

$$= \frac{1}{2\theta}A_{ij}(\boldsymbol{v}-\boldsymbol{u})\left(\partial_{\hat{x}_i}u_j + \partial_{\hat{x}_j}u_i - 2\frac{\partial_{\hat{x}_k}u_k}{D}\delta_{ij}\right) + B_i(\boldsymbol{v}-\boldsymbol{u})\frac{\partial_{\hat{x}_i}\theta}{\theta} + O(\epsilon), \tag{28b}$$

where

$$A_{ij}(\boldsymbol{v}) = v_i v_j - \frac{|\boldsymbol{v}|^2}{D}\delta_{ij}, \quad B_i(\boldsymbol{v}) = \left(\frac{|\boldsymbol{v}|^2}{2\theta} - \frac{D+2}{2}\right)v_i. \tag{29}$$

The solution to (28a) is written as

$$g_1 = \mathcal{L}_{\mathcal{M}}^{-1}(\mathrm{Id} - \Pi_{\mathscr{I}})\frac{v_k\partial_{\hat{x}_k}\mathcal{M}}{\mathcal{M}} + O(\epsilon), \tag{30}$$

where the well-posedness of (30) is guaranteed by the assumptions on $\mathcal{L}_{\mathcal{M}}$ in Sect. 2.2. Therefore, the next correction to the compressible Euler equations (24) is

$$\partial_{\hat{t}}\langle\psi_j\mathcal{M}(f_\epsilon)\rangle + \partial_{\hat{x}_i}\langle v_i\psi_j\mathcal{M}(f_\epsilon)\rangle + \epsilon\partial_{\hat{x}_i}\langle v_i\psi_j\mathcal{M}(f_\epsilon)g_1\rangle = 0 \quad \text{up to } O(\epsilon^2), \tag{31}$$

where we used (21). Neglecting higher order coefficients of ϵ, (31) reduces to the compressible Navier–Stokes–Fourier equations

$$\partial_t\rho + \partial_{x_i}(\rho u_i) = 0 \tag{32a}$$

$$\partial_t(\rho u_j) + \partial_{x_i}(\rho u_i u_j + \rho\theta\delta_{ij}) = \epsilon\partial_{x_i}\sigma_{ij} \tag{32b}$$

$$\partial_t(\rho|\boldsymbol{u}|^2 + D\rho\theta) + \partial_{x_i}(\rho|\boldsymbol{u}|^2u_i + (D+2)\rho\theta u_i) = \epsilon\partial_{x_i}(\sigma_{ij}u_j + q_i), \tag{32c}$$

To elucidate the fluid-dynamic interpretation of the first order correction g_1 we infer from (5d) and (5e) that

$$\sigma_{ij} = -\frac{\left\langle A_{ij}(\boldsymbol{v}-\boldsymbol{u})\mathcal{M}\mathcal{L}_{\mathcal{M}}^{-1}A_{kl}(\boldsymbol{v}-\boldsymbol{u})\right\rangle}{2\theta}\left(\partial_{\hat{x}_k}u_l + \partial_{\hat{x}_l}u_k - 2\frac{\partial_{\hat{x}_m}u_m}{D}\delta_{kl}\right), \tag{33a}$$

$$q_i = -\left\langle B_i(\boldsymbol{u}-\boldsymbol{v})\mathcal{M}\mathcal{L}_{\mathcal{M}}^{-1}B_j(\boldsymbol{v}-\boldsymbol{u})\right\rangle\partial_{\hat{x}_j}\theta, \tag{33b}$$

where we have used the observation that $\mathcal{L}_{\mathcal{M}}^{-1}$ preserves the even or odd symmetry of a function. This symmetry preservation follows from the special case $\boldsymbol{O} = -\mathrm{Id}$ in (13). Therefore, $\langle A_{ij}(\boldsymbol{v}-\boldsymbol{u})\mathcal{M}\mathcal{L}_{\mathcal{M}}^{-1}B_i(\boldsymbol{v}-\boldsymbol{u})\rangle$ and $\langle B_i(\boldsymbol{v}-\boldsymbol{u})\mathcal{M}\mathcal{L}_{\mathcal{M}}^{-1}A_{ij}(\boldsymbol{v}-\boldsymbol{u})\rangle$ vanish identically, since odd central moments of $\mathcal{M}$ vanish. Using the Galilean symmetry of $\mathcal{L}_{\mathcal{M}}$ and $\mathrm{tr}(A) = 0$, the relations in (33) may be simplified to read

$$\sigma_{ij} = -\omega\left(\partial_{x_j}u_i + \partial_{x_i}u_j - 2\frac{\partial_{x_k}u_k}{D}\delta_{ij}\right), \quad \omega = \frac{\left\langle A_{ij}(\boldsymbol{v}-\boldsymbol{u})\mathcal{M}\mathcal{L}_{\mathcal{M}}^{-1}A_{ij}(\boldsymbol{v}-\boldsymbol{u})\right\rangle}{(D-1)(D-2)} \tag{34a}$$

$$q_i = -\gamma\partial_{x_i}\theta, \quad \gamma = \langle B_i(\boldsymbol{u}-\boldsymbol{v})\mathcal{M}\mathcal{L}_{\mathcal{M}}^{-1}B_i(\boldsymbol{u}-\boldsymbol{v})\rangle \tag{34b}$$

where ω and γ denote the viscosity and heat conduction, respectively.

Remark 2 The closure relation in (34) represents Fourier's law of heat conduction [13] for q and the Newtonian stress tensor for a compressible fluid that satisfies Stokes hypothesis [25].

Remark 3 Higher order corrections to the compressible Navier–Stokes may be formally derived from subsequent terms in (20). However these further corrections are, in general, not well posed.

In this work, we restrict our interest to the standard BGK collision operator [5], viz.

$$\mathcal{C}(f) = \tau^{-1}(\mathcal{M}_f - f), \tag{35}$$

where $\mathcal{M}_f$ denotes the local Maxwellian (10) having the same invariant moments as f and τ is the relaxation rate. The corresponding viscosity and heat conduction are [6]

$$\omega = \tau\rho\theta, \quad \gamma = \frac{D+2}{2}\tau\rho\theta. \tag{36}$$

3 Velocity-Space-Time Galerkin Approximation

In this section we reinterpret the compressible Euler (25) and Navier–Stokes–Fourier (32) equations as Galerkin approximations, in velocity dependence, of the Boltzmann equation in renormalized form. We show that such a reinterpretation provides a natural way to stabilize the corresponding formulations and to weakly impose boundary conditions. In the spatial dependence we present an isogeometric analysis as well as a discontinuous Galerkin finite element approximation.

3.1 Velocity Discretization of a Renormalized Boltzmann Equation

The Galerkin approximation is based on a moment-system approximation in velocity dependence. In contrast to the moment system approximation in [1, 2], we consider moment-closure approximations that correspond to the Chapman-Enskog analysis (20).

Our semi-discretization of the Boltzmann equation with respect to the velocity dependence is based on velocity moments of the one-particle marginal. These velocity moments are defined over $\mathbb{R}^D$, and therefore we regard finite dimensional approximations of $f(t, \boldsymbol{x}, \boldsymbol{v})$ in (1) that are integrable over $\mathbb{R}^D$ in velocity dependence. To that end, we consider a Galerkin subspace approximation of the Boltzmann equation in renormalized form, where the renormalization maps to integrable functions. To elucidate the renormalization, let $\mathscr{M}$ denote an M-

dimensional subspace of D-variate polynomials and let $\{m_i(\boldsymbol{v})\}_{i=1}^{M}$ represent a corresponding basis. We consider the renormalization map $\beta : \mathscr{M} \to \mathscr{F}$, where

$$\mathscr{F}_{\mathscr{M}} := \left\{ f \in \mathscr{D}(C) : \quad mf \in L^1(\mathbb{R}^D),\ \boldsymbol{v}mf \in L^1(\mathbb{R}^D, \mathbb{R}^D),\ mC(f) \in L^1(\mathbb{R}^D) \quad \forall m \in \mathscr{M} \right\}. \tag{37}$$

The leading order term (23) in the Chapman-Enskog analysis (20) may be understood as a renormalization map that conforms to

$$\beta_E(v) = e^{z(\boldsymbol{v})} \tag{38}$$

such that $\beta_E : \mathscr{I} \to \mathscr{F}_{\mathscr{I}}$, where

$$z = \ln \frac{\rho}{(2\pi\theta)^{\frac{D}{2}}} - \frac{|\boldsymbol{u}|^2}{2\theta} + \frac{u_i}{\theta} v_i - \frac{|\boldsymbol{v}|^2}{2\theta}. \tag{39}$$

Therefore, the Euler equations (25) system can then be written in the Bubnov-Galerkin form:

$$\begin{aligned} \textit{Find } z \in \mathscr{L}\big((0,T) \times \Omega; \mathscr{I}\big) : \quad & \partial_t \langle m\beta_E(\boldsymbol{v})\rangle + \partial_{x_i} \langle m v_i \beta_E(\boldsymbol{v})\rangle = 0 \\ & \forall m \in \mathscr{I} \text{ a.e. } (t, \boldsymbol{x}) \in (0,T) \times \Omega. \end{aligned} \tag{40}$$

where $\mathscr{L}\big((0,T) \times \Omega; \mathscr{I}\big)$ represents a suitable vector space of functions from the considered time interval $(0,T)$ and spatial domain Ω into $\mathscr{I}$. The usual symmetry and conservation properties of the Boltzmann equation are generally retained in (40) by the choice of the subspace $\mathscr{I}$, namely that $\mathscr{I}$ constitutes the collision invariants (see (4)), and is closed under the actions of $\mathcal{T}_{\boldsymbol{u}}$ and $\mathcal{T}_{\boldsymbol{O}}$; cf. (13). Entropy conservation follows directly from Galerkin orthogonality in (40) with $\ln \beta_E(\boldsymbol{v}) = z \in \mathscr{I}$, therefore we have that

$$\partial_t \langle \eta(z) \rangle + \partial_{x_i} \langle v_i \eta(z) \rangle = 0. \tag{41}$$

where

$$\eta(z) = ze^z - e^z. \tag{42}$$

The first order distribution with (30) in the Chapman-Enskog analysis (20) may be understood as a renormalization map that conforms to

$$\beta_{NS}(\boldsymbol{v}) = e^{\Pi_{\mathscr{I}}(z(\boldsymbol{v}))}(1 + (\mathrm{Id} - \Pi_{\mathscr{I}})z(\boldsymbol{v})) \tag{43}$$

such that $\beta_{NS} : \mathscr{M} \to \mathscr{F}_{\mathscr{M}}$, where

$$z = \ln \frac{\rho}{(2\pi\theta)^{\frac{D}{2}}} - \frac{|\boldsymbol{u}|^2}{2\theta} + \frac{u_i}{\theta} v_i - \frac{|\boldsymbol{v}|^2}{2\theta} - \frac{\sigma_{ij}}{2\rho\theta^2}(v_i - u_i)(v_j - u_j) + \frac{q_i}{\rho\theta^2}\left(\frac{|\boldsymbol{v}-\boldsymbol{u}|^2}{(D+2)\theta} - 1\right)(v_i - u_i) \tag{44}$$

and $\mathscr{M} \subset \operatorname{span}\{1, v_i, v_i v_j, |\boldsymbol{v}|^2 v_i\}$ such that z only depends on ρ, $\boldsymbol{u}$, θ and their derivatives; see (20). Therefore, the Navier–Stokes–Fourier equations (32) system can then be written in the Petrov-Galerkin form:

$$\textit{Find } z \in \mathscr{L}\big((0,T)\times\Omega;\mathscr{M}\big) : \quad \partial_t \langle m\beta_{NS}(\boldsymbol{v})\rangle + \partial_{x_i}\langle m v_i \beta_{NS}(\boldsymbol{v})\rangle = 0 \quad \forall m \in \mathscr{I} \text{ a.e. } (t,\boldsymbol{x}) \in (0,T)\times\Omega. \tag{45}$$

To show entropy dissipation for the Navier–Stokes–Fourier equations we simplify the relation in (45) using (21) to read

$$\partial_t \langle m e^{\Pi_{\mathscr{I}} z}\rangle + \partial_{x_i}\langle m v_i e^{\Pi_{\mathscr{I}} z}\rangle + \partial_{x_i}\langle m v_i e^{\Pi_{\mathscr{I}} z}(1-\Pi_{\mathscr{I}})z\rangle = 0. \tag{46}$$

We use the Galerkin orthogonality with $\Pi_{\mathscr{I}}(z) \in \mathscr{I}$ to write

$$\partial_t\big\langle \eta(\Pi_{\mathscr{I}} z)\big\rangle + \partial_{x_i}\big\langle v_i \eta(\Pi_{\mathscr{I}} z)\big\rangle = -\big\langle (\Pi_{\mathscr{I}} z) v_i \partial_{x_i}\big(e^{\Pi_{\mathscr{I}} z}(1-\Pi_{\mathscr{I}})z\big)\big\rangle. \tag{47}$$

The right hand side of (47) may be rewritten using chain rule to read

$$\big\langle (\Pi_{\mathscr{I}} z) v_i \partial_{x_i}\big(e^{\Pi_{\mathscr{I}} z}(1-\Pi_{\mathscr{I}})z\big)\big\rangle = \partial_x \big\langle v_i (\Pi_{\mathscr{I}} z) e^{\Pi_{\mathscr{I}} z}(1-\Pi_{\mathscr{I}})z\big\rangle - \big\langle \big(e^{\Pi_{\mathscr{I}} z}(1-\Pi_{\mathscr{I}})z\big) v_i \partial_{x_i}\big(\Pi_{\mathscr{I}} z\big)\big\rangle. \tag{48}$$

Recalling (30), we rewrite the ultimate term of (48) as

$$\big\langle \big(e^{\Pi_{\mathscr{I}} z}(1-\Pi_{\mathscr{I}})z\big) v_i \partial_{x_i}\big(\Pi_{\mathscr{I}} z\big)\big\rangle = \Big\langle v_i \partial_{x_i}\big(\Pi_{\mathscr{I}} z\big) e^{\Pi_{\mathscr{I}} z} \mathcal{L}_{\mathcal{M}}^{-1} v_j \partial_{x_j}\big(\Pi_{\mathscr{I}} z\big)\Big\rangle \leq 0 \tag{49}$$

where the inequality follows from (15). Collecting the relations in (48) and (49) into (47) yields the entropy dissipation relation

$$\partial_t\big\langle \eta(\Pi z)\big\rangle + \partial_{x_i}\big\langle v_i \eta(\Pi_{\mathscr{I}} z)\big\rangle + \partial_{x_i}\big\langle v_i \big(\Pi_{\mathscr{I}} z\big) e^{\Pi_{\mathscr{I}} z}(1-\Pi_{\mathscr{I}})z\big\rangle \leq 0. \tag{50}$$

3.2 Space-Time Discontinuous Galerkin Approximation

In this section we present the space-time Galerkin approximation to the Euler and Navier–Stokes–Fourier equations in (40) and (45) written as velocity Galerkin approximations of a renormalized Boltzmann equation. We use the discontinuous Galerkin approximation to discretize the position and time dependence. We show that using the upwind distribution [1], and the corresponding numerical flux, the resulting DGFE moment approximation is entropy stable.

Let $\mathcal{H} := \{h_1, h_2, \ldots\} \subset \mathbb{R}_{>0}$ denote a strictly decreasing sequence of mesh parameters whose only accumulation point is 0. Consider a corresponding mesh sequence $\mathcal{T}^{\mathcal{H}}$, viz., a sequence of covers of the domain by non-overlapping element domains $\kappa \subset \Omega$. We impose on $\mathcal{T}^{\mathcal{H}}$ the standard conditions of regularity, shape-regularity and quasi-uniformity with respect to $\mathcal{H}$; see, for instance, [10] for further details. To introduce the DGFE approximation space, let $\mathcal{P}_p(\kappa)$ denote the set of D-variate polynomials of degree at most p in an element domain $\kappa \subset \mathbb{R}^D$ and by $I^n \equiv]t^n, t^{n+1}[$ the n-th time-interval in the time-domain $(0, T)$, with $n \in \mathbb{Z}_{\geq 0}$, such that $0 < t_1 < t_2 < \cdots < T$. For any $h \in \mathcal{H}$, we indicate by $V^{h,p}((0, T) \times \Omega)$ the DGFE approximation space

$$V^{h,p}((0,T) \times \Omega) = \{g \in L^2((0,T) \times \Omega) : \ g|_{\kappa \times I^n} \in \mathcal{P}_p(\kappa \times I^n), \ \forall \kappa \in \mathcal{T}^h\}, \tag{51}$$

and by $V^{h,p}(\cdot, \mathscr{M})$ the extension of $V^{h,p}(\cdot)$ to $\mathscr{M}$-valued functions.

To facilitate the presentation of the DGFE formulation, we introduce some further notational conventions. For any $h \in \mathcal{H}$, we indicate by $\mathcal{I}^h = \{\text{int}(\partial\kappa \cap \partial\hat{\kappa}) : \kappa, \hat{\kappa} \in \mathcal{T}^h, \kappa \neq \hat{\kappa}\}$ the collection of inter-element edges, by $\mathcal{B}^h = \{\text{int}(\partial\kappa \cap \partial\Omega) : \kappa \in \mathcal{T}^h\}$ the collection of boundary edges and by $\mathcal{S}^h = \mathcal{B}^h \cup \mathcal{I}^h$ their union. With every edge we associate a unit normal vector $\boldsymbol{\nu}^e$. The orientation of $\boldsymbol{\nu}^e$ is arbitrary except on boundary edges where $\boldsymbol{\nu}^e = \boldsymbol{n}|_e$. For all interior edges, let $\kappa_{\pm}^e \in \mathcal{T}^h$ be the two elements adjacent to the edge e such that the orientation of $\boldsymbol{\nu}^e$ is exterior to κ_+. We define the edge-wise jump and mean operators according to:

$$[\![c]\!] = \begin{cases} (c_+\nu_+ + c_-\nu_-) & \text{if } e \in \mathcal{I}^h \\ (c - c_B) & \text{if } e \in \mathcal{B}^h \end{cases}, \qquad \{c\} = \begin{cases} \frac{1}{2}(c_+ + c_-) & \text{if } e \in \mathcal{I}^h \\ \frac{1}{2}(c + c_B) & \text{if } e \in \mathcal{B}^h \end{cases} \tag{52}$$

where the subscripts $(\cdot)_+$ and $(\cdot)_-$ refer to the restriction of the traces of $(\cdot)|_{\kappa_+}$ and $(\cdot)|_{\kappa_-}$ to e.

3.2.1 Semi-discrete Discontinuous Galerkin Approximation

To derive the semi-discrete DGFE formulation for either of the closed moment systems (40) or (45), we note that for any $\psi \in V^{h,p}((0,T)\times\Omega,\mathscr{M})$ there holds

$$\sum_{\kappa\in\mathcal{T}^h}\int_\kappa \langle \psi\,\partial_t\beta(g)\rangle + \sum_{\kappa\in\mathcal{T}^h}\int_\kappa \langle \psi\,\partial_{x_i} v_i\beta(g)\rangle = 0, \tag{53}$$

where β may be either β_E or β_{NS} in (38) or (43), respectively. Using the product rule and integration by parts, (53) can be reformulated in weak form. The terms in the second sum in the left hand side of (53) can be recast into

$$\int_\kappa \langle \psi\,\partial_{x_i} v_i\beta(g)\rangle = \int_{\partial\kappa} \langle \psi\; v_i\; \nu_i^\kappa\,\hat{\beta}(g;v_\nu)\rangle - \int_\kappa \langle v_i\beta\,\partial_{x_i}\psi\rangle, \tag{54}$$

where $\beta(g)$ is replaced by any $\hat{\beta}(g;v_\nu)$ in compliance with the consistency condition:

$$[\![\beta(g)]\!] = 0 \quad \Rightarrow \quad \langle v_i\nu_i^\kappa\hat{\beta}(g;v_\nu)\rangle = \langle v_i\nu_i^\kappa\beta(g)\rangle. \tag{55}$$

Implicit in the identity in (54) is the assumption that β is sufficiently smooth within the elements to permit integration by parts and define traces on $\partial\kappa$. The edge distribution $\hat{\beta}(g;v_\nu)$ is defined edge-wise and on each edge e and depends on g only via $g_\pm$, viz. the restrictions of the traces of $g|_{\kappa_\pm}$ to e. The function $\psi v_i\nu_i^\kappa\hat{\beta}(g)$ in the ultimate expression in (54) can be conceived of as an upwind-flux weighted by the jump in ψ since

$$\sum_{\kappa\in\mathcal{T}^h}\int_{\partial\kappa}\langle \psi\; v_i\; \nu_i^\kappa\hat{\beta}(g)\rangle = \sum_{e\in\mathcal{S}^h}\int_e \langle \boldsymbol{v}\cdot[\![\psi\hat{\beta}(g;v_\nu)]\!]\rangle \tag{56}$$

It is to be noted that the domain of both the upwind-flux and the jump $[\![\psi]\!]$ is $e\times\mathbb{R}^D$. On boundary edges, the external distribution corresponds to exogenous data in accordance with boundary conditions. Substituting (54) and (56) into (53) yields

$$\sum_{\kappa\in\mathcal{T}^h}\int_\kappa \langle \psi\,\partial_t\beta(g)\rangle + \sum_{e\in\mathcal{S}^h}\int_e \langle \boldsymbol{v}\cdot[\![\psi\hat{\beta}(g;v_\nu)]\!]\rangle - \int_\kappa \langle v_i\beta\,\partial_{x_i}\psi\rangle = 0. \tag{57}$$

The edge distributions $\hat{\beta}$ in (71a) must be constructed such that the consistency condition (55) holds and that the formulation (72) is stable in some appropriate sense. We select the upwind edge distribution $\beta(\hat{g})$ [1] corresponding to:

$$\beta\big(\hat{g}(v_\nu)\big) = \begin{cases}\beta(g_+) & \text{if } v_{\nu_+} > 0\\ \beta(g_-) & \text{if } v_{\nu_-} > 0.\end{cases} \tag{58}$$

3.2.2 Entropy Stability

To show the entropy stability of (57) with the Euler renormalization (38) we use Galerkin orthogonality to write

$$\sum_{\kappa\in\mathcal{T}^h}\int_\kappa \langle z\,\partial_t\beta_E(z)\rangle + \sum_{e\in\mathcal{S}^h}\int_e \langle \boldsymbol{v}\cdot[\![z\beta_E(\hat{z}(v_\nu))]\!]\rangle - \sum_{\kappa\in\mathcal{T}^h}\int_\kappa \langle v_i\beta_E(z)\,\partial_{x_i}z\rangle = 0. \tag{59}$$

The last term of (59) can be recast into

$$\int_\kappa \langle v_i\beta_E(z)\,\partial_{x_i}z\rangle = \int_{\partial\kappa} \langle v_\nu z\beta_E(z)\rangle - \int_\kappa \langle v_i z\,\partial_{x_i}\beta_E(z)\rangle. \tag{60}$$

Recalling that $\eta'(z) = z\beta_E'(z)$, the last term in the right member of (60) can be reformulated as

$$\int_\kappa \langle v_i z\,\partial_{x_i}\beta_E(z)\rangle = \int_\kappa \partial_{x_i}\langle v_i\eta(z)\rangle = \int_{\partial\kappa}\langle v_\nu\eta(z)\rangle. \tag{61}$$

Collecting the results in (59)–(61)

$$d_t\int_\Omega\langle\eta(z)\rangle + \int_{\partial\Omega}\langle v_n\eta(\hat{z})\rangle = \sum_{e\in\mathcal{S}^h}\int_e \langle \boldsymbol{v}\cdot[\![z\big(\beta_E(z)-\beta_E(\hat{z})\big) - \big(\eta(z)-\eta(\hat{z})\big)]\!]\rangle \le 0, \tag{62}$$

where we used the fact that $[\![\eta(\hat{z})]\!]$ vanishes on the interior edges by the continuity of $\hat{z}$, and its aggregated contribution coincides with the boundary integral in the left member of (62). Furthermore, the last inequality follows from the convexity of $\eta(z)$ with respect to $\beta(z)$; see [1] for more details.

To show the entropy stability of (57) with the Navier–Stokes–Fourier renormalization (43) we use the Galerkin orthogonality to write

$$\sum_{\kappa\in\mathcal{T}^h}\int_\kappa \langle (\Pi_{\mathcal{I}}z)\,\partial_t e^{\Pi_{\mathcal{I}}z}\rangle + \sum_{e\in\mathcal{S}^h}\int_e \langle \boldsymbol{v}\cdot[\![(\Pi_{\mathcal{I}}z)\beta_{NS}(\hat{z})]\!]\rangle - \sum_{\kappa\in\mathcal{T}^h}\int_\kappa \langle v_i\beta_{NS}(z)\,\partial_{x_i}(\Pi_{\mathcal{I}}z)\rangle = 0. \tag{63}$$

The last term of (63) can be recast into

$$\int_\kappa \langle v_i\beta_{NS}(z)\,\partial_{x_i}(\Pi_{\mathcal{I}}z)\rangle = \int_{\partial\kappa}\langle v_\nu(\Pi_{\mathcal{I}}z)\beta_{NS}(z)\rangle - \int_\kappa \langle v_i(\Pi_{\mathcal{I}}z)\,\partial_{x_i}\beta_{NS}(z)\rangle. \tag{64}$$

The last term in the right member of (64) can be reformulated as

$$\int_{\kappa} \langle v_i(\Pi_{\mathscr{I}} z)\, \partial_{x_i} \beta_{NS}(z) \rangle = \int_{\partial\kappa} \langle v_\nu \eta(\Pi_{\mathscr{I}}) \rangle + \int_{\kappa} \langle v_i(\Pi_{\mathscr{I}} z)\, \partial_{x_i} e^{\Pi_{\mathscr{I}} z} (\mathrm{Id} - \Pi_{\mathscr{I}}) z \rangle. \tag{65}$$

The last term of (65) can we rewritten using (48) and (49) as

$$\begin{aligned}\int_{\kappa} \langle v_i(\Pi_{\mathscr{I}} z)\, \partial_{x_i} e^{\Pi_{\mathscr{I}} z} (\mathrm{Id} - \Pi_{\mathscr{I}}) z \rangle &= \int_{\partial\kappa} \langle v_\nu (\Pi_{\mathscr{I}} z) e^{\Pi_{\mathscr{I}} z} (1 - \Pi_{\mathscr{I}}) z \rangle \\ &\quad - \int_{\kappa} \left\langle v_i \partial_{x_i} (\Pi_{\mathscr{I}} z) e^{\Pi_{\mathscr{I}} z} \mathcal{L}_{\mathcal{M}}^{-1} v_j \partial_{x_j} (\Pi_{\mathscr{I}} z) \right\rangle.\end{aligned} \tag{66}$$

Collecting the results in (64)–(65) into (63)

$$\begin{aligned} d_t \int_{\Omega} \langle \eta(z) \rangle &+ \int_{\partial\Omega} \langle v_n (\eta(\hat{z}) + (\Pi_{\mathscr{I}} z) e^{\Pi_{\mathscr{I}} z} (1 - \Pi_{\mathscr{I}}) z) \rangle \\ &= \sum_{e \in \mathcal{S}^h} \int_e \langle \boldsymbol{v} \cdot [\![\Pi_{\mathscr{I}} z (e^{\Pi_{\mathscr{I}} z} - e^{\Pi_{\mathscr{I}} \hat{z}}) - (\eta(\Pi_{\mathscr{I}} z) - \eta(\Pi_{\mathscr{I}} \hat{z}))]\!] \rangle \\ &\quad + \int_{\kappa} \left\langle v_i \partial_{x_i} (\Pi_{\mathscr{I}} z) e^{\Pi_{\mathscr{I}} z} \mathcal{L}_{\mathcal{M}}^{-1} v_j \partial_{x_j} (\Pi_{\mathscr{I}} z) \right\rangle \leq 0, \end{aligned} \tag{67}$$

where we used the fact that $[\![\eta(\Pi_{\mathscr{I}} \hat{z})]\!]$ vanishes on the interior edges by the continuity of $\hat{z}$, and its aggregated contribution coincides with the boundary integral in the left member of (67). Furthermore, the last inequality follows from non-positivity of the last term of the right member of (67) (see (15)) and from the convexity of $\eta(z)$ with respect to e^z; see [1] for more details.

3.2.3 Fully-Discrete Discontinuous Galerkin Approximation

To approximate (57) in time we introduce the time integration

$$\sum_n \sum_{\kappa \in \mathcal{T}^h} \int_{\kappa \times I^n} \langle \psi\, \partial_t \beta(g) \rangle + \sum_n \sum_{e \in \mathcal{S}^h} \int_{e \times I^n} \langle \boldsymbol{v} \cdot [\![\psi \hat{\beta}(g; v_\nu)]\!] \rangle - \sum_n \int_{\kappa \times I^n} \langle v_i \beta(g)\, \partial_{x_i} \psi \rangle = 0. \tag{68}$$

The summands of the first term of the left hand side of (68) can be recast into

$$\int_{\kappa \times I^n} \langle \psi\, \partial_t \beta(g) \rangle = \int_{\kappa} \langle \psi(t_-^{n+1})\, \beta(g(t_-^{n+1})) - \psi(t_+^n)\, \beta(g(t_-^n)) \rangle - \int_{\kappa \times I^n} \langle \beta\, \partial_t \psi \rangle, \tag{69}$$

where t_+^n and t_-^n denote the one-sided limit to t^n from above and below, respectively. Implicit to the identity (69) is the assumption that β is sufficiently smooth within the time interval to permit integration by parts. Hence, any solution to (40) or (45) that is sufficiently regular in the aforementioned sense satisfies

$$a(g;\psi) = s(g;\psi) \quad \forall \psi \in V^{h,p}((0,T)\times\Omega,\mathscr{M}), \tag{70}$$

with

$$\begin{aligned} a(g;\psi) = &\sum_n \sum_{\kappa\in\mathcal{T}^h} \int_\kappa \langle \psi(t_-^{n+1})\,\beta(g(t_-^{n+1})) - \psi(t_+^n)\,\beta(g(t_-^n))\rangle \\ &+ \sum_n \sum_{\kappa\in\mathcal{T}^h} \int_{\partial\kappa\times I^n} \langle \psi\, v_i\, v_i^\kappa\, \hat{\beta}(g;v_\nu)\rangle \\ &- \sum_n \sum_{\kappa\in\mathcal{T}^h} \int_{\kappa\times I^n} \langle \beta(g)\,\partial_t\psi + v_i\beta(g)\,\partial_{x_i}\psi\rangle, \end{aligned} \tag{71a}$$

$$s(g;\psi) = \sum_n \sum_{\kappa\in\mathcal{T}^h} \int_{\kappa\times I^n} \langle \psi\mathcal{C}(\beta(g))\rangle. \tag{71b}$$

The DGFE discretization of (40) or (45) is obtained by replacing g in (70) by an approximation $g_{\mathscr{M}}^{h,p}$ in $V^{h,p}((0,T)\times\Omega,\mathscr{M})$ according to

$$\textit{Find } g_{\mathscr{M}}^{h,p} \in V^{h,p}((0,T)\times\Omega,\mathscr{M}) : a(g_{\mathscr{M}}^{h,p};\psi) = s(g_{\mathscr{M}}^{h,p};\psi) \quad \forall\psi \in V^{h,p}((0,T)\times\Omega,\mathscr{I}). \tag{72}$$

4 Numerical Results

To illustrate the approximation properties of the DGFE moment approximations (70) with the proposed renormalization map (43) for the Navier–Stokes–Fourier, we compute the transient solution of the lid-driven cavity problem [15]. We Consider a two-dimensional computational domain with length $L = 1\,\mathrm{m}$ in both the x and y directions. The top wall moving with a tangential x-direction velocity of $u_{\mathrm{lid}} = 50\,\mathrm{m\,s^{-1}}$ and the remaining three walls are stationary. All the walls have are set at a reference temperature $T_{\mathrm{ref}} = 273\,\mathrm{K}$. We set the gas constant to $R = 208\,\mathrm{J/kg\,K}$ in accordance with its value for Argon. The initial temperature is set to T_{ref}. For simplicity, we take reference values for the viscosity and heat conduction (36), ω_{ref} and γ_{ref}, corresponding to Reynolds number

$$\mathrm{Re} = \frac{\rho_{\mathrm{ref}}\,u_{\mathrm{lid}}L}{\omega_{\mathrm{ref}}} = 10000. \tag{73}$$

We use a discontinuous Galerkin finite-element approximation spaces of polynomial degree 2 in position dependence and 0 in time dependence. We solve the DGFE approximation (70) using the Newton procedure for each time-step. We use a time-step $4.8\tau \times 10^4$ and discretize the domain using a 50×50 grid. It is noteworthy that the linearization of (72) for the Newton procedure is significantly facilitated by basing the numerical flux on the upwind distribution according to (58). Traditionally, a DGFE approximation of (25a) and (32a) is equipped with an approximate Riemann solver, e.g. according to Godunov's scheme [12], Roe's scheme [23] or Osher's scheme [21], to construct a numerical flux. However, these numerical flux functions generally depend in an intricate manner on the left and right states via the eigenvalues and eigenvectors of the flux Jacobian, Riemann invariants, etc., which impedes differentiation of the resulting semi-linear form. Determining the derivative of the upwind distribution in (58) and, in turn, of the semi-linear form (72) is a straightforward operation. Moreover, the velocity integrals introduced in the DGFE moment approximation (72) does not add to the computational complexity since such integrals may be precomputed analytically.

The left and right panels of Fig. 1 show the evolution of the magnitude of the velocity and temperature contours, overlaid with velocity and heat flux streamlines, respectively. One can observe the typical roll-up of the primary vortex as time progresses, and the local temperature increase in areas of large shear deformation due to viscous dissipation. Figure 2 shows the results after the solution has reached a steady state. Noteworthy in the steady result are the secondary vortices in the bottom left and right corners and the top left corner.

5 Conclusions

In this work we have presented the compressible Euler and Navier–Stokes–Fourier equations as a Galerkin approximation of a renormalized Boltzmann equation. Such a reinterpretation allows the expression of these compressible equations as scalar kinetic systems. The scalar structure of the resulting equations allowed us to construct discontinuous Galerkin finite element methods for the compressible Euler and Navier–Stokes–Fourier equations using edge-wise upwinded distributions. We have shown that the resulting discontinuous Galerkin finite element moment method is entropy stable. Finally, we demonstrated the approximation properties of the proposed method by simulating the lid-driven cavity benchmark.

Fig. 1 Evolution of the Navier–Stokes–Fourier solution of the lid-driven cavity test case. *Left pane* shows the contours of the magnitude of the macroscopic velocity overlaid with macroscopic velocity streamlines and *right pane* shows the temperature contours overlaid with heat flux streamlines. The results are plotted at $4.8\tau \times 10^5$ (*top*), $1.4\tau \times 10^6$ (*middle*) and $2.9\tau \times 10^6$ time steps (*bottom*)

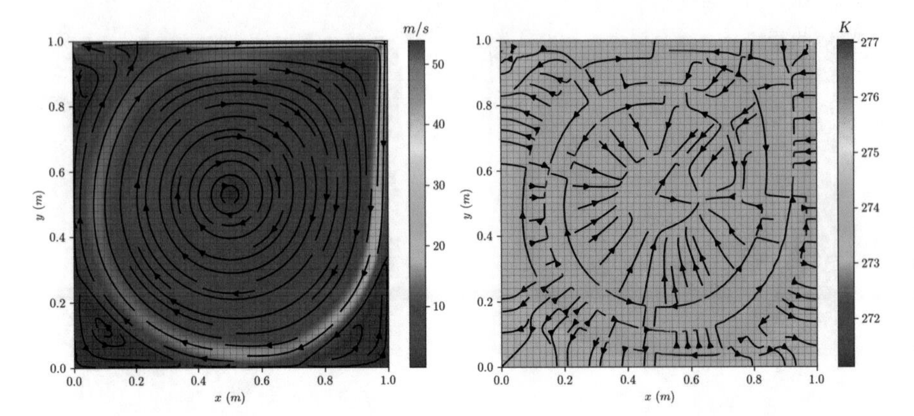

Fig. 2 Long-time solution of the Navier–Stokes–Fourier system for the lid-driven cavity test case plotted at time $6.2\tau \times 10^6$. *Left pane* shows the contours of the magnitude of the macroscopic velocity overlaid with macroscopic velocity streamlines and *right pane* shows the temperature contours overlaid with heat flux streamline

Acknowledgements This work is part of the research programme RARETRANS with project number HTSM-15376, which is (partly) financed by the Netherlands Organisation for Scientific Research (NWO). The support of ASML of the RARETRANS programme is gratefully acknowledged.

References

1. Abdelmalik, M., van Brummelen, E.: An entropy stable discontinuous Galerkin finite-element moment method for the Boltzmann equation. Comput. Math. Appl. **72**(8), 1988–1999 (2016)
2. Abdelmalik, M., van Brummelen, E.: Moment closure approximations of the Boltzmann equation based on φ-divergences. J. Stat. Phys. **164**(1), 77–104 (2016)
3. Bardos, C., Golse, F., Levermore, D.: Fluid dynamic limits of kinetic equations. I. Formal derivations. J. Stat. Phys. **63**(1), 323–344 (1991)
4. Barth, T.: On discontinuous Galerkin approximations of Boltzmann moment systems with Levermore closure. Comput. Methods Appl. Mech. Eng. **195**(25–28), 3311–3330 (2006)
5. Bhatnagar, P., Gross, E., Krook, M.: A model for collision processes in gases. I. Small amplitude processes in charged and neutral one-component systems. Phys. Rev. **94**(511—525) (1954)
6. Bouchut, F., Perthame, B.: A BGK model for small Prandtl number in the Navier-Stokes approximation. J. Stat. Phys. **71**(1–2), 191–207 (1993)
7. Cercignani, C.: The Boltzmann Equation and Its Applications. Springer, New York (1988)
8. Chandrashekar, P.: Discontinuous Galerkin method for Navier–Stokes equations using kinetic flux vector splitting. J. Comput. Phys. **233**, 527–551 (2013)
9. Chou, S.Y., Baganoff, D.: Kinetic flux–vector splitting for the Navier–Stokes equations. J. Comput. Phys. **130**(2), 217–230 (1997)
10. Di Pietro, D., Ern, A.: Mathematical Aspects of Discontinuous Galerkin Methods. Springer, Berlin (2012)
11. Esposito, R., Lebowitz, J., Marra, R.: Hydrodynamic limit of the stationary Boltzmann equation in a slab. Commun. Math. Phys. **160**(1), 49–80 (1994)

12. Godunov, S.: Finite difference method for numerical computation of discontinuous solutions of the equations of fluid dynamics. Mat. Sbornik **47**, 271–306 (1959)
13. Golse, F.: The Boltzmann equation and its hydrodynamic limits. Evol. Equ. **2**, 159–301 (2005)
14. Golse, F., Saint-Raymond, L.: The Navier–Stokes limit of the Boltzmann equation for bounded collision kernels. Invent. Math. **155**(1), 81–161 (2004)
15. John, B., Gu, X., Emerson, D.: Investigation of heat and mass transfer in a lid-driven cavity under nonequilibrium flow conditions. Numer. Heat Transfer, Part B: Fundam. **58**(5), 287–303 (2010)
16. Levermore, C.: Moment closure hierarchies for kinetic theories. J. Stat. Phys. **83**, 1021–1065 (1996)
17. Levermore, C., Masmoudi, N.: From the Boltzmann equation to an incompressible Navier–Stokes–Fourier system. Arch. Rational Mech. Anal. **196**, 753–809 (2010)
18. Lions, P.L., Masmoudi, N.: From the Boltzmann equations to the equations of incompressible fluid mechanics, I. Arch. Rational Mech. Anal. **158**(3), 173–193 (2001)
19. Lions, P.L., Masmoudi, N.: From the Boltzmann equations to the equations of incompressible fluid mechanics, II. Arch. Rational Mech. Anal. **158**(3), 195–211 (2001)
20. Liu, H., Xu, K.: A Runge–Kutta discontinuous Galerkin method for viscous flow equations. J. Comput. Phys. **224**(2), 1223–1242 (2007)
21. Osher, S., Solomon, F.: Upwind difference schemes for hyperbolic conservation laws. Math. Comput. **38**, 339–374 (1982)
22. Ren, X., Xu, K., Shyy, W., Gu, C.: A multi-dimensional high-order discontinuous Galerkin method based on gas kinetic theory for viscous flow computations. J. Comput. Phys. **292**, 176–193 (2015)
23. Roe, P.: Approximate Riemann solvers, parameter vectors, and difference schemes. J. Comput. Phys. **43**, 357–372 (1981)
24. Saint-Raymond, L.: Hydrodynamic Limits of the Boltzmann Equation. Lecture Notes in Mathematics, vol. 1971. Springer, Berlin (2009)
25. Temam, R., Miranville, A.: Mathematical modeling in continuum mechanics. Cambridge University Press, Cambridge (2005)

Space-Time NURBS-Enhanced Finite Elements for Solving the Compressible Navier–Stokes Equations

Michel Make, Norbert Hosters, Marek Behr, and Stefanie Elgeti

Abstract This article considers the NURBS-Enhanced Finite Element Method (NEFEM) applied to the compressible Navier–Stokes equations. NEFEM, in contrast to conventional finite element formulations, utilizes a NURBS-based computational domain representation. Such representations are typically available from Computer-Aided-Design tools. Within the NEFEM, the NURBS boundary definition is utilized only for elements that are touching the domain boundaries. The remaining interior of the domain is discretized using standard finite elements. Contrary to isogeometric analysis, no volume splines are necessary.

The key technical features of NEFEM will be discussed in detail, followed by a set of numerical examples that are used to compare NEFEM against conventional finite element methods with the focus on compressible flow.

Keywords Spline-based methods · NURBS-enhanced finite elements · Stabilized space-time finite elements · Compressible Navier–Stokes equations

1 Introduction

Geometries in engineering applications are commonly designed with the use of Computer-Aided-Design (CAD) tools. In general, these tools utilize Non-Uniform Rational B-Splines (NURBS) to accurately represent complex geometric domains by means of surface splines. When performing numerical analysis on such domains, it is common practice to first discretize the domain into finite sub-domains or elements. This discretization process, typically results in loss of the exact geometry.

M. Make (✉) · N. Hosters · M. Behr · S. Elgeti
RWTH-Aachen University, Aachen, Germany
e-mail: make@cats.rwth-aachen.de; hosters@cats.rwth-aachen.de; behr@cats.rwth-aachen.de; elgeti@cats.rwth-aachen.de

H. van Brummelen et al. (eds.), *Numerical Methods for Flows*,
Lecture Notes in Computational Science and Engineering 132,
https://doi.org/10.1007/978-3-030-30705-9_9

An alternative approach, known as Isogeometric Analysis (IGA), was proposed in [6]. The key idea of IGA is to use the NURBS basis functions not only for the geometric representation, but also for the numerical solution itself. By doing so, numerical analysis can be applied to the CAD model directly without the loss of geometric accuracy caused by discretizing the computational domain. Numerical analysis of fluid flow problems, however, commonly involves complex three-dimensional volume domains. Parametrizing such domains using closed volume splines can be challenging.

An alternative was proposed in [9], and further extended for space-time finite elements and free-surface flows in [10]. This method was then modified for interface-coupled problems in [4]. This approach suggests to use standard finite elements in the interior of the computational domain supplemented with so-called NURBS-enhanced finite elements along domain boundaries. These elements make use of NURBS to accurately represent complex geometries. The NURBS-Enhanced Finite Element Method (NEFEM) allows for maintaining as much as possible the proven computational efficiency of standard finite element methods, while utilizing the accurate geometric representation provided by the NURBS.

In this work, we apply NEFEM to supersonic flow problems. For this type of problems, accurate geometry representation can be important, especially due to the presence of shock waves and their interaction with solid walls.

2 Quasi-Linear Form of the Navier–Stokes Equations

Before presenting the NEFEM concept, first the governing Navier–Stokes equations are presented. For this, let $\Omega_t \subset \mathbb{R}^{n_{sd}}$ and $t \in (0, T)$ be the spatial and temporal domains respectively, and let Γ_t denote the boundary of Ω_t. Here, n_{sd} represents the number of spatial dimensions. The model problem can now be written as a generalized advective-diffusive system:

$$\frac{\partial \mathbf{U}}{\partial t} + \frac{\partial \mathbf{F}_i}{\partial x_i} - \frac{\partial \mathbf{E}_i}{\partial x_i} = \mathbf{0} \qquad \text{on } \Omega_t \quad \forall\, t \in (0, T), \tag{1}$$

where $\mathbf{U} = (\rho, \rho u_1, \rho u_2, \rho u_3, \rho e)^T$ is the solution vector. ρ, u_i, and e represent the density, velocity components, and total energy per unit mass respectively. For the three-dimensional case, the Euler and viscous flux vectors $\mathbf{F}_i$ and $\mathbf{E}_i$ are defined as:

$$\mathbf{F}_i = \begin{pmatrix} u_i \rho \\ u_i \rho u_1 + \delta_{i1} p \\ u_i \rho u_2 + \delta_{i2} p \\ u_i \rho u_3 + \delta_{i3} p \\ u_i(\rho e + p) \end{pmatrix}, \qquad \mathbf{E}_i = \begin{pmatrix} 0 \\ \tau_{i1} \\ \tau_{i2} \\ \tau_{i3} \\ -q_i + \tau_{ik} u_k \end{pmatrix}, \tag{2}$$

where q_i, τ_{ik}, p, and δ represent the heat flux, viscous stress tensor, pressure, and the Kronecker delta respectively. The boundary and initial conditions are given by:

$$\mathbf{U} \cdot \mathbf{e}_d = g_d \quad \text{on } (\Gamma_t)_{g_d}, \quad d = 1 \dots n_{dof}\ \forall t \in (0, T), \tag{3}$$

$$(n_i \mathbf{E}_i) \cdot \mathbf{e}_d = h_d \quad \text{on } (\Gamma_t)_{h_d}, \quad d = 1 \dots n_{dof}\ \forall t \in (0, T), \tag{4}$$

$$\mathbf{U}(\mathbf{x}, 0) = \mathbf{U}_0(\mathbf{x}) \quad \text{on } \Omega_0, \tag{5}$$

where $(\Gamma_t)_{g_d}$ and $(\Gamma_t)_{h_d}$ are the subsets of Γ_t, $\mathbf{e}_d$ is a basis in $\mathbb{R}^{n_{dof}}$, and n_{dof} is the number of degrees of freedom. The quasi-linear form of Eq. (1) is written as:

$$\frac{\partial \mathbf{U}}{\partial t} + \mathbf{A}_i \frac{\partial \mathbf{U}}{\partial x_i} - \frac{\partial}{\partial x_i}\left(\mathbf{K}^h_{ij} \frac{\partial \mathbf{U}}{\partial x_j}\right) = \mathbf{0} \quad \text{on } \Omega_t \quad \forall t \in (0, T), \tag{6}$$

where $\mathbf{A}_i = \frac{\partial \mathbf{F}_i}{\partial \mathbf{U}}$ represent the Euler Jacobians, and $\mathbf{K}^h_{ij} \frac{\partial \mathbf{U}}{\partial x_j} = \mathbf{E}_i$ the diffusivity matrices. $\mathbf{A}_i$ and $\mathbf{K}^h_{ij}$ are defined according to the set of solution variables (conservation variables in this case). For a detailed discussion on the various variable sets and corresponding $\mathbf{A}_i$ and $\mathbf{K}^h_{ij}$ matrices see, e.g. [3].

3 Stabilized Space-Time Finite Element Formulation

Following [1], the deformable spatial domain/stabilized space-time (DSD/SST) finite element formulation is derived for the quasi-linear form in Eq. (6).

In order to construct the finite element function spaces for the DSD/SST formulation, the time interval $(0, T)$ is decomposed into subintervals $I_n = (t_n, t_{n+1})$, where t_n and t_{n+1} are part of the ordered series: $0 = t_0 < t_1 < \cdots < t_N = T$. Additionally, we define $\Omega_n = \Omega_{t_n}$, and $\Gamma_n = \Gamma_{t_n}$. Q_n now represents a so-called *space-time slab*, which is the domain enclosed by Ω_n, Ω_{n+1} and P_n. Here, P_n is the surface described by the boundary Γ_t along interval I_n (see Fig. 1).

Similar to Γ_t given in Sect. 2, P_n can be decomposed into $(P_n)_{g_d}$ and $(P_n)_{h_d}$. The discrete finite element space-time function spaces for the trial and weighting

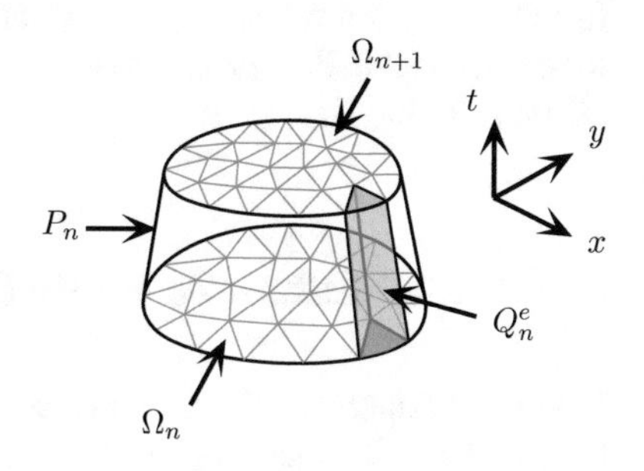

Fig. 1 Space-time slab with space-time element Q^e_n. The space-time slab is enclosed by spatial domains Ω_n and Ω_{n+1} together with P_n

functions $\mathbf{U}$ and $\mathbf{W}$ are given by:

$$(\mathscr{S}_U^h)_n = \left\{ \mathbf{U}^h | \mathbf{U}^h \in \left[H^{1h}(Q_n) \right]^{n_{dof}}, \ \mathbf{U}^h \cdot \mathbf{e}_d = g_d^h \text{ on } (P_n)_{g,d}, d = 1 \ldots n_{dof} \right\}, \tag{7}$$

$$(\mathscr{V}_W^h)_n = \left\{ \mathbf{W}^h | \mathbf{W}^h \in \left[H^{1h}(Q_n) \right]^{n_{dof}}, \ \mathbf{W}^h \cdot \mathbf{e}_d = 0 \text{ on } (P_n)_{h,d}, d = 1 \ldots n_{dof} \right\}. \tag{8}$$

Using the Streamline-Upwind Petrov–Galerkin (SUPG) formulation, the weak form of Eq. (6) states: given $(\mathbf{U}^h)_n^-$, find $\mathbf{U}^h \in (\mathscr{S}_U^h)_n$, such that $\forall\, \mathbf{W}^h \in (\mathscr{V}_W^h)_n$:

$$\begin{aligned}
&\int_{Q_n} \mathbf{W}^h \cdot \left(\frac{\partial \mathbf{U}^h}{\partial t} + \mathbf{A}_i^h \cdot \frac{\partial \mathbf{U}^h}{\partial x_i} \right) dQ + \int_{Q_n} \left(\frac{\partial \mathbf{W}^h}{\partial x_i} \right) \cdot \left(\mathbf{K}_{ij}^h \frac{\partial \mathbf{U}^h}{\partial x_j} \right) dQ + \\
&\quad \sum_{e=1}^{(n_{el})_n} \int_{Q_n^e} \boldsymbol{\tau}_{mom} \left[(\mathbf{A}_k^h)^T \frac{\partial \mathbf{W}^h}{\partial x_k} \right] \cdot \left[\frac{\partial \mathbf{U}^h}{\partial t} + \mathbf{A}_i^h \frac{\partial \mathbf{U}^h}{\partial x_i} - \frac{\partial}{\partial x_i} \left(\mathbf{K}_{ij}^h \frac{\partial \mathbf{U}^h}{\partial x_j} \right) \right] dQ + \\
&\quad \sum_{e=1}^{(n_{el})_n} \int_{Q_n^e} \boldsymbol{\tau}_{DC} \left(\frac{\partial \mathbf{W}^h}{\partial x_i} \right) \cdot \left(\frac{\partial \mathbf{U}^h}{\partial x_i} \right) dQ + \\
&\int_{\Omega_n} (\mathbf{W}^h)_n^+ \cdot \left((\mathbf{U}^h)_n^+ - (\mathbf{U}^h)_n^- \right) d\Omega = \int_{(P_n)_h} \mathbf{W} \cdot \mathbf{h}^h dP.
\end{aligned} \tag{9}$$

Here, the first two integrals on the left-hand side and the integral on the right-hand side represent the standard Galerkin form. The third and fourth left-hand side integrals represent the SUPG stabilisation and shock-capturing terms respectively. Continuity over the time-slab interface Ω_n is weakly imposed by the jump term, i.e., the fifth left-hand side integral in Eq. (9). Here, the $\pm$ subscripts refer to the upper and lower time-slab solutions at time n. Equation (9) is solved sequentially for all space-time slabs $Q_1, Q_2, \ldots, Q_{N-1}$ with initial condition $(\mathbf{U}^h)_0^- = \mathbf{U}_0$.

For $\boldsymbol{\tau}_{mom}$ in the SUPG stabilization term in Eq. (9), the formulation proposed in [5] was used. The shock-capturing parameter $\boldsymbol{\tau}_{DC}$ used in this work, is similar to that given in [8], which is a modification of the original definition presented in [5]. For brevity, a detailed discussion on the stabilization and shock-capturing formulations is omitted. For an extensive discussion on DSD/SST finite elements for compressible flow problems including SUPG-stabilization and shock-capturing, please refer to [7].

4 NURBS-Enhanced Finite Elements

In this section, the NURBS-enhanced finite element method as proposed in [4] will be presented. The key idea of this method, is to use a NURBS definition of the computational domain to *enhance* the finite elements along the domain boundary.

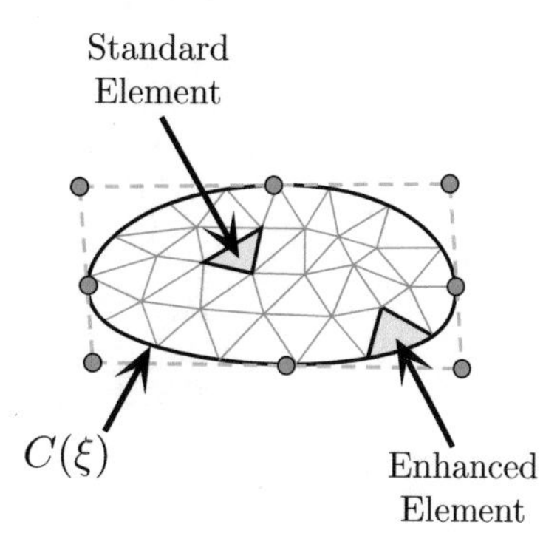

Fig. 2 The computational domain defined by a NURBS curve $\mathbf{C}(\xi)$ expressed by means of parametric coordinate ξ and a control polygon. The NURBS-enhanced elements are located along the NURBS boundary. The standard elements are located in the remaining interior part of the domain

On all remaining elements in the interior of the domain a standard finite element formulation is used (cf. Fig. 2).

Before discussing how the boundary elements make use of the NURBS boundary, let us first define a NURBS-curve of degree p. Such a curve is composed of piecewise rational basis functions $\mathbf{R}_i^p(\xi)$, and control points $\mathbf{B}_i$. The curve is then expressed by means of parametric coordinate $\xi \in (0, 1)$ as follows:

$$\mathbf{C}(\xi) = \sum_{i=1}^{n} \mathbf{R}_i^p(\xi)\mathbf{B}_i, \tag{10}$$

where n denotes the total number of control points.

The elements that touch the NURBS domain boundary make use of a non-linear mapping between a reference element and the element in physical coordinates. This mapping, was proposed in [4] as Triangle-Rectangle-Triangle (TRT) mapping. The mapping $\boldsymbol{\Phi}(s, r)$ is given by:

$$\boldsymbol{\Phi}(s, r) = (1 - s - r)x_3 + (s + r)\mathbf{C}\left(\frac{s\,\xi_1 + r\,\xi_2}{s + r}\right). \tag{11}$$

Here, s and r are the parametric coordinates of the triangular reference element, x_3 denotes the physical coordinate of the interior node, and ξ_1 and ξ_2 are the parametric coordinates of the NURBS curve at which the element boundary nodes are located. A graphical representation of this mapping is shown in Fig. 3.

By using the TRT mapping, the NURBS definition can be incorporated into the numerical analysis. As a result, the distribution of the integration points is determined from the exact geometry and not the erroneous discretized geometry (cf. Fig. 4).

Furthermore, the shape functions corresponding to the interior nodes of the boundary elements remain zero along the NURBS (cf. Fig. 4). This has the advantage that there is no contribution of the interior nodes when considering Dirichlet boundaries or boundary integrals. Especially for interface-coupled problems this can be important, where Dirichlet boundaries and boundary integrals are used to compute the coupling conditions [4].

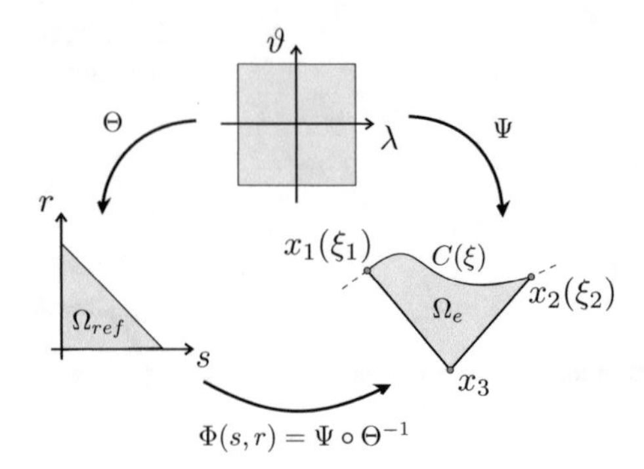

Fig. 3 Triangle-Rectangle-Triangle (TRT) mapping, as used in the NEFEM formulation

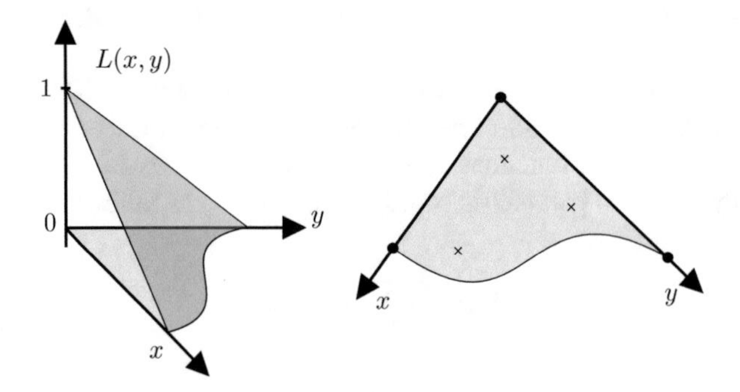

Fig. 4 **Left:** interior shape function, $L(x, y)$, on an NEFEM element using TRT mapping. **Right:** distribution of integration points on an NEFEM element

For the numerical integration additional quadrature points are used for the NEFEM elements. This is necessary in order to properly capture the geometric description provided by the NURBS. Since only a small portion of the computational domain is discretized using the enhanced elements the additional computational effort is kept to a minimum [9].

5 Numerical Examples

To demonstrate the performance of the NEFEM in comparison to standard finite elements (SFEM), two test cases are considered next: (1) 2D supersonic viscous flow around a cylinder; (2) 2D transonic inviscid flow around a NACA0012 airfoil.

5.1 Cylinder Flow

The supersonic flow around a 2D cylinder is computed using the NEFEM and the SFEM. The flow conditions and the computational domain are shown in Table 1 and

Table 1 Flow conditions for the supersonic flow around the cylinder

Flow conditions	
Mach	1.7
Re	2.0×10^5
ρ_{in}	1.0
u_{in}	1.0
v_{in}	0.0
e_{in}	1.1179

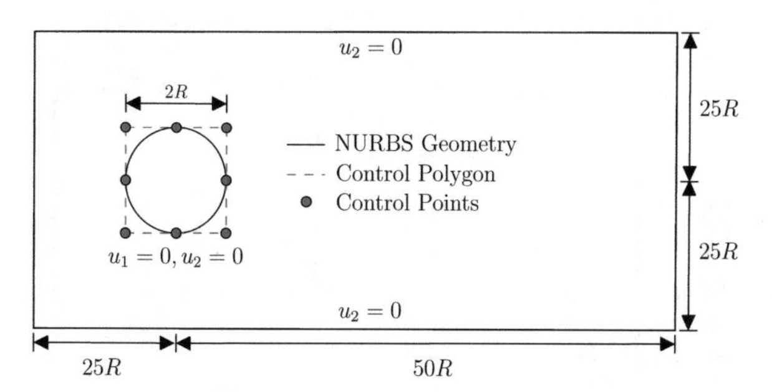

Fig. 5 Computational domain and boundary conditions for a supersonic flow around a 2D cylinder

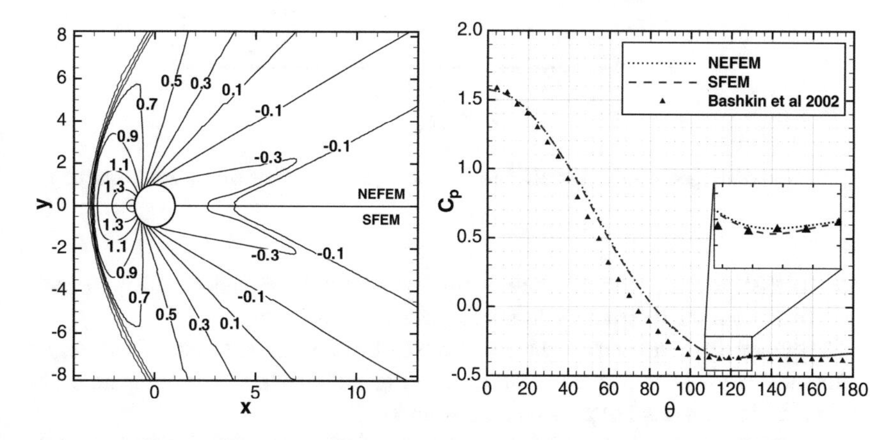

Fig. 6 **Left:** contour lines of the pressure coefficient for NEFEM and SFEM. **Right:** pressure coefficient along the cylinder wall with angular coordinate θ

Fig. 5 respectively. For the NEFEM computations, the cylinder is represented by a second order NURBS-curve (cf. Fig. 5).

The flow solution for both methods, presented by means of the pressure coefficient, C_p, is given in Fig. 6. In this figure, it can be seen that the NEFEM and the SFEM result in similar flow solutions. However, small differences can be observed when looking at the pressure coefficients along the cylinder wall. These differences could be a result of the improved geometry representation within

Table 2 Grids used for the grid refinement study

Grid #	n_{en}	$n_{en_{wall}}$
0	6.72K	64
1	26.88K	128
2	107.52K	256
3	430.08K	512
4	1.72M	1028

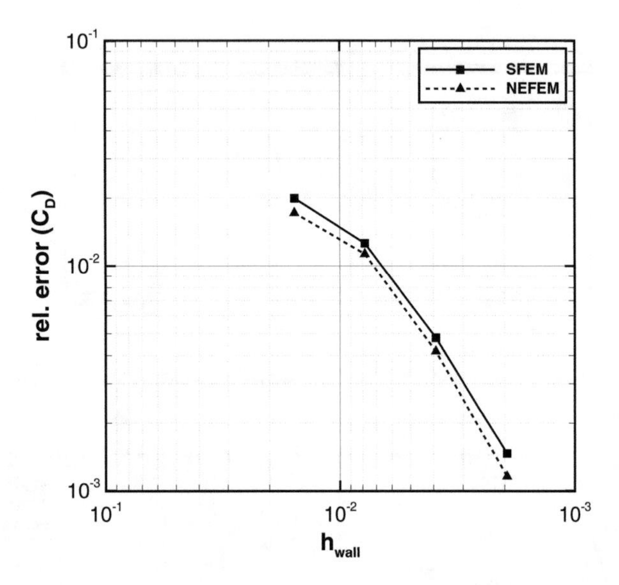

Fig. 7 Grid convergence for supersonic flow around a cylinder. Results are relative to that of grid 4 in Table 2

the NEFEM. Overall, the pressure coefficient along the cylinder wall is in good agreement with the reference solution [2], as shown in Fig. 6.

To demonstrate the performance of the NEFEM compared to the SFEM, a grid refinement study is performed in which the drag coefficient, C_D, is compared. The grids used in the study are presented in Table 2.

The relative error in C_D in Fig. 7 shows a similar convergence rate for both methods. It can be seen, however, that the NEFEM has a reduced error for all grids.

5.2 *NACA0012 Airfoil*

The transonic inviscid flow around a 2D NACA0012 airfoil is computed using the NEFEM and the SFEM. The flow conditions and computational domain are shown in Table 3 and Fig. 8 respectively. The airfoil is represented by a fourth-order NURBS curve and is positioned with a zero degree angle of attack (cf. Fig. 8).

Table 3 Flow conditions for the transonic flow around the NACA0012 airfoil

Flow conditions	
Mach	0.8
ρ	1.0
u	1.0
v	0.0
e	3.29

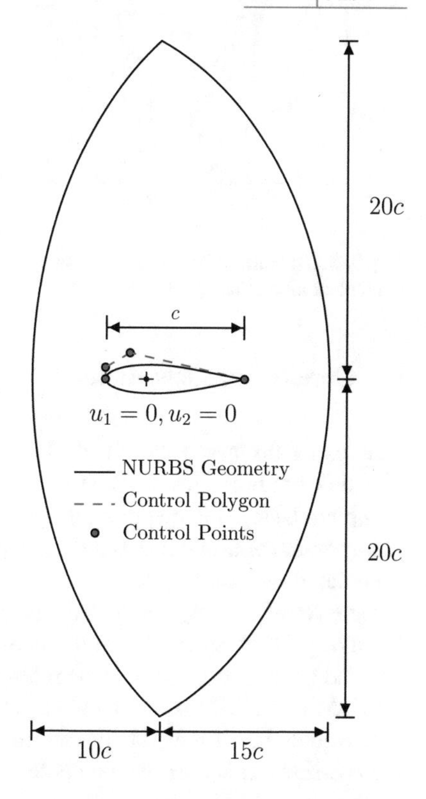

Fig. 8 Computational domain and boundary conditions for a supersonic flow around a NACA0012 airfoil

The flow solutions obtained with the NEFEM and the SFEM, are presented by means of the pressure coefficient C_p in Fig. 9. Again, it can be observed that both methods result in similar flow solutions. Additionally the results are in close agreement with the reference solution provided by Vassberg and Jameson [11].

In Fig. 9, the jump in C_p along the airfoil wall is a result of the transition from supersonic to subsonic flow conditions. It can be seen that the location of this jump slightly differs between the NEFEM and the SFEM results. As for the cylinder test case, the differences between the NEFEM and the SFEM could potentially be a result of the improved geometry representation within the NEFEM.

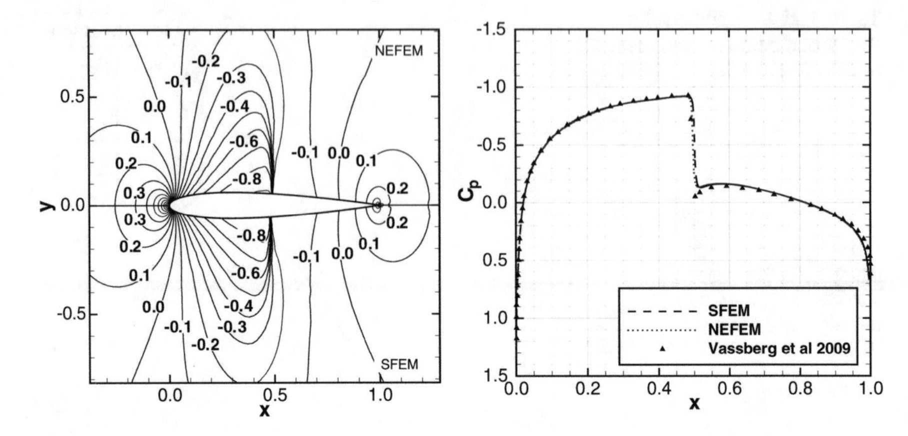

Fig. 9 **Left:** contour lines of the pressure coefficient C_p for NEFEM and SFEM. **Right:** pressure coefficient along the airfoil wall

6 Concluding Remarks

This paper discussed the NEFEM in the context of compressible flow problems. For this method, the DSD/SST formulation was used together with the TRT mapping. Using this mapping, the NURBS definition of the domain boundary was incorporated within the shape functions and numerical integration for the elements touching these boundaries.

The NEFEM was then tested and compared against the SFEM. Two test cases involving 2D viscous and inviscid flows were considered. Overall the methods showed good agreement with the reference solutions. Small differences between the NEFEM and SFEM solutions along solid walls in the flow domain were observed. These could be attributed to the improved geometry representation accounted for in the NEFEM. This observation could be a suitable starting point for future research on the benefits of the NEFEM over the conventional methods.

References

1. Aliabadi, S.K., Tezduyar, T.E.: Parallel fluid dynamics computations in aerospace applications. Int. J. Numer. Methods Fluids **21(10)**, 783–805 (1995)
2. Bashkin, V.A., Vaganov, A.V., Egorov, I.V., Ivanov, D.V., Ignatova, G.A.: Comparison of calculated and experimental data on supersonic flow past a circular cylinder. Fluid Dyn. **37(3)**, 473–483 (2002). https://doi.org/10.1023/A:1019675027402
3. Hauke, G., Hughes, T.J.R.: A Comparative study of different sets of variables for solving compressible and incompressible flows. Comput. Methods Appl. Mech. Eng. **153(1–2)**, 1–44 (1998)
4. Hosters, N., Helmig, J., Stavrev, A., Behr, M., Elgeti, S.: Fluid–structure interaction With NURBS-based coupling. Comput. Methods Appl. Mech. Eng. **332**, 520–539 (2018). https://doi.org/10.1016/j.cma.2018.01.003

5. Hughes, T.J.R., Mallet, M.: A new finite element formulation for computational fluid dynamics: IV. Discontinuity-capturing operator for multidimensional advective-diffusive systems. Comput. Methods Appl. Mech. Eng. **58**, 329–339 (1986)
6. Hughes, T.J.R., Cottrell, J.A., Bazilevs, Y.: Isogeometric analysis: CAD, finite elements, NURBS, exact geometry and mesh refinement. Comput. Methods Appl. Mech. Eng. **194**, 4135–4195 (2005)
7. Hughes, T.J.R., Scovazzi, G., Tezduyar, T.E.: Stabilized methods for compressible flows. SIAM J. Sci. Comput. **43(3)**, 343–368 (2010). https://doi.org/10.1007/s10915-008-9233-5
8. Kirk, B.J.: Adaptive Finite Element Simulation of Flow and Transport Applications on Parallel Computers. The University of Texas, Austin (2009)
9. Sevilla, R., Fernández-Méndez, S., Huerta, A.: NURBS-enhanced finite element method (NEFEM). Int. J. Numer. Methods Eng. **76(1)**, 56–83 (2008)
10. Stavrev, A., Knechtges, P., Elgeti, S., Huerta, A.: Space-time NURBS-enhanced finite elements for free-surface flows. 2D. Int. J. Num. Methods Fluids **81**, 426–450 (2016)
11. Vassberg, J.C., Jameson, A.: In pursuit of grid convergence for two-dimensional Euler solutions. J. Aircr. **47(4)**, 1152–1166 (2010). https://doi.org/10.2514/1.46737

Fluid Flow Simulation from Geometry Data Based on Point Clouds

Simon Santoso, Hassan Bouchiba, Luisa Silva, Francois Goulette, and Thierry Coupez

Abstract It is nowadays a real challenge to perform fluid flow simulation from human-acquired data, in particular when the geometries are reduced to a set of 3D points without any connectivity. Assuming that a sufficient set of points can represent the underlying geometries with a high level of details, we present in this paper a method to perform CFD computations directly from the point cloud raw data. Using an error estimator, a level-set function is built directly from the point cloud, which bypass the explicit surface reconstruction step. The level-set is then used in a mesh-adaptation procedure to optimize the representation of the distance field near the its zero value. Secondly, we use the immersion volume method to define the boundary conditions at nodes. Finally, we used a VMS finite element solver to perform the fluid flow calculation. We finally present computations on 3D point clouds self-acquired in urban environments.

Keywords Point cloud geometry · Finite element for flow · Anisotropic adaptive meshing · Immersed boundary method

1 Introduction

Due to the ongoing recent progress of the 3D scanning technologies, real world geometries can be described with a higher and higher level of details [7]. Nevertheless, it is still a challenge to use the point clouds provided by these instruments

S. Santoso · L. Silva
Ecole Centrale de Nantes, High Performance Computing Institute, Nantes, France

H. Bouchiba · F. Goulette
MINES ParisTech, CAOR, Paris, France

T. Coupez (✉)
MINES ParisTech, CEMEF, Sophia-Antipolis, France
e-mail: thierry.coupez@mines-paristech.fr

H. van Brummelen et al. (eds.), *Numerical Methods for Flows*,
Lecture Notes in Computational Science and Engineering 132,
https://doi.org/10.1007/978-3-030-30705-9_10

to perform fluid flow computations. Indeed, the surface definition and the domain representation for a suitable discretization are at least technical and often unfeasible. The common way to create a volume mesh from a given surface is the body-fitted mesh approach which requires in a first step the reconstruction of the manifold's surface. A number of techniques have been developed to reconstruct the surface of the manifold are for instance in [11] and [8]. The next step consists in the generation of a surface mesh which must be used as a boundary constraint of a volume mesh [6, 9]. The body-fitted mesh approach from the only data of point clouds is difficult to implement and rarely free of human interaction. Indeed, points clouds are usually massive, polluted with statistical noise, without information on the topology of the manifold, moreover, the surface reconstruction and mesh generation steps often relies on a slow optimization process. We explore here another route, by combining the embedded boundary techniques with an anisotropic adaptive meshing process. This approach helps to simplify the mesh generation problem by avoiding the need of an explicit reconstruction of the surfaces [3]. It is based on the implicit representation of the interfaces with the help of a Level-Set function whose zero represents the surfaces to be reconstructed [12]. The Level-Set function is defined by the Euclidean distance to a boundary. The distance is then signed whether one point is inside or outside the domain, allowing to define the physical parameters and also to impose boundary conditions. Embedded boundary approaches rely on the continuity of the surface. One of the difficulties to overcome, is that the distance to the points clouds is not defining a zero level that could represent a continuous manifold. The proposed solution is to thicken the point cloud in an anisotropic way in order to avoid holes and to smooth asperities. In this paper the point clouds are immersed directly in the computational domain and the distance function calculation is designed in a non euclidean way in order to account for the discrete nature of the sampling of the surfaces. A local metric is introduced in order to reduce the apparent distance between neighbouring points of the sampling and the desired thickness of the surface in a pseudo normal direction. The proposed strategy bypasses the explicit reconstruction of the surface and allows computations to be performed on real objects by using an adaptive anisotropic mesh procedure and a stabilized finite element method to solve the incompressible Navier–Stokes equations.

The paper is structured as follows. Section 2 is a review of the immersion volume method. Section 3 highlights the immersion method for point clouds and introduced the metric used in the distance function calculation. Section 4 describes an error estimation to compute the metric. Last section provides numerical examples.

2 Volume Immersion

This section describes the volume immersion method and introduces some useful notations.

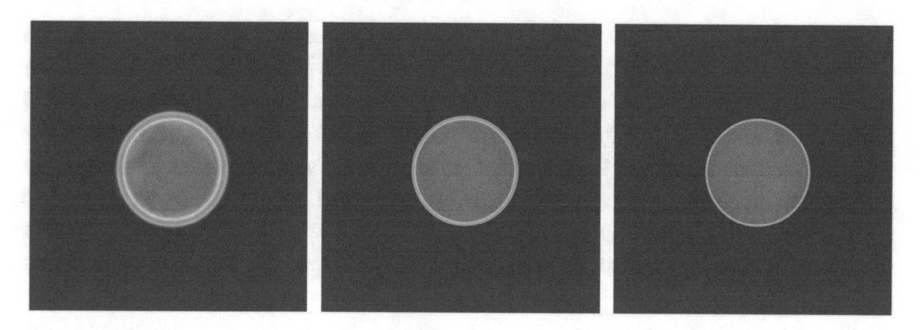

Fig. 1 Phase function for $\epsilon = 0.005, 0.002, 0.001$

Let $\Omega \in \mathbb{R}^D$. We denote his frontier as $\partial\Omega$. We define the signed distance function such as:

$$\forall \mathbf{x} \in \mathbb{R}^D, \alpha(\mathbf{x}) = \begin{cases} d(x, \partial\Omega) & \text{if } \mathbf{x} \in \Omega \\ -d(x, \partial\Omega) & \text{otherwise} \end{cases} \tag{1}$$

where d is the Euclidean distance. In the scope of [3] the phase function is defined by:

$$\forall \mathbf{x} \in \mathbb{R}^D, u_\epsilon(\mathbf{x}) = \epsilon \tanh\left(\frac{\alpha(\mathbf{x})}{\epsilon}\right) \tag{2}$$

The phase function varies between $+\epsilon$ and $-\epsilon$ in the vicinity of the immersed boundary and Fig. 1 shows how it depends on ϵ. From the phase function we define a smooth Heaviside function by:

$$\forall \mathbf{x} \in \mathbb{R}^D, H_\epsilon(\mathbf{x}) = \frac{1}{2}\left(1 + \frac{u_\epsilon(\mathbf{x})}{\epsilon}\right) \tag{3}$$

H_ϵ is equal to 0 outside of Ω and equal to 1 inside with a smooth transition. The physical parameters entering in the flow calculation are defined with the help of H_ϵ by simple geometrical mixing. A physical parameter is defined as a continuous field q in the whole domain from the two values it can takes on each side of the interface by:

$$\forall \mathbf{x} \in \mathbb{R}^D, q(\mathbf{x}) = q_1 H_\epsilon(\mathbf{x}) + q_2\left(1 - H_\epsilon(\mathbf{x})\right) \tag{4}$$

ϵ is thus a regularization parameter that controls the discontinuity at the interface between domains. In the same manner it enables to enforce weakly the boundary conditions.

The efficiency of IVM has already been proven [5] but the approach is limited to a continuous definition of the geometries for which it is possible to determine the interior and the exterior of a domain without ambiguity. A point cloud can represent complex geometries with high degree of precision, but without connectivity of the points it defines a surface from a point-wise point of view. In order to extend the embedded approach to a point-wise surface definition, we introduce a *distance to a point cloud*. The next sections highlight the issue due to the immersion of points clouds with Euclidean distance and give a definition of the *distance to a points cloud.*

3 Immersion of Point Clouds

This section gives some details on the immersion of point clouds. First, we highlight the issues caused by the use of the Euclidean distance applied to point clouds. Then, we give a solution to bypass those issues by introducing a metric field. This metric field is computed thanks to the neighbourhood of each point of $\mathscr{P}$. It also defines an approximated tangent plane and orthogonal direction. Finally, we will give a method to adapt this neighborhood to homogenize the error made on the approximation of the orthogonal direction.

3.1 Distance Function to a Set of Points

A straightforward use of the Euclidean distance to a set of points shows limitations. Let $\mathscr{P} \in (\mathbb{R}^D)^N$ be a points cloud.

$$\mathscr{P} = \{\mathbf{p}_i \in \mathbb{R}^D, i = 1, .., N)\}$$

We denote as $\mathbf{p}_{ij}$ the vector $\overrightarrow{\mathbf{p}_i\mathbf{p}_j} = \mathbf{p}_j - \mathbf{p}_i$ and ϵ_c the sampling distance of $\mathscr{P}$:

$$\epsilon_c = \max_{\mathbf{p}_i} \min_{\mathbf{p}_j, j \neq i} |\mathbf{p}_{ij}| \tag{5}$$

The following example (Fig. 2) shows that if $\epsilon < \frac{\epsilon_c}{2}$ then holes appears in the phase function. On the contrary, if $\epsilon > \frac{\epsilon_c}{2}$, then asperities are created. The fact that we want to build a closed frontier makes the condition $\epsilon > \frac{\epsilon_c}{2}$ mandatory. It is then needed to reduce asperities in the orthogonal direction and a normal direction at each point $\mathbf{p}_i$ is required to control its contribution in the distance function. The main idea here is to change the distance calculation by taking into account the direction of a target point to the sampling.

Let us consider the point cloud such as a Riemannian object equipped with a pointwise metric $\mathbb{M}_i$ associated to each point $\mathbf{p}_i$. As $\mathbb{M}_i$ is a symmetric positive-

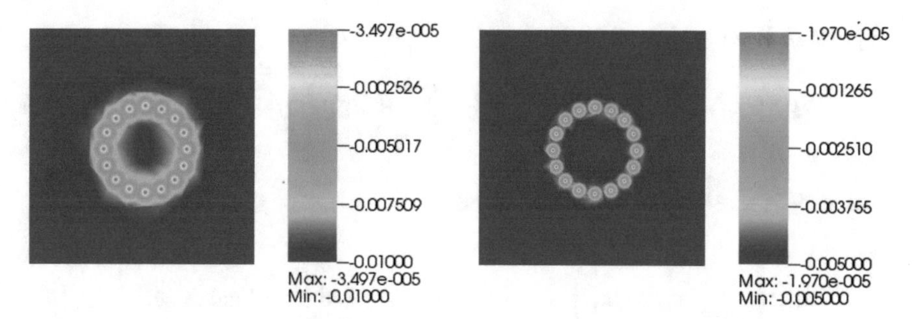

Fig. 2 Function phase of a 16 points-discretized circle of radius $R = 0.05\,\text{m}$ for $\epsilon > \frac{\epsilon_c}{2}$ and $\epsilon < \frac{\epsilon_c}{2}$. $\epsilon_c \simeq 0.0195\,\text{m}$

definite tensor, it contains several useful information. Indeed, the eigen-direction of the smallest eigenvalue gives the minor axe of the quadric associated to $\mathbb{M}_i$. It then gives the orthogonal direction at point $\mathbf{p}_i$. The space spanned by the other directions gives the tangent plane. In a Riemannian space, the distance function is given by:

$$\forall \mathbf{x} \in \mathbb{R}^D d(\mathbf{x}, \mathscr{P}) = (\mathbb{M}_i \mathbf{x}\mathbf{p}_i, \mathbf{x}\mathbf{p}_i)^{\frac{1}{2}} \text{ where } \mathbf{p}_i \text{ is the closest point of } \mathbf{x} \tag{6}$$

In the euclidean geometry context, every $\mathbb{M}_i$ is equal to the identity tensor $\mathbb{I}_D$. The isovalue curve associated to $\mathbb{I}_D$ are spheres. Minor and major axes are associated to the same eigenvalue equal to 1. Reducing asperities is equivalent to reducing the length of the minor axe. As the inside and the outside of $\mathscr{P}$ must be defined, we are unable to sign the distance function. That is why, we introduce the smooth Dirac function which can be seen as a spatial derivative of the smooth Heaviside function.

$$\forall \mathbf{x} \in \mathbb{R}^D, \delta_\epsilon(\mathbf{x}) = \frac{1}{2\epsilon}(1 - (\frac{u_\epsilon(\mathbf{x})}{\epsilon})^2) \tag{7}$$

δ_ϵ tends to $\frac{1}{2\epsilon}$ near the points clouds and is equal to 0 anywhere else. As a result, this function is a good representation of a surface. The boundary conditions can now be applied by considering a threshold $0 < s < \frac{1}{2\epsilon}$ and a mixing of the fluid parameters. In practice, we use $s = \frac{1}{4\epsilon}$.

3.2 Unit Metric Construction

We follow the work introduced in [2]. In this paper, the author builds a unit metric field for each point of a connected mesh. We extend this method to the point clouds by introducing an arbitrary neighbourhood of the points. The unit metric field, $\mathbb{M}_i$, at point p_i is set by:

$$\forall \mathbf{p}_j \in \mathscr{P}, (\mathbb{M}_i \mathbf{p}_{ij}, \mathbf{p}_{ij}) = 1 \implies \sum_{\mathbf{p}_j \in \mathscr{N}_i} (\mathbb{M}_i \mathbf{p}_{ij}, \mathbf{p}_{ij}) = \sum_{\mathbf{p}_j \in \mathscr{N}_i} 1 \tag{8}$$

where $\mathscr{N}_i$ is a neighbourhood of $\mathbf{p}_i$. Then:

$$\mathbb{M}_i : \left(\sum_{\mathbf{p}_j \in \mathscr{N}_i} \mathbf{p}_{ij} \otimes \mathbf{p}_{ij} \right) = |\mathscr{N}_i| \text{ where } |\mathscr{N}_i| \text{ is the cardinal of } \mathscr{N}_i \tag{9}$$

The length distribution tensor at each point $\mathbf{p}_i$ and denoted as $\mathbf{X}_i$ is then defined by:

$$\mathbb{X}_i = \frac{1}{|\mathscr{N}_i|} \sum_{\mathbf{p}_j \in \mathscr{N}_i} \mathbf{p}_{ij} \otimes \mathbf{p}_{ij} \tag{10}$$

When every $\mathbf{p}_{ij}$ are aligned then, $\det(\mathbb{X}_i) = 0$, we can define the metric at the point $\mathbf{p}_i$ by:

$$\mathbb{M}_i = \frac{|\mathscr{N}_i|}{D} (\mathbb{X} + \epsilon^2 n \times n)_i^{-1} \tag{11}$$

where the pseudo normal n is associated with the direction of the eigenvector of the smallest eigenvalue. Indeed, it is shown in [8] that the tangent plane associated to $\mathbb{X}_i$ is the best plane that fits the neighborhood of $\mathbf{p}_i$. As $\mathbb{X}_i$ and $\mathbb{M}_i$ have the same eigen-directions, we are ensured $\mathbb{M}_i$ is well oriented.

3.3 Definition of the Neighbourhood

The construction of the field metric is depending on the choice of the neighborhood. The literature exhibits two mains methods: Choose the K-nearest neighbors or every points contained in a ball of radius R to be defined. This two parameters are arbitrary defined and are the same for every point of $\mathscr{P}$. Based on the work of [10], this subsection gives a method to choose a ball of radius R to minimize the error made on the computation of the normal $\mathbf{n}$. For the sake of clarity, we will develop the construction of the error estimator only in 2D. We denote as $\mathbf{n}^*$ the real normal at $\mathbf{p}_i$ of the sampled surface. We can choose a base where $\mathbf{n}^* = [0\ 1]^T$. We introduce λ^- as the smallest eigenvalue and $\mathbf{n} = [e\ 1]^T$ the associated vector and e is the error to be evaluated. We have:

$$\mathbb{X}\mathbf{n} = \begin{pmatrix} A & B \\ B & C \end{pmatrix} \begin{bmatrix} e \\ 1 \end{bmatrix} = \lambda^- \begin{bmatrix} e \\ 1 \end{bmatrix} \tag{12}$$

We assume that $A >> C$ such that $\mathrm{Tr}(\mathbb{X}) \simeq A$. Mitra et al. [10] show that:

$$e = -\frac{(A + C - 2\lambda^-)B}{(A - \lambda^-)^2 + B^2} \tag{13}$$

Using the trace $\mathrm{Tr}(\mathbb{X})$ and the discriminant Δ of the characteristic polynomial of $\mathbb{X}$, λ^- can be written:

$$\lambda^- = \frac{\mathrm{Tr}(\mathbb{X}) - \sqrt{\Delta}}{2} \tag{14}$$

The numerator of e can be written:

$$A + C - 2\lambda^- = \mathrm{Tr}(\mathbb{X}) - 2 \times \frac{\mathrm{Tr}(\mathbb{X}) - \sqrt{\Delta}}{2} = \sqrt{\Delta} \tag{15}$$

The denominator of e can be written:

$$(A-\lambda^-)^2+B^2 = A^2-A\,\mathrm{Tr}(\mathbb{X})+A\sqrt{\Delta}+\frac{\mathrm{Tr}(\mathbb{X})^2}{4}-\frac{1}{2}\,\mathrm{Tr}(\mathbb{X})\sqrt{\Delta}+\frac{\Delta}{4}+B^2 \tag{16}$$

As $\mathrm{Tr}(\mathbb{X}) \simeq A$, several approximations can be done:

$$A^2 - A\,\mathrm{Tr}(\mathbb{X}) \simeq 0 \tag{17}$$

$$\Delta = (A - C)^2 + 4B^2 \simeq \mathrm{Tr}(\mathbb{X})^2 + B^2 \tag{18}$$

$$A\sqrt{\Delta} \simeq \mathrm{Tr}(\mathbb{X})\sqrt{\Delta} \tag{19}$$

Finally, the denominator of e can be approximated by:

$$(A - \lambda)^2 + B^2 \simeq \frac{\mathrm{Tr}(\mathbb{X})\sqrt{\Delta} + \Delta}{2} \tag{20}$$

The error e is then majored by:

$$|e| \leq \frac{2}{\mathrm{Tr}(\mathbb{X}) + \sqrt{\Delta}}|B| \tag{21}$$

We now have to major B to complete the error estimation. As $\mathscr{P}$ is a discretization of a $(D-1)$-manifold, we assume that it exists a $\mathscr{C}^2$ function g such as $y = g(\mathbf{x})$ in the neighborhood of $\mathbf{p}_i$. A Taylor-expansion at $\mathbf{p}_i$ gives us:

$$\forall \mathbf{p}_j \in \mathscr{N}_i,\, y_j = g(\mathbf{p}_i + \mathbf{p}_{ij}) = g(\mathbf{p}_i) + \underbrace{\nabla g(\mathbf{p}_i)}_{=0}\,\mathbf{p}_{ij} + \frac{1}{2}(\mathbb{H}(\mathbf{p}_i)\mathbf{p}_{ij}, \mathbf{p}_{ij}) + o(||\mathbf{p}_{ij}||^2) \tag{22}$$

where $\mathbb{H}(\mathbf{p}_i)$ is the Hessian at $\mathbf{p}_i$ of the sampled surface.

Since all $\mathbf{p}_j$ are contained in a ball of center $\mathbf{p}_i$ of radius R then $||\mathbf{p}_{ij}|| \leq R$. We finally have:

$$|B| \leq R \times \frac{R^2}{2}||\mathbb{H}(\mathbf{p}_i)|| \leq ||\mathbb{H}(\mathbf{p}_i)||\frac{R^3}{2} \tag{23}$$

The error e can finally be bounded by:

$$\boxed{e \leq M(R)||\mathbb{H}(\mathbf{p}_i)||\frac{R^3}{2}}$$

where $M(R) = \frac{2}{\mathrm{Tr}(\mathbb{X})+\sqrt{\Delta}}$ in 2D and $M(R) = \sqrt{\frac{\mathrm{Tr}(\mathbb{X})^2+2I_2}{I_2^2}}$ in 3D (I_2 is the second invariant of the length distribution tensor). We estimate the Hessian tensor with the following relation:

$$\mathbb{H}(\mathbf{p}_i) = \frac{2}{D-1}\left(\frac{\sum\limits_{p_j \in \mathcal{N}_i} \mathbf{p}_{ij} \otimes \mathbf{p}_{ij}}{\sum\limits_{p_j \in \mathcal{N}_i} y_j}\right)^{-1} \tag{24}$$

This estimation is subject to discussion. Indeed, this relation tends to mean curvatures in each eigen-direction of $\mathbb{H}(\mathbf{p}_i)$. As the curvature in 2D is a single scalar, this estimation is efficient, but there are 2 curvatures in 3D, and the norm can be ill-estimated if one curvature is negligible compared to the other one. The construction of the previous error estimator allows to build an iterative scheme to minimize the error for every $\mathbf{p}_i$. Starting from an initial radius R^0 and a targeted error e, the following iterative scheme adapt the radius to minimize the error.

$$R^{k+1} = \left(\frac{e}{||\mathbb{H}^k||M(R^k)}\right)^{\frac{1}{3}} \tag{25}$$

If the error e is too small for the sampling distance, we choose to define the neighborhood as the K-nearest neighbor of $\mathbf{p}_i$ where K is user defined.

4 A Numerical Example

We show in this section the capability of our method to perform CFD on a complex point sampled surface. The following point cloud has been acquired with the help of a LIDAR and represents a part of the Rue Madame in Paris. Figure 3 shows the acquired point cloud: The presence of holes in the point cloud is due to objects in the

Fig. 3 Acquired point cloud of the Madame Street

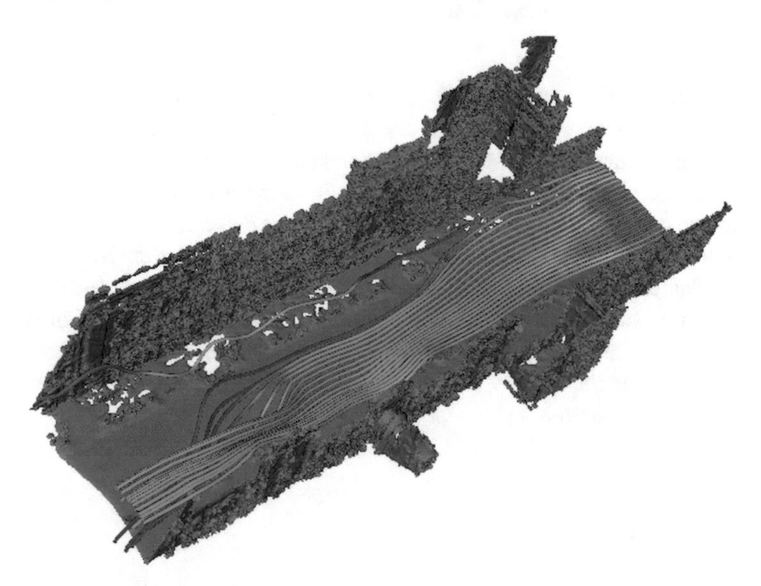

Fig. 4 Computed Streamlines using our method of surface reconstruction

street such as cars, trees or bystanders. We reconstructed the Dirac function with the help of our algorithm described in the last section. The targeted error is equal to $e = 0.2$. A wind comes from the right side of the street with a velocity of $0.1\ \mathrm{m/s}$ and we study the streamlines of the flow. We use the mesh adaptation method [2] in order to adapt the mesh around the iso-zero of the distance function. The computation of the flow is made with the use of the Variational Multiscale Method [1]. Figure 4 shows

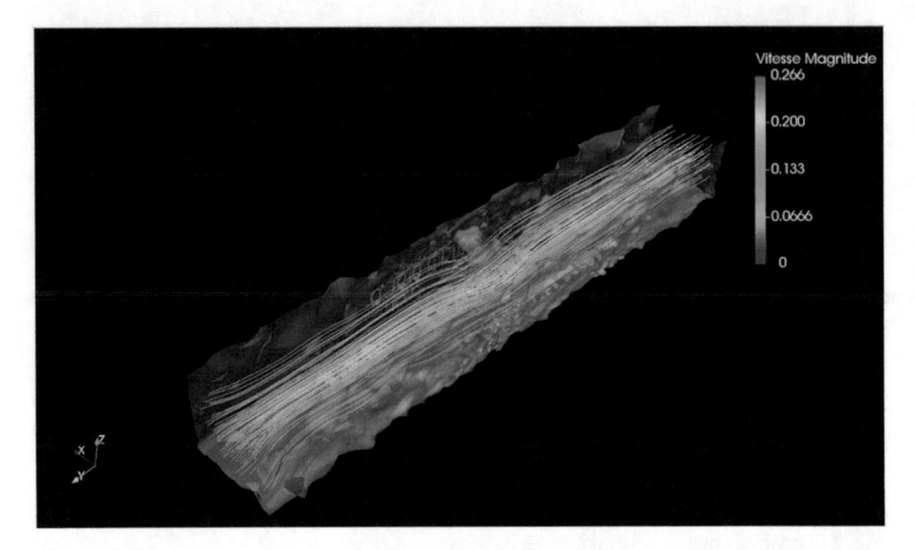

Fig. 5 Computed Streamlines using IMLS reconstruction method

those streamlines. We can see that the Dirac function is a good representation of the sampled surface. The quality of the computation we may proceed will be altered by the presence of holes. The streamlines can deviate, especially on the left side, due to the presence of holes. We can consider two methods to fill the holes. The first one is to use a higher ϵ but we noticed the fact that it can create asperities. It is also possible to use another reconstruction method that fill the holes such as the Implicit Moving Least Square Method (IMLS) [4]. The same computation is performed with the previous reconstruction technique. Figure 5 shows the streamlines of the wind when holes are filled. The IMLS method is a time-consuming method. Our method is more direct but the quality of the computation is extremely dependent of the quality of the point cloud.

5 Conclusion

In this paper, we propose to extend the embedded boundary approaches (Immersed boundary or immersed volume techniques) directly to a point cloud. This approach bypasses the need of the surface mesh reconstruction from the raw data of the 3D points. A length distribution tensor field is associated with the point cloud and calculated with the help of an error estimator giving rise to a pseudo normal direction. This estimator is based on an invariant metric and on the computation of the norm of the estimated Hessian matrix. CFD simulation can be achieved directly from the point cloud by combining the proposed approach to a mesh adaptation process and a finite element solver using the Variational Multiscale Method to perform CFD. The process is easy to implement and it opens the door to fluid flow calculation from human-acquired point cloud data.

References

1. Codina, R.: Stabilization of incompressibility and convection through orthogonal sub-scales in finite element methods. Comput. Methods Appl. Mech. Eng. **190**(13), 1579–1599 (2000)
2. Coupez, T.: Metric construction by length distribution tensor and edge based error for anisotropic adaptive meshing. J. Comput. Phys. **230**(7), 2391 – 2405 (2011)
3. Coupez, T., Silva, L., Hachem, E.: Implicit Boundary and Adaptive Anisotropic Meshing, pp. 1–18. Springer, Cham, (2015).
4. Cheng, Z.Q., Wang, Y.Z., Li, B., Xu, K., Dang, G., Jin, S.Y.: A survey of methods for moving least squares surfaces. In: Proceedings of the Fifth Eurographics/IEEE VGTC Conference on Point-Based Graphics, SPBG'08, pp. 9–23. Eurographics Association, Aire-la-Ville (2008)
5. El Jannoun, G.: Adaptation anisotrope précise en espace et temps et méthodes d'éléments finis stabilisées pour la résolution de problèmes de mécanique des fluides instationnaires. PhD thesis, 2014. Thèse de doctorat dirigée par Coupez, Thierry et Hachem, Elie Mécanique numérique Paris, ENMP 2014
6. George, P.L., Hecht, F., Saltel, E.: Automatic mesh generator with specified boundary. Compu. Methods Appl. Mech. Eng. **92**(3), 269–288 (1991)
7. Goulette, F., Nashashibi, F., Abuhadrous, I., Ammoun, S., Laurgeau, C.: An integrated on-board laser range sensing system for on-the-way city and road modelling. Int. Arch. Photogramm. Remote. Sens. Spat. Inf. Sci. **36**(Part 1) (2006) [on CDROM]
8. Hoppe, H., DeRose, T., Duchamp, T., McDonald, J., Stuetzle, W.: Surface reconstruction from unorganized points. SIGGRAPH Comput. Graph. **26**(2), 71–78 (1992)
9. Mavriplis, D.J.: Unstructured mesh generation and adaptivity. Technical report, 1995
10. Mitra, N.J., Nguyen, A., Guibas, L.: Estimating surface normals in noisy point cloud data. Int. J. Comput. Geom. Appl. **14**, 261–276 (2004)
11. Muraki, S.: Volumetric shape description of range data using "blobby model". SIGGRAPH Comput. Graph. **25**(4), 227–235 (1991)
12. Peskin, C.S.: Numerical analysis of blood flow in the heart. J. Comput. Phys. **25**(3), 220 – 252 (1977)

Thermomechanically-Consistent Phase-Field Modeling of Thin Film Flows

Christopher Miles, Kristoffer G. van der Zee, Matthew E. Hubbard, and Roderick MacKenzie

Abstract We use phase-field techniques coupled with a Coleman–Noll type procedure to derive a family of thermomechanically consistent models for predicting the evolution of a non-volatile thin liquid film on a flat substrate starting from mass conservation laws and the second law of thermodynamics, and provide constraints which must be met when modeling the dependent variables within a constitutive class to ensure dissipation of the free energy. We show that existing models derived using different techniques and starting points fit within this family. We regularise a classical model derived using asymptotic techniques to obtain a model which better handles film rupture, and perform numerical simulations in 2 and 3 dimensions using linear finite elements in space and a convex splitting method in time to investigate the evolution of a flat thin film undergoing rupture and dewetting on a flat solid substrate.

Keywords Thin films · Thermomechanical consistency · Coleman–Noll procedure · Phase-field model · Thin-film rupture · Dewetting · Free-energy dissipation · Rational mechanics

1 Introduction

The ability to accurately predict the evolution of the morphology of a thin film has a large range of applications, from the linings of mammalian lungs in biophysics and lava flows in geology [11] to the fabrication of thin-film solar cells [6]. In the last

C. Miles · K. G. van der Zee (✉) · M. E. Hubbard
School of Mathematical Sciences, University of Nottingham, Nottingham, UK
e-mail: Christopher.Miles@nottingham.ac.uk; kg.vanderzee@nottingham.ac.uk; Matthew.Hubbard@nottingham.ac.uk

R. MacKenzie
Faculty of Engineering, University of Nottingham, Nottingham, UK
e-mail: Roderick.MacKenzie@nottingham.ac.uk

H. van Brummelen et al. (eds.), *Numerical Methods for Flows*, Lecture Notes in Computational Science and Engineering 132,
https://doi.org/10.1007/978-3-030-30705-9_11

case, it is particularly key to know the final morphology of the film, as dewetting of the film on the substrate, driven by a combination of evaporation and interaction energies such as disjoining pressures can cause poor surface coverage, resulting in low device efficiency [3].

There are two main methods for developing a model to describe thin film evolution. The first method is an asymptotic approach which assumes density, viscosity and thermal conductivity are negligible in the vapour phase of the system, and employs a long-wave approximation where it is assumed that the gradients of the height and temperature functions are small in the area considered. This methodology is demonstrated in full complexity by Burelbach et al. [1]. This method is rigorous in its derivation in the sense that the resulting thin-film equations are obtained from the bulk fluid equations. However, the resulting thin-film model may have difficulty handling film rupture, when the height of the film becomes zero at a point, and a hole forms.

The second method is an energy-gradient dynamics approach directly applied to the film height, in which it is postulated that the energy of the system dissipates according to gradient dynamics, and that the film grows towards an equilibrium which is achieved at the minimal energy. Thiele [14] uses this method to build on the work in [12] to describe phenomena such as dewetting and evaporation in thin films. While this model is less rigorously derived, it is able to naturally cope with film rupture.

In this paper, we introduce a general derivation for a family of thin-film flow models based on the classical theory of thermomechanics and the Coleman–Noll procedure [2, 7]. We derive the family of models using as a starting point the fundamental axioms of conservation of mass and the second law of thermodynamics. Following [5], we stipulate that the free energy of the film Ψ depends on the film's height and its gradient, $\Psi = \widehat{\Psi}(h, \nabla h)$. We then derive constraints on the remaining constitutive variables, and allow these to additionally depend on the variational derivative μ of the total free energy and its gradient, $\nabla\mu$. This ensures energy dissipation for allowed choices made by the modeler. We show that the above-mentioned models developed in [14] and [1] fit into our framework despite having been derived from different starting points and with different techniques. Finally, numerical simulations show how a small perturbation in a flat film evolves. In order to enable the more rigorously derived asymptotic model to progress past the point of film rupture, we use phase-field techniques to regularise this model within our framework, and thereby guarantee consistency with the second law of thermodynamics, i.e. free energy dissipation.

This paper is structured as follows. In Sect. 2 we derive a basic framework for thin film models. We follow phase-field arguments to derive a family of simple models to describe the evolution of thin films while ensuring energy dissipation in Sects. 3 and 4. In Sect. 5, we show that the model derived in [1] fits into this family when evaporation is not taken into consideration. In Sect. 6, the model is regularised and discretised using linear finite elements and numerical experiments are carried out. Finally, concluding remarks are made in Sect. 7.

2 Axioms

There are two key principles behind our derivation of a model for describing the dynamics of a thin film on a flat substrate. First, by considering the conservation of mass of a thin liquid film on a horizontal substrate $\mathfrak{D} \subset \mathbb{R}^{n-1}$ where $n = 2$ or 3, we derive an equation for the height function of the film $h(\mathbf{x}, t)$ for $\mathbf{x} \in \mathfrak{D},\ t \geq 0$.

We follow the standard argument of considering the horizontal flux $\mathbf{j}$ across an arbitrary sub-domain of the thin film, $\Omega \subset \mathfrak{D}$, such as presented in [8]. We consider the film to have constant density $\rho = 1$, and that rate of mass lost across the interface of the film (due to evaporation for example) is given by R. From this, we obtain the conservation of mass equation

$$\frac{\partial h}{\partial t} + \nabla \cdot \mathbf{j} = -R. \tag{1}$$

The constitutive classes of $\mathbf{j}$ and R are chosen below.

The second key principle is a mechanical version of the second law of thermodynamics. This states that the increase in free energy of an arbitrary control volume Ω increases at a rate no greater than the work done on the region [7]. For some energy functional $\mathscr{F}(\Omega) = \int_\Omega \mathscr{E}\, dx$, this can be written as

$$\frac{d}{dt}\mathscr{F}(\Omega) = \mathscr{W}(\Omega) - \mathscr{D}(\Omega), \tag{2}$$

where $\mathscr{W}(\Omega)$ contains the work done on Ω and the free energy flux through the boundary $\partial\Omega$, and $\mathscr{D}(\Omega) \geq 0$ is the dissipation of the free energy.

The total energy density is given by $\mathscr{E} = \Psi + \Xi$, where Ψ is the Helmholtz free energy density of the system, and Ξ is a function encapsulating energies from other sources, including kinetic energy, energy from magnetic fields and thermal energy [16]. The constitutive class of Ψ, and the composition of Ξ are a choice to be made by the modeler. In this work, we take $\Xi \equiv 0$, since non-zero Ξ requires independent study.

3 Constitutive Dependence

Phase-field type models are driven by the variational derivative of the free energy functional. We follow arguments made in [5] and consider the Helmholtz free energy density Ψ to depend on h and its gradient, that is to say

$$\Psi = \widehat{\Psi}\,(h, \nabla h), \tag{3}$$

and the total energy functional is given by

$$\mathscr{F}(\Omega)[h] := \int_{\Omega} \widehat{\Psi}\,(h, \nabla h)\;dx. \tag{4}$$

The variational derivative μ of $\mathscr{F}$ is defined as

$$\mu = \frac{\delta \mathscr{F}}{\delta h} = \partial_h \widehat{\Psi} - \nabla \cdot \left(\partial_{\nabla h} \widehat{\Psi}\right). \tag{5}$$

An example of a classical choice for this energy which applies here is

$$\widehat{\Psi}(h, \nabla h) = W(h) + \frac{\sigma^2}{2}\,|\nabla h|^2\,, \tag{6}$$

with corresponding variational derivative

$$\mu = W'(h) - \sigma^2 \Delta h, \tag{7}$$

where $W(h)$ is a free energy function depending only on the height h, and the second term is a surface energy contribution.

We now define a constitutive class for $\mathbf{j}$ and R in Eq. (1) by postulating that these variables are dependent on h, the variational derivative μ, and the gradients of these variables, that is to say

$$\mathbf{j} = \widehat{\mathbf{j}}\,(h, \nabla h, \mu, \nabla \mu)\,, \tag{8}$$

$$R = \widehat{R}\,(h, \nabla h, \mu, \nabla \mu)\,. \tag{9}$$

Having set up the constituent classes for the dependent variables in the model, we now derive constraints such that the second law of thermodynamics (2) holds.

4 Deriving Constraints

We follow the procedure outlined in [5]. Using that $\Psi = \widehat{\Psi}(h, \nabla h)$ the left hand side of (2) equals:

$$\frac{d}{dt}\left(\int_{\Omega} \widehat{\Psi}\,(h, \nabla h)\,dx\right) = \int_{\Omega} \left(\partial_h \widehat{\Psi}\,\partial_t h + \partial_{\nabla h} \widehat{\Psi} \cdot \partial_t\,(\nabla h)\right) dx, \tag{10}$$

where ∂_x is the partial derivative with respect to x. Switching the time and space derivatives in the last term of (10), integrating by parts, and using (5) we obtain

$$\frac{d}{dt}\left(\int_{\Omega}\widehat{\Psi}(h,\nabla h)dx\right)=\int_{\Omega}\mu\partial_t h\,dx+\int_{\partial\Omega}\partial_t h\partial_{\nabla h}\widehat{\Psi}\cdot\mathbf{n}\,ds. \tag{11}$$

We can now substitute (1) into (11), and integrating by parts the term involving $\mu\nabla\cdot\mathbf{j}$ gives

$$\frac{d}{dt}\int_{\Omega}\Psi\,dx=-\int_{\Omega}(\mu R-\mathbf{j}.\nabla\mu)\,dx+\int_{\partial\Omega}\left(-\mu\mathbf{j}+\partial_t h\partial_{\nabla h}\widehat{\Psi}\right)\cdot\mathbf{n}\,ds. \tag{12}$$

Comparing (12) to (2), we identify the domain integral to be the dissipation $\mathscr{D}(\Omega)$ and the boundary integral to be $\mathscr{W}(\Omega)$, which are natural identifications, similar as in earlier work [7].

Thus, a family of models that suitably describes the evolution of a thin film on a solid substrate while ensuring energy dissipation is given by

$$\frac{\partial h}{\partial t}+\nabla\cdot\widehat{\mathbf{j}}=-\widehat{R}, \tag{13}$$

where $\widehat{\mathbf{j}}$ and $\widehat{R}$ are chosen to be as in (8) and (9) and

$$\mu\widehat{R}-\widehat{\mathbf{j}}\cdot\nabla\mu\geq 0, \tag{14}$$

with $\mu=\partial_h\widehat{\Psi}-\nabla\cdot\left(\partial_{\nabla h}\widehat{\Psi}\right)$.

5 Choices and Connections

In this section, we show that the family of models described above is consistent with existing models for thin film evolution when the modeler makes particular choices for the constitutive relations.

Thiele's model [14] for a non-volatile case is given by

$$\frac{\partial h}{\partial t}=\nabla\cdot\left[M_c(h)\nabla\frac{\delta\mathscr{F}}{\delta h}\right], \tag{15}$$

where $M_c(h)\geq 0$ is the mobility function for the thin film and $\delta\mathscr{F}/\delta h$ is given in (5), with $\mathscr{F}$ given in (4). It is clear that this model fits into the framework (13) with $\widehat{R}(h,\nabla h,\mu,\nabla\mu)=0$ and $\widehat{\mathbf{j}}(h,\nabla h,\mu,\nabla\mu)=-M_c(h)\nabla\mu$, and with these choices it is also clear that constraint (14) is satisfied, implying the dissipation $\mathscr{D}(\Omega)=\int_{\Omega}M_c(h)|\nabla\mu|^2\geq 0$.

We now show that the model derived using asymptotic approaches by Burelbach et al. [1] also satisfies these requirements. Equation for a non-volatile case ($R = 0$) given in [1] is

$$\frac{\partial h}{\partial t} + S\nabla \cdot \left(h^3 \nabla \Delta h\right) + \nabla \cdot \left(\left[Ah^{-1}\right] \nabla h\right) = 0, \tag{16}$$

where $A \geq 0$ is a non-dimensionalised version of the Hamaker constant, $S \geq 0$ is the non-dimensionalised surface tension, and $\Delta = \nabla \cdot \nabla$.

Using the chain rule $\nabla f(h) = f'(h) \nabla h$, with $f(h) = h^{-3}$, (16) can be rewritten as

$$\frac{\partial h}{\partial t} + \nabla \cdot \left[-\frac{Sh^3}{\sigma^2} \nabla \left(\frac{A\sigma^2}{3S} h^{-3} - \sigma^2 \Delta h \right) \right] = 0. \tag{17}$$

By considering $\widehat{\Psi}(h, \nabla h)$ as in (6), and choosing

$$W(h) = -\frac{A\sigma^2}{6S} h^{-2}, \tag{18}$$

we observe that (17) can be rewritten in terms of the variational derivative μ given in (7), as

$$\frac{\partial h}{\partial t} + \nabla \cdot \left[-\frac{Sh^3}{\sigma^2} \nabla \mu \right] = 0. \tag{19}$$

Hence, by choosing $\widehat{R}(h, \nabla h, \mu, \nabla \mu) = 0$ and $\widehat{\mathbf{j}}(h, \nabla h, \mu, \nabla \mu) = -Sh^3\sigma^{-2}\nabla\mu$ we see that the model fits the family defined in (13), with dissipation

$$\mathscr{D}(\Omega) = \int_{\Omega} \frac{Sh^3}{\sigma^2} |\nabla \mu|^2 \, dx \geq 0, \tag{20}$$

for $h \geq 0$.

A point of interest here is that the so-called disjoining pressure $\Pi(h)$ chosen in the derivation of the model (16) is given by $\Pi(h) = -kW'(h) = Ah^{-3}$ for constant $k = \frac{3S}{\sigma^2}$, and so is directly proportional to $-W'(h)$, see also [15].

6 Regularisation of the Asymptotic Model

A characteristic of model (16) is that it breaks down as the film ruptures since $h^{-1} \to \infty$. In typical numerical simulations this breakdown is observed by h becoming negative. To enable simulations to continue past the point of rupture one

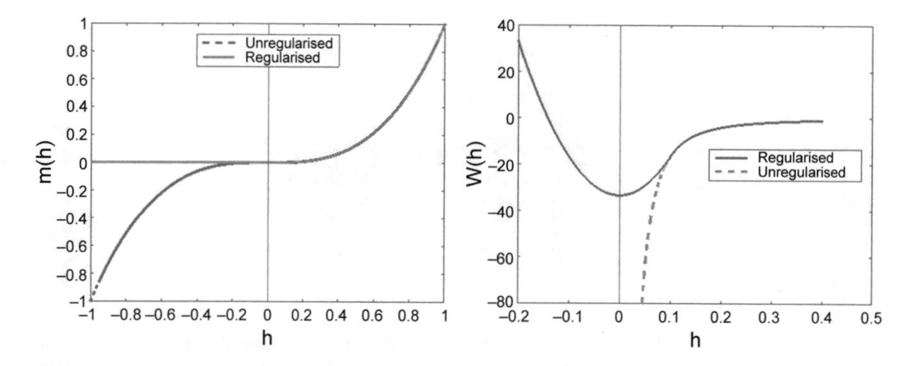

Fig. 1 Graphs of the original mobility $m(h)$ (left) and free energy $W(h)$ (right), along with the regularised versions of these functions

can regularise the bulk free energy $W(h)$ and the mobility function $m(h) = Sh^3\sigma^{-2}$ as follows.

The dotted lines in Fig. 1 show the non-regularised $m(h)$ and $W(h)$. To regularise the mobility, we force $m(h) = 0$ for $h \leq 0$ (Fig. 1, left). To handle $W(h)$, we choose a small $\epsilon > 0$ and construct $W(h)$ to be quadratic for $h < \epsilon$, and remain as given in (18) for $h \geq \epsilon$. We require the minimum of $W(h)$ to be at $h = 0$ and for the function to be continuous with a continuous derivative. The regularised function is given by

$$W(h) = \begin{cases} \dfrac{1}{6\epsilon^4}h^2 - \dfrac{1}{3\epsilon^2} & \text{if } h < \epsilon, \\ -\dfrac{1}{6}h^{-2} & \text{if } h \geq \epsilon. \end{cases} \tag{21}$$

and is shown in Fig. 1, right. This regularization leads to a potential $W(h)$ that is similar to those used in thin-film models with so-called pre-cursor films (although in our case the minimum of $W(h)$ is located at $h = 0$ instead of the pre-cursor film thickness); see for more details, e.g. [13].

To perform numerical simulations we use a linear finite element discretisation in space for h and μ in (19) and (7), employing homogeneous Neumann boundary conditions and triangular elements for the case of $n = 3$. For the time discretisation we use a convex splitting method in which the non-linear term is split as $W(h) = W_+(h) + W_-(h)$ with $W_+(h)$ being convex and $W_-(h)$ being concave. It is shown in [5] that if $W_+(h)$ is treated implicitly and $W_-(h)$ explicitly then the method is energy stable. In addition, if $\exists\, L_W > 0$ such that $|W''(h)| \leq L_W\ \forall h$ then there exists a convex split with $W_+(h) = L_W h^2/2$. This is a useful property as it results in the implicit terms being linear, removing the need to use a non-linear solver. Also, we use a semi-implicit treatment of the mobility term $m(h)$.

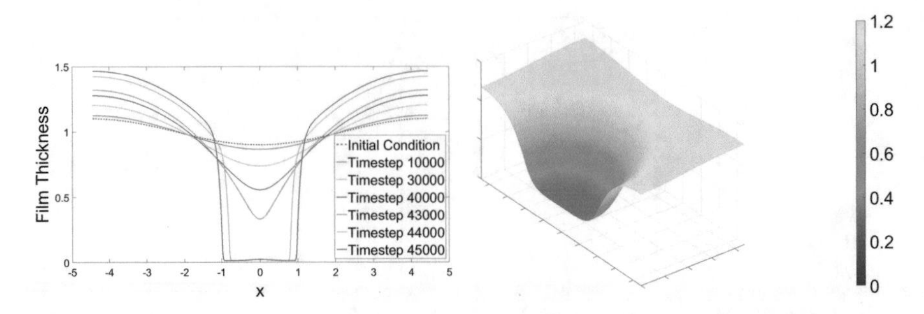

Fig. 2 **Left:** Simulation with $n = 2$ of the regularised asymptotic model showing film rupture in the domain $\mathfrak{D} = [-\pi\sqrt{2}, \pi\sqrt{2}]$. **Right:** Simulation with $n = 3$ showing how a small perturbation in a flat thin film can result in a hole forming. Only half the domain $\mathfrak{D} = [-\pi\sqrt{2}, \pi\sqrt{2}] \times [0, \pi\sqrt{2}]$ is shown to visualise the dewetted area and the final time of $T = 6.25$

Figure 2 shows examples of numerical solutions for $n = 2$ (left) and $n = 3$ (right). σ, S and A are taken to be 1. For $n = 2$, $\epsilon = 0.1$ and $\Delta t = 0.00032$, with an initial condition of $h(x, 0) = 1 - 0.1\cos(x/\sqrt{2})$. For $n = 3$, $\epsilon = 0.5$, $\Delta t = 0.025$ and $h(x, y, 0) = 1 - 0.05(\cos(x/\sqrt{2}) + \cos(y/\sqrt{2}))$. The chosen initial conditions represent a small perturbation in a flat film.

It is clear that the small perturbation in the film grows until the film ruptures, at which point a hole forms and grows via dewetting.

7 Conclusion

In this work a family of thermomechanically consistent models for predicting the evolution of a non-volatile thin liquid film on a flat substrate was derived from mass conservation laws and the second law of thermodynamics, and it was shown that existing models fit within this family. In particular, this allows for regularisations that can be applied to modeling choices to better handle film rupture and dewetting.

In [13–15] more complex thin-film processes are described that require a change in the energy functional $W(h)$, but the general form of the equation remains unchanged. Similarly, Lyushnin et al. [10] postulate a different choice of $W(h)$ to simulate fingering instabilities. Further, it can be shown that other existing models, such as those developed in [4, 9] fit the framework, covering a wide range of applications from introducing a regime to account for slip to the growth of dry regions.

Current work being undertaken is directed at investigating volatile thin films, where $R \not\equiv 0$ in (1), as well as multiphase extensions, which can be used to simulate the evolution of a substance mixed with a volatile solvent, as in the fabrication of thin-film solar cells.

Acknowledgements This work was funded by the Leverhulme Trust Modeling and Analytics for a Sustainable Society Grant. The contribution of the second author was partially supported by the Engineering and Physical Sciences Research Council (EPSRC) under grant EP/I036427/1.

References

1. Burelbach, J.P., Bankoff, S.G., Davis, S.H.: Nonlinear stability of evaporating/condensing liquid films. J. Fluid Mech **195**, 463–494 (1988)
2. Coleman, B.D., Noll, W.: The thermodynamics of elastic materials with heat conductivity and viscosity. Arch. Ration. Mech. Anal. **13**, 167–178 (1963)
3. Eperon, G.E., Burlakov, V.M., Docampo, P., Goriely, A, Snaith, H.J.: Morphological control for high performance, solution-processed planar heterojunction Perovskite solar cells. Adv. Funct. Mater **24**, 151–157 (2014)
4. Fetzer, R., Jacobs, K., Munch, A., Wagner, B., Witelski, T.P.: New slip regimes and the shape of dewetting thin liquid films. Phys. Rev. Lett. **95**(12), 127801 (2005). https://doi.org/10.1103/PhysRevLett.95.127801
5. Gomez, H., van der Zee, K. G.: Computational phase-field modeling. In: Encyclopedia of Computational Mechanics, 2nd edn. Wiley, London (2017). ISBN 978-1-119-00379-3
6. Gratzel, M.: The light and shade of perovskite solar cells. Nat. Mater. **13**, 838–842 (2014)
7. Gurtin, M.E.: Generalized Ginzburg-Landau and Cahn-Hilliard equations based on microforce balance. Phys. D. **92**, 178–192 (1996)
8. Gurtin, M. E., Fried, E., Anand, L.: The Mechanics and Thermodynamics of Continua. Cambridge University Press, Cambridge (2010)
9. Kheshgi, H.S., Scriven, L.E.: Dewetting: nucleation and growth of dry regions. Chem. Eng. Sci **46**, 519–526 (1991)
10. Lyushnin, A.V., Golovin, A.A., Pisman, L.M.: Fingering instability of thin evaporating liquid films. Phys. Rev. E **65**(2), 021602 (2002). https://doi.org/10.1103/PhysRevE.65.021602
11. Oron, A., Davis, S.H., Bankoff, S.G.: Long-scale evolution of liquid thin films. Rev. Mod. Phys. **69**, 931–980 (1997)
12. Pismen, L.M., Pomeau, Y.: Disjoining potential and spreading of thin liquid layers in the diffuse-interface model coupled to hydrodynamics. Phys. Rev. E. **62**, 2480–2492 (2000)
13. Sibley, D.N., Nold, A., Savva, N., Kalliadasis, S.: A comparison of slip, disjoining pressure, and interface formation models for contact line motion through asymptotic analysis of thin two-dimensional droplet spreading. J. Eng. Math. **94**, 19–41 (2015)
14. Thiele, U: Thin film evolution equations from (evaporating) dewetting liquid layers to epitaxial growth. J. Phys. Condens. Matter. **22**, 1–11 (2010)
15. Thiele, U: Note on thin film equations for solutions and suspensions. Eur. Phys. J. Special Topics **197**, 213–220 (2011)
16. Wodo, O., Ganapathysubramanian, B.: Modeling morphology evolution during solvent-based fabrication of organic solar cells. Comput. Mater. Sci. **55**, 113–126 (2012)

On the Sensitivity to Model Parameters in a Filter Stabilization Technique for Advection Dominated Advection-Diffusion-Reaction Problems

Kayla Bicol and Annalisa Quaini

Abstract We consider a filter stabilization technique with a deconvolution-based indicator function for the simulation of advection dominated advection-diffusion-reaction (ADR) problems with under-refined meshes. The proposed technique has been previously applied to the incompressible Navier-Stokes equations and has been successfully validated against experimental data. However, it was found that some key parameters in this approach have a strong impact on the solution. To better understand the role of these parameters, we consider ADR problems, which are simpler than incompressible flow problems. For the implementation of the filter stabilization technique to ADR problems we adopt a three-step algorithm that requires (1) the solution of the given problem on an under-refined mesh, (2) the application of a filter to the computed solution, and (3) a relaxation step. We compare our deconvolution-based approach to classical stabilization methods and test its sensitivity to model parameters on a 2D benchmark problem.

Keywords Advection-diffusion-reaction problems · Nonlinear filtering · Approximate deconvolution

1 Introduction

We adapt to time-dependent advection-diffusion-reaction (ADR) problems a filter stabilization technique proposed in [12] for evolution equations and mostly developed for the Navier-Stokes equations [4, 12, 22]. This technique applied to the Navier-Stokes equations has been extensively tested on both academic problems [4, 12, 22] and realistic applications [2, 26]. It was found in [2] that key parameters in this approach have a strong impact on the solution. In order to understand the role

---

K. Bicol · A. Quaini (✉)
University of Houston, Houston, TX, USA
e-mail: kmbicol@math.uh.edu; quaini@math.uh.edu

H. van Brummelen et al. (eds.), *Numerical Methods for Flows*,
Lecture Notes in Computational Science and Engineering 132,
https://doi.org/10.1007/978-3-030-30705-9_12

of these key parameters, we apply the filter stabilization technique in the simplified context of ADR problems.

It is well known that the Galerkin method for advection dominated ADR problems can lead to unstable solutions with spurious oscillations [6, 8, 10, 15, 16, 18, 24]. The proposed stabilization technique cures these oscillations by using an indicator function to tune the amount and location of artificial viscosity. The main advantage of this technique is that it can be easily implemented in legacy solvers.

2 Problem Definition

We consider a time-dependent advection-diffusion-reaction problem defined on a bounded domain $\Omega \in \mathbb{R}^d$, with $d = 2, 3$, over a time interval of interest $(0, T]$:

$$\partial_t u - \mu \Delta u + \nabla \cdot (\mathbf{b} u) + \sigma u = f \quad \text{in } \Omega \times (0, T], \tag{1}$$

endowed with boundary conditions:

$$u = u_D \qquad \text{on } \partial\Omega_D \times (0, T], \tag{2}$$

$$\mu \nabla u \cdot \mathbf{n} = g \quad \text{on } \partial\Omega_N \times (0, T], \tag{3}$$

and initial condition $u = u_0$ in $\Omega \times \{0\}$. Here $\overline{\partial\Omega_D} \cup \overline{\partial\Omega_N} = \overline{\partial\Omega}$ and $\partial\Omega_D \cap \partial\Omega_N = \emptyset$ and u_D, g, and u_0 are given. In (1)–(3), μ is a diffusion coefficient, $\mathbf{b}$ is an advection field, σ is a reaction coefficient, and f is the forcing term. For the sake of simplicity, we consider μ, $\mathbf{b}$, and σ constant.

Let L be a characteristic macroscopic length for problem (1)–(3). To characterize the solution of problem (1)–(3), we introduce the Péclet number: $\mathbb{P}\mathrm{e} = ||\mathbf{b}||_\infty L/(2\mu)$. We will assume that the problem is convection dominated, i.e. $||\mathbf{b}||_\infty \gg \mu$, which implies large Péclet numbers. Notice that the role of the $\mathbb{P}\mathrm{e}$ for advection-diffusion-reaction problems is similar to the role played by the Reynolds number for the Navier-Stokes equations.

In order to write the variational formulation of problem (1)–(3), we define the following spaces:

$$V = \left\{ v : \Omega \to \mathbb{R},\ v \in H^1(\Omega),\ v = u_D \text{ on } \partial\Omega_D \right\},$$

$$V_0 = \left\{ v : \Omega \to \mathbb{R},\ v \in H^1(\Omega),\ v = 0 \text{ on } \partial\Omega_D \right\}.$$

and bilinear form

$$b(u, w) = (\mu \nabla u, \nabla w)_\Omega + (\nabla \cdot (\mathbf{b} u), w)_\Omega + (\sigma u, w)_\Omega. \tag{4}$$

The variational form of problem (1)–(3) reads: Find $u \in V$ such that

$$(\partial_t u, w)_\Omega + b(u, w) = (f, w)_\Omega + (g, w)_{\partial\Omega_N}, \quad \forall w \in V_0. \tag{5}$$

The conditions for existence and unicity of the solution of problem (5) can be found, e.g., in [24, Chapter 12].

For the time discretization of problem (5), we consider the backward Euler scheme for simplicity. Let $\Delta t \in \mathbb{R}$, $t^n = n\Delta t$, with $n = 0, \ldots, N_T$ and $T = N_T \Delta t$. We denote by y^n the approximation of a generic quantity y at the time t^n. Problem (5) discretized in time reads: given $u^0 = u_0$, for $n \geq 0$ find $u^{n+1} \in V$ such that

$$\frac{1}{\Delta t}(u^{n+1}, w)_\Omega + b(u^{n+1}, w) = \frac{1}{\Delta t}(u^n, w)_\Omega + (f^{n+1}, w)_\Omega + (g^{n+1}, w)_{\partial\Omega_N}, \tag{6}$$

$\forall w \in V_0$. We remark that time discretization approximates problem (5) by a sequence of quasi-static problems (6). In fact, we can think of Eq. (6) as the variational formulation of problem:

$$L_t u^{n+1} = f_t^{n+1}, \tag{7}$$

where

$$L_t u^{n+1} = -\mu \Delta u^{n+1} + \nabla \cdot (\mathbf{b} u^{n+1}) + \left(\sigma + \frac{1}{\Delta t}\right) u^{n+1}, \quad f_t^{n+1} = \frac{u^n}{\Delta t} + f^{n+1}.$$

With regard to space discretization, we use the Finite Element method. Let $\mathscr{T}_h = \{K\}$ be a generic, regular Finite Element triangulation of the domain Ω composed by a set of finite elements, indicated by K. As usual, h refers to the largest diameter of the elements of $\mathscr{T}_h$. Let V_h and $V_{0,h}$ be the finite element spaces approximating V and V_0, respectively. The fully discrete problem reads: given u_h^0, for $n \geq 0$ find $u_h^{n+1} \in V_h$ such that

$$\frac{1}{\Delta t}(u_h^{n+1}, w)_\Omega + b(u_h^{n+1}, w) = \frac{1}{\Delta t}(u_h^n, w)_\Omega + (f_h^{n+1}, w)_\Omega + (g_h^{n+1}, w)_{\partial\Omega_N}, \tag{8}$$

$\forall w \in V_{0,h}$, where u_h^0, f_h^{n+1}, and g_h^{n+1} are appropriate finite element approximations of u_0, f^{n+1}, and g^{n+1}, respectively. It is known that the solution of problem (8) converges optimally to the solution of problem (5). However, the Finite Element method can perform poorly if the coerciveness constant of the bilinear form (4) is small in comparison with its continuity constant. In particular, the error estimate can have a very large multiplicative constant if μ is small with respect to $||\mathbf{b}||_\infty$, i.e. when $\mathbb{P}\mathrm{e}$ is large. In those cases, the finite element solution u_h can be globally polluted with strong spurious oscillations. To characterize the solution of problem (8), we introduce the local counterpart of the Péclet number: $\mathbb{P}\mathrm{e}_h = ||\mathbf{b}||_\infty h/(2\mu)$.

Several stabilization techniques have been proposed to eliminate, or at least reduce, the numerical oscillations produced by the standard Galerkin method in case of large $\mathbb{P}\mathrm{e}$. In the next section, we will go over a short review of these stabilization techniques before introducing our filter stabilization method in Sect. 3.

2.1 Overview of Stabilization Techniques

We will restrict our attention to stabilization techniques that consists of adding a stabilization term $b_s(u_h^{n+1}, w)$ to the left-hand side of time-discrete problem (6). In the following, we will use τ to denote a stabilization parameter that can depend on the element size h and the equation coefficients. Parameter τ takes different values for the different stabilization schemes. We will use the broken inner product $(\cdot,\cdot)_K = \sum_K(\cdot,\cdot)$, where $\sum_K$ denotes summation over all the finite elements.

Perhaps the easiest way to stabilize problem (6) is by introducing artificial viscosity either in the whole domain, leading to $b_s(u_h^{n+1}, w) = (\tau\nabla u_h^{n+1}, \nabla w)_K$, or streamwise, leading to $b_s(u_h^{n+1}, w) = (\tau\mathbf{b}\cdot\nabla u_h^{n+1}, \mathbf{b}\cdot\nabla w)_K$. See, e.g., [16, 24]. In this way, the effective $\mathbb{P}\mathrm{e}_h$ becomes smaller. The artificial viscosity τ in these schemes is proportional to h. The drawbacks of these schemes are that they are first order accurate only and not strongly consistent. An improvement over the artificial viscosity schemes is given by the strongly consistent stabilization methods. In fact, strong consistency allows the stabilized method to maintain the optimal accuracy.

Let us introduce the residual for problem (7) and the skew-symmetric part of operator L_t:

$$R(u_h^{n+1}) = f_t^{n+1} - L_t u_h^{n+1}, \quad L_{SS}v = \frac{1}{2}\nabla\cdot(\mathbf{b}v) + \frac{1}{2}\mathbf{b}\cdot\nabla v.$$

One of the most popular strongly consistent stabilized finite element methods is the Streamline Upwind Petrov-Galerkin (SUPG) method [6], for which $b_s(u_h^{n+1}, w) = -(\tau R(u_h^{n+1}), L_{SS}w)_K$. The Galerkin Least Squares (GLS) method [18] is a generalization of the SUPG method: $b_s(u_h^{n+1}, w) = -(\tau R(u_h^{n+1}), L_t w)_K$. The Douglas-Wang method [10] replaces $L_t w$ in the GLS method with $-L_t^* w$, where L_t^* is the adjoint of operator L_t. Thus, we have: $b_s(u_h^{n+1}, w) = (\tau R(u_h^{n+1}), L_t^* w)_K$. For the SUPG, GLS, and Douglas-Wang methods, $\tau = \delta h_K/|\mathbf{b}|$ for $\delta > 0$. Finally, we mention a method based on the Variational Multiscale approach [15] called algebraic subgrid scale (ASGS), for which $b_s(u_h^{n+1}, w) = (\tau R(u_h^{n+1}), L_t^* w)_K$. The difference between the Douglas-Wang and ASGS method consists in the choice for parameter τ. One possibility for τ in the ASGS method is $\tau = [4\mu/h_K^2 + 2|\mathbf{b}|/h_K + \sigma]^{-1}$ (see [8]).

Table 1 Stabilization term $b_s(u_h^{n+1}, w)$ for some stabilization methods

	Stabilization method	$b_s(u_h^{n+1}, w)$
Not strongly consistent	Artificial viscosity	$(\tau \nabla u_h^{n+1}, \nabla w)_K$
	Streamline upwind	$(\tau \mathbf{b} \cdot \nabla u_h^{n+1}, \mathbf{b} \cdot \nabla w)_K$
Strongly consistent	Streamline upwind Petrov-Galerkin	$-(\tau R(u_h^{n+1}), L_{SS} w)_K$
	Galerkin least-squares	$-(\tau R(u_h^{n+1}), L_t w)_K$
	Douglas-Wang	$(\tau R(u_h^{n+1}), L_t^* w)_K$
	Variational multiscale, ASGS	$(\tau R(u_h^{n+1}), L_t^* w)_K$

We report in Table 1 a summary of the methods in this overview. All of the strongly consistent stabilization techniques come with stability estimates that improve the one that can be obtained for the Galerkin method. See [6, 8, 10, 15, 18, 24].

3 A Filter Stabilization Technique

We adapt to the time-dependent advection-diffusion-reaction problem defined in Sect. 2 a filter stabilization technique proposed in [12]. For the implementation of this stabilization technique we adopt an algorithm called *evolve-filter-relax* (EFR) that was first presented in [22]. The EFR algorithm applied to problem (7) with boundary conditions (2)–(3) reads: given u^n

(i) *Evolve*: find the intermediate solution v^{n+1} such that

$$L_t v^{n+1} = f_t^{n+1} \quad \text{in } \Omega, \tag{9}$$

$$v^{n+1} = u_D \quad \text{on } \partial\Omega_D, \tag{10}$$

$$\mu \nabla v^{n+1} \cdot \mathbf{n} = g \quad \text{on } \partial\Omega_N. \tag{11}$$

(ii) *Filter*: find $\overline{v}^{n+1}$ such that

$$\overline{v}^{n+1} - \delta^2 \nabla \cdot (a(v^{n+1}) \nabla \overline{v}^{n+1}) = v^{n+1} \quad \text{in } \Omega, \tag{12}$$

$$\overline{v}^{n+1} = u_D \quad \text{on } \partial\Omega_D, \tag{13}$$

$$\mu \nabla \overline{v}^{n+1} \cdot \mathbf{n} = 0 \quad \text{on } \partial\Omega_N. \tag{14}$$

Here, δ can be interpreted as the *filtering radius* (that is, the radius of the neighborhood were the filter extracts information) and $a(\cdot) \in (0, 1]$ is a scalar function called *indicator function*. The indicator function has to be such that

$a(v^{n+1}) \simeq 1$ where v^{n+1} does need to be filtered from spurious oscillations, and $a(v^{n+1}) \simeq 0$ where v^{n+1} does not need to be filtered.

(iii) *Relax*: set

$$u^{n+1} = (1 - \chi)v^{n+1} + \chi\overline{v}^{n+1}, \tag{15}$$

where $\chi \in (0, 1]$ is a relaxation parameter.

The EFR algorithm has the advantage of modularity: since the problems at steps (i) and (ii) are numerically standard, they can be solved with legacy solvers without a considerable implementation effort. Algorithm (9)–(15) is sensitive to the choice of key parameters δ and χ [2, 3]. A common choice for δ is $\delta = h$. However, in [2] it is suggested that taking $\delta = h$ might lead to excessive numerical diffusion and it is proposed to set $\delta = h_{min}$, where h_{min} is the length of the shortest edge in the mesh. As for χ, in [22] the authors support the choice $\chi = O(\Delta t)$ because it guarantees that the numerical dissipation vanishes as $h \to 0$ regardless of Δt. In [2] the value of χ is set with a heuristic formula which depends on both physics and discretization parameters.

Remark 3.1 The choice of the filtering radius δ is still an open problem when dealing with non-uniform grids. As we mentioned before, it is a common choice to set $\delta = h$, where h usually refers to the largest diameter of the elements of the mesh. When using this choice with non-uniform grids, the region where the filter has a significant effect would not be guaranteed to be confined within a single element and may include also one (or more) neighboring elements for the smallest elements. This could lead to excessive numerical diffusion. In order to avoid this issue, in Sect. 4 we use uniform, structured grids.

Different choices of $a(\cdot)$ for the Navier-Stokes equations have been proposed and compared in [5, 19, 22, 25]. Here, we focus on a class of deconvolution-based indicator functions:

$$a(u) = a_D(u) = |u - D(F(u))|, \tag{16}$$

where F is a linear, invertible, self-adjoint, compact operator from a Hilbert space V to itself, and D is a bounded regularized approximation of F^{-1}. In fact, since F is compact, the inverse operator F^{-1} is unbounded. The composition of the two operators F and D can be interpreted as a low-pass filter.

A possible choice for D is the Van Cittert deconvolution operator D_N, defined as

$$D_N = \sum_{n=0}^{N}(I - F)^n.$$

The evaluation of a_D with $D = D_N$ (deconvolution of order N) requires then to apply the filter F a total of $N + 1$ times. Since F^{-1} is not bounded, in practice N

is chosen to be small, as the result of a trade-off between accuracy (for a regular solution) and filtering (for a non-regular one).

We select F to be the linear Helmholtz filter operator F_H [14] defined by

$$F = F_H \equiv \left(I - \delta^2 \Delta\right)^{-1}.$$

It is possible to prove [11] that

$$u - D_N(F_H(u)) = (-1)^{N+1} \delta^{2N+2} \Delta^{N+1} F_H^{N+1} u. \tag{17}$$

Therefore, $a_{D_N}(u)$ is close to zero in the regions of the domain where u is smooth. Indicator function (16) with $D = D_N$ and $F = F_H$ has been proposed in [4] for the Navier-Stokes equations. Algorithm (9)–(15) with indicator function (16) is also sensitive to the choice of N [2, 3].

In order to compare our approach with the stabilization techniques reported in Sect. 2.1, let us assume that problem (1) is supplemented with homogeneous Dirichlet boundary conditions on the entire boundary, i.e. $\partial\Omega_D = \partial\Omega$ and $u_D = 0$ in (2). Let us start by writing the weak form of Eq. (9):

$$(L_t v^{n+1}, w) = \left(f_t^{n+1}, w\right)_{\Omega}. \tag{18}$$

Next, we apply operator L_t to Eq. (12) and write the corresponding weak form, using also Eq. (18):

$$(L_t \overline{v}^{n+1}, w) - (\nabla \cdot (\overline{\mu} \nabla \overline{v}^{n+1}), L_t^* w) = \left(f_t^{n+1}, w\right)_{\Omega}, \quad \overline{\mu} = \delta^2 a(v^{n+1}). \tag{19}$$

Here, $\overline{\mu}$ is the artificial viscosity introduced by our stabilization method. Now, we multiply Eq. (18) by $(1 - \chi)$ and add it to Eq. (19) multiplied by χ. Using the relaxation step (15), we obtain:

$$(L_t u^{n+1}, w) - \chi(\nabla \cdot (\overline{\mu} \nabla \overline{v}^{n+1})), L_t^* w) = \left(f_t^{n+1}, w\right)_{\Omega}, \tag{20}$$

The second term at the left-hand side in (20) is the stabilization term added by the filter stabilization technique under consideration. Notice that Eq. (20) is a consistent perturbation of the original advection-diffusion-reaction problem. The perturbation vanishes with coefficient χ, which goes to zero with the discretization parameters (recall that a possible choice is $\chi = O(\Delta t)$). Using (15) once more, we can rewrite the stabilization term as:

$$b_s(u^{n+1}, w) = -(\nabla \cdot (\overline{\mu} \nabla u^{n+1}), L_t^* w) + (1 - \chi)(\nabla \cdot (\overline{\mu} \nabla v^{n+1}), L_t^* w). \tag{21}$$

We see that the stabilization term here does not depend only on the end-of-step solution u^{n+1}. We remind that usually $\delta = h$. Thus, as $h \to 0$ the artificial viscosity $\overline{\mu}$ in (19) vanishes. It is then easy to see that the filter stabilization technique we consider is consistent, although not strongly.

4 Numerical Results

We consider a benchmark test proposed in [20]. The prescribed solution is given by:

$$\begin{aligned} u(x, y, t) =& 16 \sin(\pi t) x(1-x) y(1-y) \\ & \cdot \left[\frac{1}{2} + \frac{\arctan(2\mu^{-1/2}(0.25^2 - (x-0.5)^2 - (y-0.5^2)))}{\pi} \right], \end{aligned} \tag{22}$$

in $\Omega = (0, 1) \times (0, 1)$ and in time interval $(0, 0.5]$. We set $\sigma = 1$, $\mu = 10^{-5}$, and $\mathbf{b} = [2, 3]^T$, which yield $\mathbb{P}\mathrm{e} = 1.5 \cdot 10^5$. Solution (22) is a hump changing its height in the course of the time. The internal layer in solution (22) has size $O(\sqrt{\mu})$. The forcing term f in (1) and the initial condition u_0 follow from (22). We impose boundary condition (2) with $u_D = 0$ on the entire boundary, which is consistent with exact solution (22). We use this test to compare the solution computed by the EFR method with the solution given by other methods, and to show the sensitivity of the solution computed by the EFR method to parameters N and δ. The sensitivity to χ will be object of future work. All the computational results have been obtained with FEniCS [1, 13, 23].

We take $\Delta t = 10^{-3}$. We consider structured meshes with 5 different refinement levels $\ell = 0, \cdots, 4$ and $\mathbb{P}_2$ finite elements. Triangulation $\mathcal{T}_{h_\ell}$ of Ω consists of n_ℓ^2 sub-squares, each of which is further divided into 2 triangles. The associated mesh size is $h_\ell = \sqrt{2}/n_\ell$. In Table 2, we report n_ℓ and $\mathbb{P}\mathrm{e}_h$ for each of the meshes under consideration. We see that even on the finest mesh, the local Péclet number is much larger than 1. Table 2 gives also the value of χ used for the EFR method on the different meshes.

Since the problem is convection-dominated and the solution has a (internal) layer, the use of a stabilization method is necessary. See Fig. 1(left) for a comparison of the solution at $t = 0.5$ computed on mesh $\ell = 0$ with the standard Galerkin element method, the SUPG method, and the EFR method with $\delta = 1/n_\ell$ and $N = 0$.We

Table 2 Number of partitions n_ℓ for each side, local Péclet number $\mathbb{P}\mathrm{e}_h$, and value of χ for the EFR method for the meshes associated to 5 different refinement levels

Refinement level	$\ell = 0$	$\ell = 1$	$\ell = 2$	$\ell = 3$	$\ell = 4$
n_ℓ	25	50	100	200	400
$\mathbb{P}\mathrm{e}_h$	8485.3	4242.6	2121.3	1060.7	530.3
χ	1	1/2	1/4	1/16	1/256

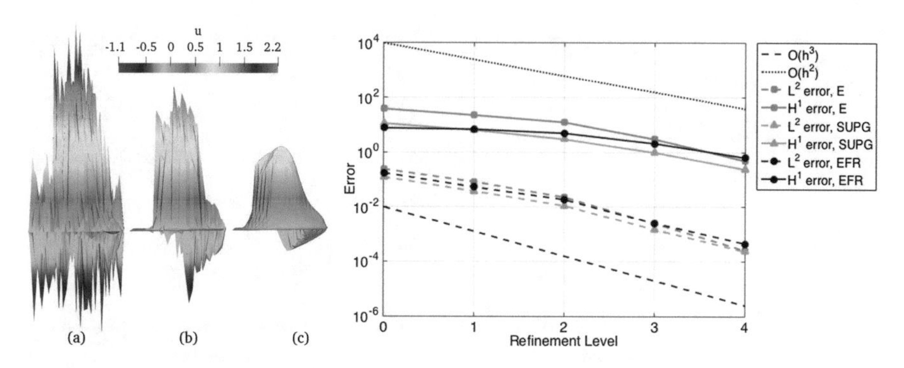

Fig. 1 Left: Solution at $t = 0.5$ computed on mesh $\ell = 0$ with (a) the standard Galerkin method, (b) the SUPG method, and (c) the EFR method with $\delta = 1/n_\ell$ and $N = 0$. Right: L^2 and H^1 norms of the error for u at $t = 0.5$ given by the standard Galerkin method (E), the SUPG method, and the EFR method plotted against the refinement level

see that the solution obtained with the non-stabilized method is globally polluted with spurious oscillations. Oscillations are still present in SUPG method (mainly at the top of the hump and in the right upper part of the domain), but they are reduced in amplitude. The amplitude of the oscillations is further reduced in the solution computed with the EFR method. The other strongly consistent stabilization methods reported in Sect. 2.1 give results very similar to the SUPG methods. For this reason, those results are omitted. Figure 1(right) shows the L^2 and H^1 norms of the error for the solution at $t = 0.5$ plotted against the mesh refinement level ℓ. We observe that the EFR method gives errors comparable to those given by the SUPG method on the coarser meshes. When the standard Galerkin method gives a smooth approximation of the solution, e.g. with mesh $\ell = 4$, the errors given by the EFR method are comparable to the errors given by the standard Galerkin method. In fact although mesh $\ell = 4$ is characterized by $\mathbb{P}e_h = 530.3$, the L^2-error is $4.5 \cdot 10^{-4}$ and the solution (not shown for space constraints) does not display oscillations.

In Fig. 2, we report the minimum value (left) and maximum value (right) of the solution at $t = 0.5$ computed by the standard Galerkin method, the SUPG method, and EFR method with $\delta = 1/n_\ell$ and $N = 0$ plotted against the refinement level ℓ. The SUPG method does not eliminate the under- and over-shoots given by a standard Galerkin method but it reduces their amplitude. It is well-known that all finite element methods that rely on streamline diffusion stabilization produce under- and over-shoots in regions where the solution gradients are steep and not aligned with the direction of **b**. From Fig. 2, we see that the EFR method gives under- and overs-shoots of smaller or comparable amplitude when compared to the SUPG method. In some practical applications, such imperfections are small in magnitude and can be tolerated. In other cases, it is essential to ensure that the numerical solution remains nonnegative and/or devoid of spurious oscillations. This can be achieved with, e.g., discontinuity-capturing or shock-capturing techniques [7, 9, 17, 21]. However, this is outside the scope of the present work.

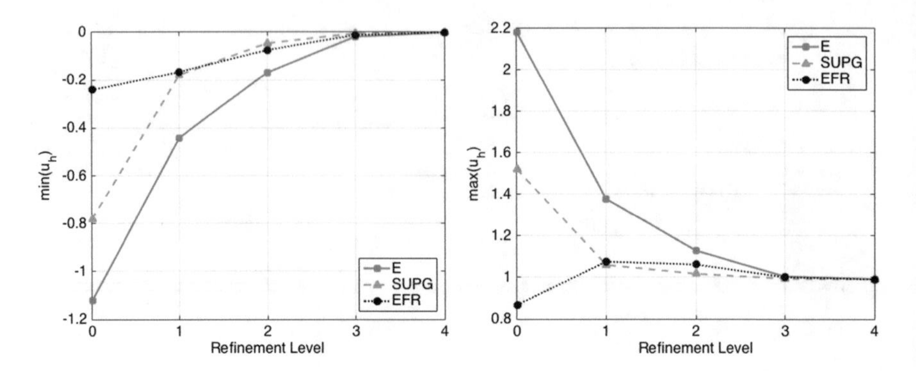

Fig. 2 Minimum value (left) and maximum value (right) of the solution at $t = 0.5$ computed by the standard Galerkin method (E), the SUPG method, and the EFR method with $\delta = 1/n_\ell$, $N = 0$, and $\chi = 1$ plotted against the refinement level

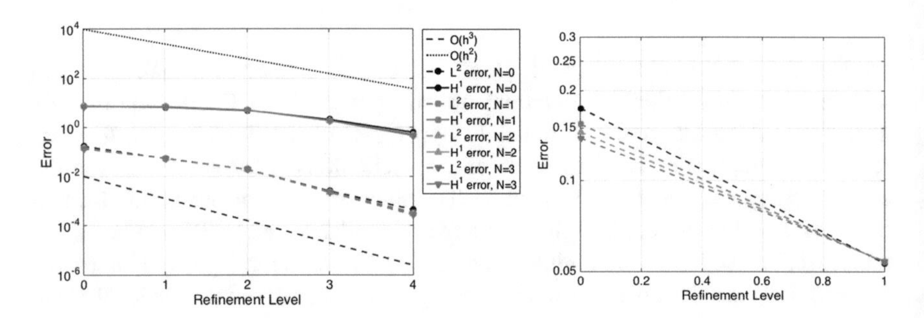

Fig. 3 Left: L^2 and H^1 norms of the error for u at $t = 0.5$ given by the EFR method with $\delta = 1/n_\ell$ and $N = 0, 1, 2, 3$ plotted against the refinement level. Right: zoomed-in view around $\ell = 0, 1$

Next, we focus on the EFR algorithm and vary the order of the deconvolution N. In Fig. 3(left) we show L^2 and H^1 norms of the error for u at $t = 0.5$ given by the EFR method with $\delta = 1/n_\ell$ and $N = 0, 1, 2, 3$ plotted against the refinement level. The only visible difference when N varies is for the finer meshes, with both errors slightly decreasing as N is increased. Figure 3(right) displays a zoomed-in view of Fig. 3(left) around $\ell = 0, 1$. It shows that also for the coarser meshes the errors get slightly smaller when N increases. We recall that indicator function (16) with $D = D_N$ requires to apply the Helmholtz filter $N + 1$ times. So, the slightly smaller errors for large N come with an increased computational time.

In Fig. 4, we report the indicator function at $t = 0.5$ for $\delta = 1/n_\ell$ and $N = 0, 1, 2, 3$. In all the cases, the largest values of the indicator function are around the edge of the hump. Moderate values are aligned with the direction of $\mathbf{b}$. While we see some differences in the indicator functions for $N = 0$ and $N = 1$, for $N > 1$ the indicator function does not seem to change substantially.

Finally, we fix $N = 0$ and vary δ. In Fig. 5(left) we show L^2 and H^1 norms of the error for u at $t = 0.5$ given by the EFR method with $\delta = c/n_\ell$, $c = 1, \sqrt{2}, 2, 5,$

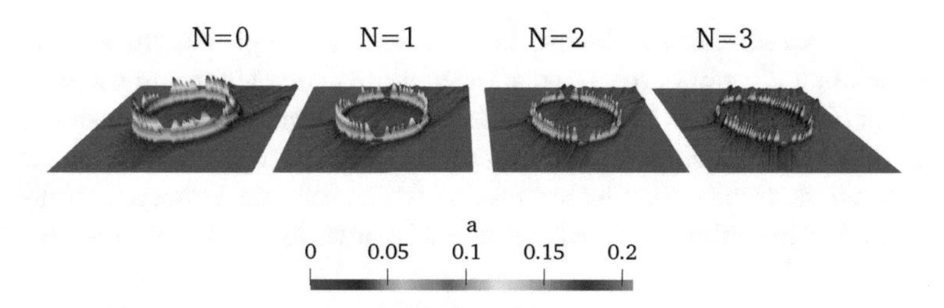

Fig. 4 Indicator function on mesh $\ell = 2$ at $t = 0.5$ for $\delta = 1/n_\ell$ and $N = 0, 1, 2, 3$

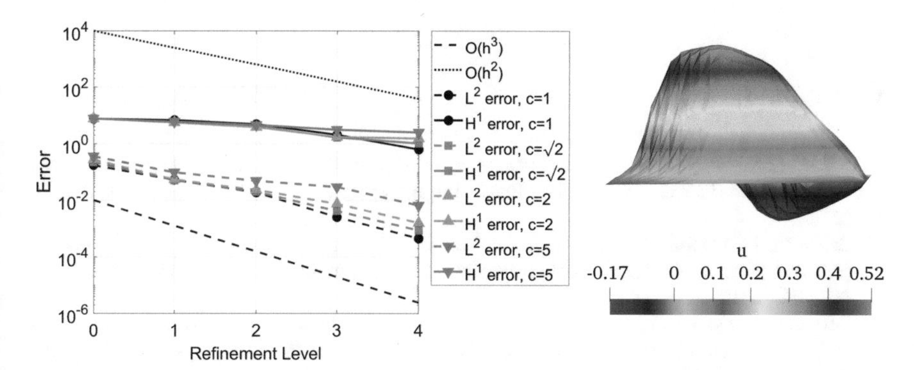

Fig. 5 Left: L^2 and H^1 norms of the error for u at $t = 0.5$ given by the EFR method with $N = 0$ and $\delta = c/n_\ell$, $c = 1, \sqrt{2}, 2, 5$, plotted against the refinement level. Right: solution at $t = 0.5$ computed by the EFR method with $\delta = 5/n_\ell$ on mesh $\ell = 0$

plotted against the refinement level. Notice that $c = \sqrt{2}$ corresponds to the choice $\delta = h$, while $c = 1$ corresponds to $\delta = h_{min}$. In [2], it was found that $\delta = h_{min}$ makes the numerical results (for a Navier-Stokes problem on unstructured meshes) in better agreement with experimental data. Our results confirm that $\delta = h_{min}$ is the best choice. In fact, it minimizes the error and gives optimal convergence rates. Higher values of c, i.e. $c > 1$, seem to spoil the convergence rate of the EFR method. From Figs. 3(left) and 5(left) we see that the computed solution is much more sensitive to δ than it is to N. Figure 5(right) shows the solution at time $t = 0.5$ computed by the EFR method with $\delta = 5/n_\ell$ on mesh $\ell = 0$. Remember that δ is the filtering radius, i.e. the radius of the circle over which we average (in some sense) the solution. Thus, it is not surprising that for a large value of δ the EFR method has an over-smoothing effect.

5 Conclusions

We considered a deconvolution-based filter stabilization technique recently proposed for the Navier-Stokes equations and adapted it to the numerical solution of advection dominated advection-diffusion-reaction problems with under-refined meshes. Our stabilization technique is consistent, although not strongly. For the implementation of our approach we adopted a three-step algorithm called evolve-filter-relax (EFR) that can be easily realized within a legacy solver. We showed that the EFR algorithm is competitive when compared to classical stabilization methods on a benchmark problem that features an analytical solution. However, special care has to be taken in setting the filtering radius δ in order to avoid over-smoothing.

Acknowledgement This research has been supported in part by the NSF under grants DMS-1620384.

References

1. Alnæs, M.S., Blechta, J., Hake, J., Johansson, A., Kehlet, B., Logg, A., Richardson, C., Ring, J., Rognes, M.E., Wells, G.N.: The FEniCS Project version 1.5. Arch. Numer. Softw. **3**(100), (2015)
2. Bertagna, L., Quaini, A., Veneziani, A.: Deconvolution-based nonlinear filtering for incompressible flows at moderately large Reynolds numbers. Int. J. Numer. Methods Fluids **81**(8), 463–488 (2016)
3. Bertagna, L., Quaini, A., Rebholz, L.G., Veneziani, A.: On the sensitivity to the filtering radius in Leray models of incompressible flow. In: Computational Methods in Applied Sciences. Springer-ECCOMAS series, pp. 111–130. Springer, Cham (2019)
4. Bowers, A.L., Rebholz, L.G.: Numerical study of a regularization model for incompressible flow with deconvolution-based adaptive nonlinear filtering. Comput. Methods Appl. Mech. Eng. **258**, 1–12 (2013)
5. Bowers, A.L., Rebholz, L.G., Takhirov, A., Trenchea, C.: Improved accuracy in regularization models of incompressible flow via adaptive nonlinear filtering. Int. J. Numer. Methods Fluids **70**(7), 805–828 (2012)
6. Brooks, A.N., Hughes, T.J.R.: Streamline upwind/Petrov-Galerkin formulations for convection dominated flows with particular emphasis on the incompressible Navier-Stokes equation. Comput. Methods Appl. Mech. Eng. **32**, 199–259 (1982)
7. Codina, R.: A discontinuity-capturing crosswind-dissipation for the finite element solution of the convection-diffusion equation. Comput. Methods Appl. Mech. Eng. **110**(3), 325–342 (1993)
8. Codina, R.: Comparison of some finite element methods for solving the diffusion-convection-reaction equation. Comput. Methods Appl. Mech. Eng. **156**, 185–210 (1998)
9. de Sampaio, P.A.B., Coutinho, A.L.G.A.: A natural derivation of discontinuity capturing operator for convection-diffusion problems. Comput. Methods Appl. Mech. Eng. **190**(46), 6291–6308 (2001)
10. Douglas, J., Wang, J.: An absolutely stabilized finite element method for the Stokes problem. Math. Comp. **52**, 495–508 (1989)
11. Dunca, A., Epshteyn, Y.: On the Stolz-Adams deconvolution model for the large-eddy simulation of turbulent flows. SIAM J. Math. Anal. **37**(6), 1980–1902 (2005)
12. Ervin, V.J., Layton, W.J., Neda, M.: Numerical analysis of filter-based stabilization for evolution equations. SIAM J. Numer. Anal. **50**(5), 2307–2335 (2012)
13. FEniCS Project. https://fenicsproject.org
14. Germano, M.: Differential filters of elliptic type. Phys. Fluids **29**, 1757–1758 (1986)
15. Hughes, T.J.R.: Multiscale phenomena: green's function, the Dirichlet-to-Neumann formulation, subgrid scale models, bubbles and the origins of stabilized formulations. Comput. Methods Appl. Mech. Eng. **127**, 387–401 (1995)

16. Hughes, T.J.R., Brooks, A.N.: A multidimensional upwind scheme with no crosswind diffusion. In: Hughes, T.J.R. (ed.), FEM for Convection Dominated Flows. ASME, New York (1979)
17. Hughes, T.J.R., Mallet, M.: A new finite element formulation for computational fluid dynamics: IV. A discontinuity-capturing operator for multidimensional advective-diffusive systems. Comput. Methods Appl. Mech. Eng. **58**(3), 329–336 (1986)
18. Hughes, T.J.R., Franca, L.P., Hulbert, G.M.: A new finite element formulation for computational fluid dynamics: VIII. The Galerkin/least-squares method for advective-diffusive equations. Comput. Methods Appl. Mech. Eng. **73**, 173–189 (1989)
19. Hunt, J.C., Wray, A.A., Moin, P.: Eddies Stream and Convergence Zones in Turbulent Flows. Technical Report CTR-S88, CTR report (1988)
20. John, V., Schmeyer, E.: Finite element methods for time-dependent convection-diffusion-reaction equations with small diffusion. Comput. Methods Appl. Mech. Eng. **198**(3), 475–494 (2008)
21. Kuzmin, D.: A Guide to Numerical Methods for Transport Equations. University Erlangen-Nuremberg, Erlangen (2010)
22. Layton, W., Rebholz, L.G., Trenchea, C.: Modular nonlinear filter stabilization of methods for higher Reynolds numbers flow. J. Math. Fluid Mech. **14**, 325–354 (2012)
23. Logg, A., Mardal, K.-A., Wells, G.N., et al.: Automated Solution of Differential Equations by the Finite Element Method. Springer, Heidelberg (2012)
24. Quarteroni, A., Valli, A.: Numerical Approximation of Partial Differential Equations. Springer, Heidelberg (1994)
25. Vreman, A.W.: An eddy-viscosity subgrid-scale model for turbulent shear flow: algebraic theory and applications. Phys. Fluids **16**(10), 3670–3681 (2004)
26. Xu, H., Piccinelli, M., Leshnower, B.G., Lefieux, A., Taylor, W.R., Veneziani, A.: Coupled morphological–hemodynamic computational analysis of type B aortic dissection: a longitudinal study. Ann. Biomed. Eng. **46**(7), 927–939 (2018)

One-Dimensional Line SIAC Filtering for Multi-Dimensions: Applications to Streamline Visualization

Jennifer K. Ryan and Julia Docampo-Sanchez

Abstract Smoothness-Increasing Accuracy-Conserving (SIAC) filters for Discontinuous Galerkin (DG) methods are designed to increase the smoothness and improve the convergence rate of the DG solution through post-processing. These advantages can be exploited during flow visualization, for example by applying the SIAC filter to DG data before streamline computations. However, introducing these filters in engineering applications can be challenging since the filter is based on a convolution over an area of $[(r + \ell + 1)h]^d$, where d is the dimension, h is the uniform element length, and r and ℓ depend on the construction of the filter. This can become computationally prohibitive as the dimension increases. However, by exploiting the underlying mathematical framework, this problem can be overcome in order to realize a technique that allows for appropriate filtering along a streamline curve. Numerical experiments of such an idea were proposed in Walfisch et al. (J Sci Comput 38(2):164–184, 2009). Here, we review the introduction of the Line SIAC post-processing filter by Docampo et al. (SIAM J Sci Comput 39(5):A2179–A2200, 2017), which showed how the underlying mathematics can be exploited to make the SIAC filter more tractable and illustrate the promise of LSIAC in assisting in streamline visualization.

Keywords Discontinuous Galerkin · Post-processing · Filtering · SIAC filtering · Streamlines

---

J. K. Ryan (✉)
Applied Mathematics & Statistics, Colorado School of Mines, Golden, CO, USA
e-mail: jkryan@mines.edu

J. Docampo-Sanchez
Aerospace Computational Design Lab, Department of Aeronautics and Astronautics, Massachusetts Institute of Technology, Cambridge, MA, USA
e-mail: docampo@mit.edu

H. van Brummelen et al. (eds.), *Numerical Methods for Flows*, Lecture Notes in Computational Science and Engineering 132,
https://doi.org/10.1007/978-3-030-30705-9_13

1 Introduction and Motivation

The goal of a numerical simulation is to provide an approximate solution of a model designed to understand a physical problem such as flow past an aircraft or weather forecasting. Hence, it is necessary to apply visualisation techniques that extract and evaluate the information from the numerical solution. Vector field visualisation through streamlines is a popular post-processing technique employed to understand fluid flow behaviour. Streamlines, curves everywhere tangent to the velocity field, are described by an Ordinary Differential Equation (ODE) which can be solved using many techniques including Runge-Kutta methods [1]. However, the theoretical error estimates of these methods rely on Taylor series and therefore assume smooth field conditions. Vector fields obtained through a DG method present constraints since the solution is only continuous inside each element. A suitable solver for computing streamlines over non-smooth fields has to be able to detect, locate and effectively step over a discontinuity [6]. This can be achieved through a Predictor-Corrector method [7, 10] or by controlling the error through adaptive step size methods such as the Runge-Kutta-Fehlberg solvers [4, 5]. The downside of these methods is that they require intense computations since detecting and passing over a discontinuity implies increasing the number of evaluations per iteration.

Alternatively, Smoothness-Increasing Accuracy-Conserving (SIAC) filters can be applied to obtain a local smooth solution where a relatively simple ODE solver can be implemented. Furthermore, since the filtered solution usually reduces the error from the DG approximation, the new filtered velocity field should lead to more accurate field lines [17, 18]. An example streamline calculation is illustrated in Fig. 1a. In this figure, three different streamlines are shown, with the exact streamline in black, the streamline obtained using DG in blue, and the SIAC filtered DG solution in red. For the top two streamlines, all three calculations coincide. For the bottom streamline, the calculation using DG data without post-processing deviates from the exact streamline. However, the SIAC post-processed streamline remains close to the exact streamline. Traditionally, the post-processor for a $2D$ field is a tensor product of one-dimensional functions. This can pose computational challenges as the area required to post-process one point in d-dimensions is $[(r + \ell + 1)h]^d$, where r and ℓ depend on the construction of the filter and h is the uniform mesh size. This is illustrated in Fig. 1b for $d = 2$. Each colored quadrilateral represents a section where Gaussian quadrature needs to be performed. Line SIAC significantly reduces the computational support as it is performed along a one-dimensional line. In addition to visualizing streamlines, Line SIAC post-processing can be of assistance when visualizing vorticity [8].

Applying SIAC filters for flow visualisation implies combining different kernel types. For example, during streamline computations, since particles can move across the entire field, the filter has to be able to post-process points at the boundaries of the computational domain. In previous experiments, numerical results were given for the one dimensional boundary filters. The error plots suggested that these filters are not as effective as the symmetric filters when reducing the error from the

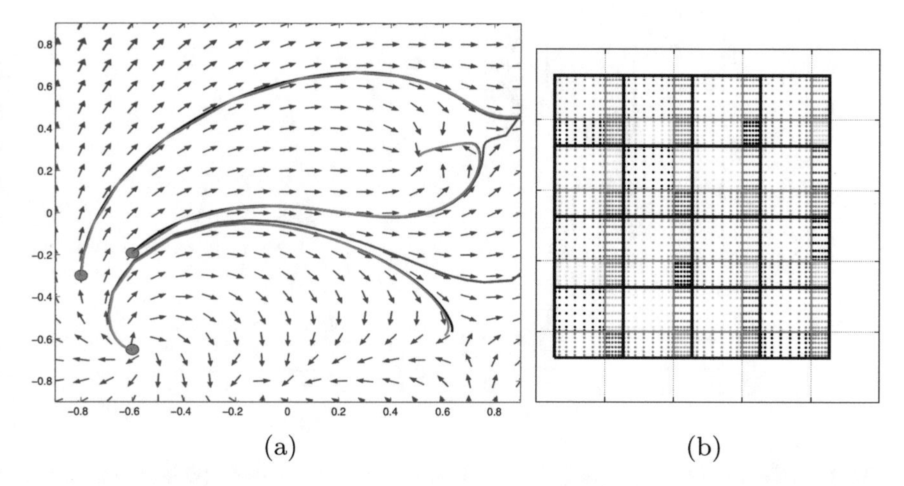

Fig. 1 (**a**) Plot of streamlines. (**b**) An illustration of the computational support size of the 2D kernel in order to post-process one point. Each colored quadrilateral represents a section where Gaussian quadrature needs to be performed

DG approximation [2]. Numerical results for Line filters suggest that alternative orientations to those for which superconvergence can be proven still lead to error reduction. This was observed for the cases where the filter was oriented using the flow direction or when symmetries from the initial condition were used to choose the orientation [3]. The results showed that for such alignments, the filtered solution presented lower errors than the original one. This can be exploited further near the boundaries, rotating the filter conveniently to fit a symmetric kernel, thus avoiding shifting its support. Here, we present one investigation of the potential of Line SIAC filters for accuracy enhancement during flow visualisation. Further results are presented in [2].

2 Background

Traditionally, Smoothness-Increasing Accuracy-Conserving filtering has been designed to recover smoothness and extract up to $2p+1$ order accuracy [9, 12, 14–16]. The post-processing filter is applied to the given data at the final time,

$$u_h^\star(\overline{\mathbf{x}}, T) = \frac{1}{H^d}\int_{\mathbb{R}^d} \underbrace{K^{(r+1,\ell)}\left(\frac{\overline{x}_1 - y_1}{H}\right)\cdots K^{(r+1,\ell)}\left(\frac{\overline{x}_d - y_d}{H}\right)}_{\text{Tensor product kernel}} \underbrace{u_h(\mathbf{y}, T)}_{\text{Input data}}\, d\mathbf{y}, \tag{1}$$

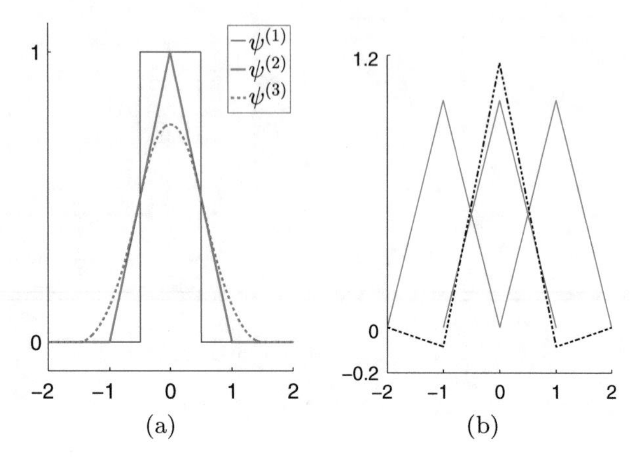

Fig. 2 (**a**) The first three B-splines ($\psi^{(1)}(x)$ in blue, $\psi^{(2)}(x)$ in red, and $\psi^{(3)}(x)$ in green). (**b**) The SIAC kernel for piecewise linear data using three B-splines of order two

where $\mathbf{y} = (y_1, y_2, \ldots, y_d)$ represents the cartesian coordinate system. The convolution kernel, $K^{(r+1,\ell)}(x)$, consists of a linear combination of $r + 1$ B-Splines of order ℓ :

$$K^{(r+1,\ell)}(\eta) = \sum_{\gamma=0}^{r} \underbrace{c_\gamma}_{\text{weights}} \underbrace{\psi^{(\ell)}(\eta - x_\gamma)}_{\text{B-Splines}}. \tag{2}$$

The weights in the kernel are designed to ensure that the post-processor is consistent and maintains the first r moments. The order of the B-spline, ℓ, effectively determines the amount of oscillations allowed in the error as well as allowing for continuity of $\ell - 2$. This leads to a kernel support size of $[(r + \ell + 1)H]^d$, where d is the dimension, and H is the filter scaling which is typically the uniform mesh size (cf. Fig. 1b). In Fig. 2a, an illustration of the first three B-splines is given. In Fig. 2b, the symmetric kernel for piecewise linear data using three B-splines of order two is given.

3 Line SIAC Filters: Reducing the Filter Dimension

Although the tensor product kernel has been shown to aid in reducing the errors and improve the ability to visualize streamlines resulting from DG solutions [17], the computational cost can be prohibitive with dimension. Instead, a Line SIAC filter is proposed to handle such applications.

The $2D$ Line SIAC filter is essentially a rotated 1D SIAC filter applied along a prescribed direction, $\Gamma(t) = t(\cos\theta, \sin\theta) + (\overline{x}, \overline{y})$, where θ is a fixed angle and $(\overline{x}, \overline{y})$ is the evaluation point. The Line SIAC kernel is constructed as a linear combination of these (scaled) B-Splines:

$$K_{\Gamma,H}^{(r+1,\ell)}(t) = \sum_{\gamma=-\lceil \frac{r}{2} \rceil}^{\lceil \frac{r}{2} \rceil} c_\gamma \psi_H^{(\ell)}(t-\gamma), \qquad \psi_H^{(\ell)}(t-\gamma) = \frac{1}{H}\psi^{(\ell)}\left(\frac{t}{H}-\gamma\right). \tag{3}$$

The 2D convolution for the Line SIAC filter in cartesian coordinate is given by:

$$u^\star(\overline{x}, \overline{y}) = \frac{1}{H}\int_\Gamma K_{\Gamma,H}\left(\frac{t}{H}\right) u_h(\Gamma(t))dt, \tag{4}$$

where, again, we have used that $\Gamma(t) = t(\cos\theta, \sin\theta) + (\overline{x}, \overline{y})$ and $||\Gamma'(t)|| = 1$.

4 Numerical Results

In the following experiments, the potential of Line filters for accuracy enhancement is demonstrated. More complete results that include comparisons to tensor product and boundary filters are given in [2].

We present the potential of the Line SIAC filter by applying it to the following field:

$$z = x + iy, \qquad u = Re(r), \qquad v = -Im(r),$$

where the field, CF1, was given by:

$$r = (z - (0.74 + 0.35i))(z - (0.68 - 0.59i))(z - (-0.11 - 0.72i)) \tag{5}$$

$$(\overline{z} - (-0.58 + 0.64i))(\overline{z} - (0.51 - 0.27i))(\overline{z} - (-0.12 + 0.84))^2. \tag{6}$$

This field has been studied before for 2D symmetric filtering in [17] and for more general filters in [9, 11]. The computational domain used for the simulations corresponded to $\Omega = [-1, 1] \times [-1, 1]$, using two uniform quadrilateral meshes made of 40×40 and 80×80 elements respectively. The unfiltered solutions were obtained by performing the L^2-projection of the function onto a piecewise polynomial basis which mimics a DG solution at the initial time.

This experiment was done over a DG approximation using $\mathbb{P}^1$ polynomials. The filtered solutions were obtained using a $K_H^{(3,2)}$ symmetric Line filter with scaling $H = \sqrt{2}h$, h being the DG mesh size. The flow based filters were calculated in the following way: the first orientation (at streamline seed) was chosen to be $3\pi/4$. The rest of the points were post-processed using the direction given by the last two computed streamline points. Both the unfiltered and filtered streamlines were computed using the RK2 method with time step $dt = 0.01$. The final time was determined by the exact streamline, corresponding to the last point inside the computational domain or when a streamline reached zero velocity. The exact streamlines were obtained by implementing the RK4 method with time step $dt = 1e^{-5}$ directly on the analytic velocity fields. The choice for the angle relies on the fact that for the proof of superconvergence, the rotation angle is $\theta = \arctan\left(\frac{h_x}{h_y}\right)$. For a mesh made of uniform square elements, this implies $\theta = \pi/4$ or $\theta = 3\pi/4$. In addition, since other orientations also allowed for error reduction, filters oriented using flow information have also been implemented.

Figure 3 shows streamlines belonging to the velocity field CF1 (Eq. (5)) using three different filter orientations based on the underlying mesh ($\theta = \pi/4, 3\pi/4$) and the flow direction. Observe how the flow based filters produced a diverging streamline for the lower seed (starting at a critical point) even after mesh refinement. The $\pi/4$ and $3\pi/4$ filtered streamlines converge towards the exact curve. Note that the $\pi/4$ Line filter performs better since for a coarse mesh (40×40 elements), the filtered streamline moved away from the exact solution initially and eventually converged towards the exact streamline.

Table 1 shows two error estimates; the first is a local error computed as a maximum error from each iteration,

$$\max_{n=0:N} e_n = \max_{n=0:N} d(p_n, \tilde{p}_n), \tag{7}$$

where $d(p, \tilde{p})$ denotes the $\mathbb{E}$uclidean distance, p_n and $\tilde{p}_n$ the exact and approximate solutions respectively. The global error corresponds to the difference between the solutions at final time. The errors in this table show that even when both the filtered and unfiltered streamlines converge, the filtered solution has generally lower values. Regarding the differences between both filters, the numbers are very similar, especially after mesh refinement. The results suggest that both the $\pi/4$ and $3\pi/4$ orientations are suitable for effective post-processing. These orientations should be chosen (whenever possible) instead of flow aligned filters since the latter ones require longer and more complicated computations and do not seem to produce more accurate streamlines.

5 Conclusions and Future Work

In this article, a small exploration into the applications of Line filtering during flow visualisation. A more detailed exploration is given in [2]. The Line SIAC filter has demonstrated the ability of these filters to recover smoothness and increase the accuracy from the original DG solution. Here, the filters have been applied to more general vector fields (including singularities) and the results suggest that this post-processor enhances the accuracy from the original solution, leading in this case to more accurate streamlines.

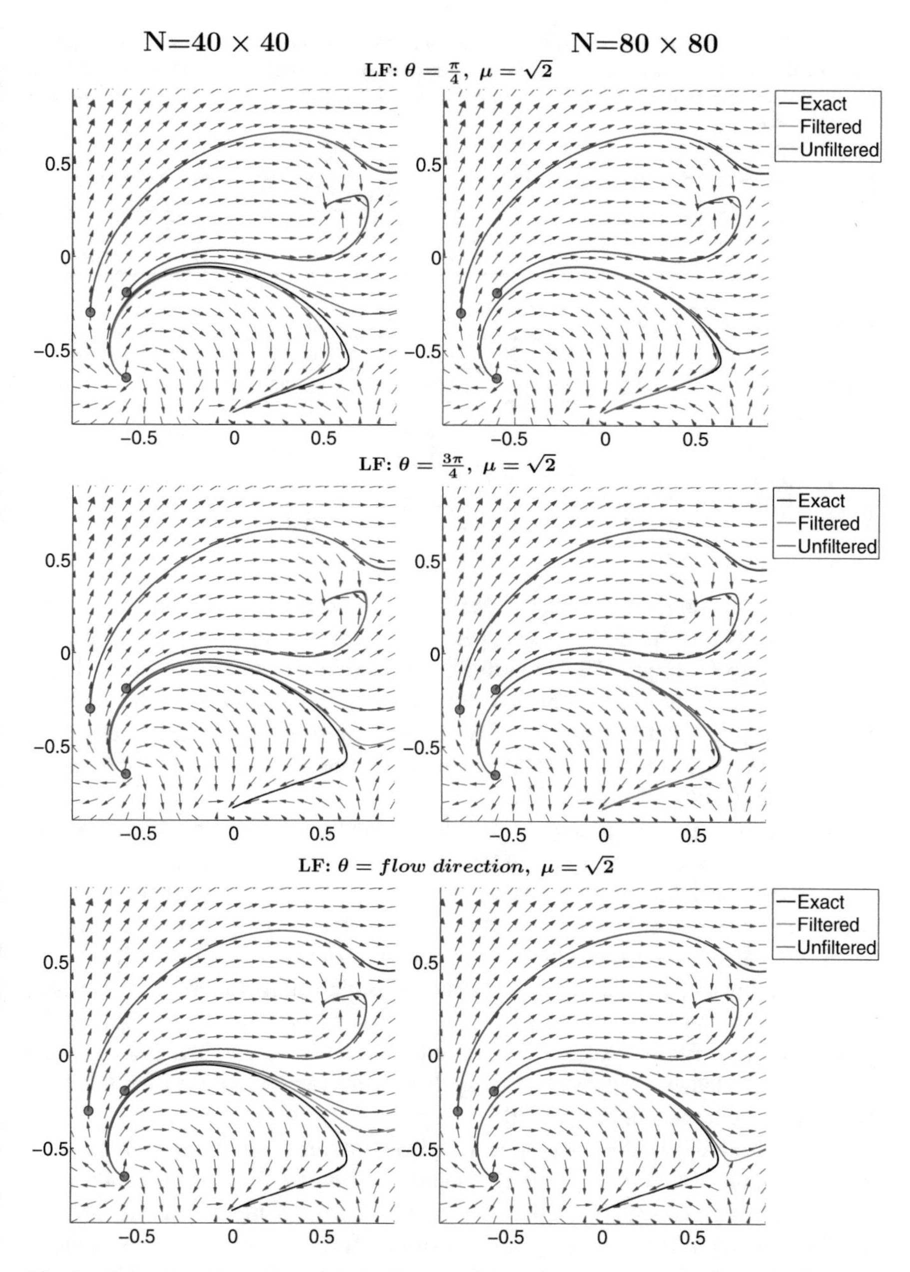

Fig. 3 Streamlines along CF1 (Eq. (5)) for two meshes ($N = 40 \times 40$ and $N = 80 \times 80$) before and after applying different symmetric Line Filters (LFs) using the RK2 solver with $dt = 0.01$. The plots where the exact curve cannot be seen is because it overlaps with the filtered streamline

Table 1 Maximum Distance (MD) taken as the greatest point distance between each iteration and Global Errors (GE) measuring the point distance at final time comparing unfiltered and filtered streamlines for Line filters over the CF1 velocity field and two different meshes ($N = 40 \times 40$ and $N = 80 \times 80$)

CF1						
			Line filtering: $H = \sqrt{2}h$			
	Unfiltered		$\theta = \pi/4$		$\theta = \frac{3\pi}{4}$	
Seed	MD	GE	MD	GE	MD	GE
$N = 40 \times 40$						
$(-0.6, -0.651)$	Diverged		2.1e−01	3.0e−02	DIV	DIV
$(-0.6, -0.192)$	4.0e−03	9.7e−04	6.1e−03	1.6e−04	3.1e−02	8.9e−05
$(-0.8, -0.3)$	4.1e−02	4.1e−02	3.8e−02	3.8e−02	2.2e−02	2.2e−02
$N= 80 \times 80$						
$(-0.6, -0.651)$	Diverged		4.3e−02	7.3e−03	4.1e−02	7.7e−03
$(-0.6, -0.192)$	2.6e−03	1.2e−04	5.5e−04	9.6e−06	1.3e−03	5.2e−06
$(-0.8, -0.3)$	3.6e−02	3.6e−02	2.7e−02	2.7e−02	2.9e−02	2.9e−02

Line filters using the symmetric kernel show excellent performance and the low computational times associated with them make them great candidates for engineering applications. The experiments suggested that mesh resolution has a strong effect on Line filters. For coarse meshes, this limitation can be overcome by increasing the order and number of B-Splines employed to build the Line kernel, producing a solution that matches the quality of the one obtained through the Tensor Product filter results obtained in [2].

Further, the experiments suggest that these filters (in particular filters along the streamline) can be suitable for accuracy enhancement during flow visualisation. In many cases where the unfiltered streamline diverged from the exact solution, the filtered curve converged back towards the exact curve. From this study it was not possible to conclude whether a particular orientation or type of boundary filter could give optimal results because there are many possible configurations. For instance, there is no guarantee that the kernel scalings used here ($h = h,\ H = \sqrt{2}h$) are the appropriate ones for flow aligned filters. In fact, it could be possible that since the rotation angle changes, the kernel scaling should also vary within the filter location. This will be addressed in future work.

To make LSIAC feasible for practical applications, the methodology for selecting both the kernel rotation and scaling should be robust enough so that the filter can be applied in general settings including non-linear solutions and unstructured meshes. Although SIAC filtering has shown to improve results over unstructured grids [8, 13], developing a solid theoretical background for such problems is an ongoing work. Furthermore, the reliance of the kernel scaling and rotation on the flux function needs to be established. Finally, the appropriate pairing between theory and efficient implementation should include a study of different implementations in order to ensure efficient computation.

Acknowledgements This research was sponsored by the European Office of Aerospace Research and Development (EOARD) under the U.S. Air Force Office of Scientific Research (AFOSR) under grant number FA8655-13-1-3017.

References

1. Butcher, J.C., Johnston, P.B.: Estimating local truncation errors for Runge-Kutta methods. J. Comput. Appl. Math. **45**(1), 203–212 (1993)
2. Docampo Sánchez, J.: Smoothness-Increasing Accuracy-Conserving (SIAC) Line Filtering: Effective Rotation for Multidimensional Fields. University of East Anglia, East Anglia (2017). Ph.D. thesis
3. Docampo Sánchez, J., Ryan, J.K., Mirzargar, M., Kirby, R.M.: Multi-dimensional filtering: reducing the dimension through rotation. SIAM J. Sci. Comput. **39**(5), A2179–A2200 (2017)
4. Fehlberg, E.: New high-order Runge-Kutta Formulas with step size control for systems of first-and second-order differential equations. In: Zeitschrift fur Angewandte Mathematik und Mechanik, vol. 44, p. T17. Wiley-VCH Verlag Gmbh Muhlenstrasse 33-34, Berlin (1964)
5. Fehlberg, E.: Low-Order Classical Runge-Kutta Formulas with Stepsize Control and their Application to Some Heat Transfer Problems. Tech. rep., NASA (1969)
6. Gear, C.W., Osterby, O.: Solving ordinary differential equations with discontinuities. ACM Trans. Math. Softw. **10**(1), 23–44 (1984)
7. Hull, T.E., Creemer, A.L.: Efficiency of predictor-corrector procedures. J. Assoc. Comput. Mach. **10**(3), 291–301 (1963)
8. Jallepalli, A., Docampo-Sánchez, J., Ryan, J.K., Haimes, R., Kirby, R.M.: On the treatment of field quantities and elemental continuity in FEM solutions. IEEE Trans. Vis. Comput. Graph. **24**(1), 903–912 (2018)
9. Ji, L., Van Slingerland, P., Ryan, J.K., Vuik, K.C.: Superconvergent error estimates for position-dependent smoothness-increasing accuracy-conserving (SIAC) post-processing of discontinuous Galerkin solutions. Math. Comput. **83**(289), 2239–2262 (2014)
10. Klopfenstein, R.W., Millman, R.S.: Numerical stability of a one-evaluation predictor-corrector algorithm for numerical solution of ordinary differential equations. Math. Comput. **22**(103), 557–564 (1968)
11. Li, X.: Smoothness-Increasing Accuracy-Conserving Filters for Discontinuous Galerkin Methods: Challenging the Assumptions of Symmetry and Uniformity. Delft University of Technology, Delft (2015). Ph.D. thesis
12. Mirzaee, H., Ryan, J.K., Kirby, R.M.: Efficient implementation of smoothness-increasing accuracy-conserving (SIAC) filters for discontinuous Galerkin solutions. J. Sci. Comput. **52**, 85–112 (2012)
13. Mirzaee, H., King, J., Ryan, J.K., Kirby, R.M.: Smoothness-increasing accuracy-conserving (SIAC) filters for discontinuous Galerkin solutions over unstructured triangular meshes. SIAM J. Sci. Comput. **35**, A212–A230 (2013)
14. Mirzargar, M., Ryan, J.K., Kirby, R.M.: Smoothness-increasing accuracy-conserving (SIAC) filtering and quasi-interpolation: a unified view. J. Sci. Comput. **67**(1), 237–261 (2016)
15. Ryan, J.K.: Exploiting Superconvergence Through Smoothness-Increasing Accuracy-Conserving (SIAC) Filtering, pp. 87–102. Springer, Cham (2015)
16. Ryan, J.K., Shu, C-W., Atkins, H.: Extension of a postprocessing technique for the discontinuous Galerkin method for Hyperbolic equations with application to an Aeroacoustic problem. SIAM J. Sci. Comput. **26**(3), pp. 821–843 (2005)

17. Steffen, M., Curtis, S., Kirby, R.M., Ryan, J.K.: Investigation of smoothness-increasing accuracy-conserving filters for improving streamline integration through discontinuous fields. IEEE Trans. Vis. Comput. Graph. **14**(3), pp. 680–692 (2008)
18. Walfisch, D., Ryan, J.K., Kirby, R.M., Haimes, R.: One-sided smoothness-increasing accuracy-conserving filtering for enhanced streamline integration through discontinuous fields. J. Sci. Comput. **38**(2), pp. 164–184 (2009)

A High Performance Computing Framework for Finite Element Simulation of Blood Flow in the Left Ventricle of the Human Heart

Jeannette Hiromi Spühler, Johan Jansson, Niclas Jansson, and Johan Hoffman

Abstract We present a high performance computing framework for finite element simulation of blood flow in the left ventricle of the human heart. The mathematical model is described together with the discretization method and the parallel implementation in Unicorn which is part of the open source software framework FEniCS-HPC. We show results based on patient-specific data that capture essential features observed with other computational models and imaging techniques, and thus indicate that our framework possesses the potential to provide relevant clinical information for diagnosis and medical treatment. Several other studies have been conducted to simulate the three dimensional blood flow in the left ventricle of the human heart with prescribed wall movement. Our contribution to the field of cardiac research lies in establishing an open source framework modular both in modelling and numerical algorithms.

Keywords Finite element method · Arbitrary Lagrangian–Eulerian method · Parallel algorithm · Blood flow · Left ventricle · Patient-specific heart model

J. H. Spühler (✉) · N. Jansson
Department of Computational Science and Technology, CSC, KTH Royal Institute of Technology, Stockholm, Sweden
e-mail: spuhler@kth.se; njansson@kth.se

J. Jansson · J. Hoffman
Department of Computational Science and Technology, CSC, KTH Royal Institute of Technology, Stockholm, Sweden

BCAM-Basque Center for Applied Mathematics, Bilbao Basque Country, Spain
e-mail: jjan@kth.se; jhoffman@kth.se

H. van Brummelen et al. (eds.), *Numerical Methods for Flows*, Lecture Notes in Computational Science and Engineering 132,
https://doi.org/10.1007/978-3-030-30705-9_14

1 Introduction

Cardiac disease is the number one cause of death in the world [28], and therefore the understanding of normal cardiac function and diseases is vital. Today, computer simulation is emerging as an important tool in enhancing our understanding of the heart and offers the potential to serve as decision support in diagnostics and treatment.

It may be desirable to embed the various micro- and macro-scopic properties of the heart in one computational framework as e.g. the electrical excitation, the cardiac muscle contraction or fluid mechanics of the blood flow. Coupling several parts into one model is associated with challenges as e.g. handling different scales [21] or high computational costs. However, depending on the problem, we do not need such an extensive simulation model, but instead one specific aspect of the heart function can be identified and examined, separately.

For cardiologists it is a highly aspired goal to non-invasively detect cardiac dysfunction. Analyzing the ventricular *wall motion* using echocardiography is a common way, but abnormalities may not be detectable in an early stage of pathology. Therefore, metrics based on intraventricular *blood flow* have come into focus, and their potential as an alternative approach is investigated and debated (see e.g. [2, 3, 13]).

In this paper, we present an Arbitrary Lagrangian-Eulerian (ALE) finite element framework for patient-specific simulation of the blood flow in the left ventricle of a human heart, embedded in the modular framework offered by the open source software Unicorn [11, 12] for high performance finite element simulations. Preliminary work has been presented in conferences and workshops [25]. We use an approach with prescribed wall motion where the input is a set of snapshots in time of a deforming surface mesh that describes the dynamics of the endocardium. This set of surface meshes can come from a cardiac wall model, or patient-specific measurements. Here, we illustrate the framework for a dataset developed at Umeå University [1, 25]. The model geometry in the form of a surface triangulation is generated from ultrasound measurements of the position of the endocardial wall of the left ventricle (LV) at a number of snapshots in time during the cardiac cycle, from which intermediate states are constructed by interpolation. A three dimensional volume mesh is deformed in time to fit these LV surface meshes using mesh smoothing algorithms. Finally, an ALE space-time finite element method is used to simulate the blood flow by solving the incompressible Navier-Stokes equations. Velocity and pressure boundary conditions are set to model the inflow from the mitral valve and outflow through the aortic valve.

Several studies have been conducted to simulate the three dimensional blood flow in the left ventricle of the human heart based on medical imaging. Our contribution to the field of cardiac research lies in combining patient-specific measurements and parallel computing in a framework which allows a thorough validation based on an extensive number of individuals. To our knowledge, the current image based models only have been studied for single, or maximal 30 patients [6]. This paper presents the mathematical and numerical approach using one geometry as example, but the workflow from clinical image acquisition based on echocardiography, surface segmentation, volume mesh generation and numerical simulation has been automated and applied on more than 100 test cases [19, 20]. Furthermore, since the open-source software FEniCS-HPC provides a problem solving environment where partial differential equations (PDE) are automatically converted to low level source code, complex multiphysics problems can be implemented by users without profound knowledge of HPC concepts or coding experience. Thus, our model can be extended in a modular way, modular in terms of simple implementation of mathematical and numerical methods due to automatic code generation as well as combination of different models such as including fluid-structure interaction (FSI) [10] or adaptivity [4].

We structure the rest of the paper as follows. Section 2 specifies our Heart Solver in terms of the basic mathematical equations and the finite element method including the parallel implementation. The results of our simulations are presented in Sect. 3. We conclude the paper with a summary and outlook towards future work.

2 Computational Model

2.1 The Mathematical Model

The human heart is a muscular organ on the scale of a fist and consists of four chambers, the two atria and the two ventricles [17]. The left ventricle possesses the mitral and the aortic valves which ensure unidirectional flow. The opening and closing of the valves are mainly controlled by the pressure gradient between the ventricle and the adjacent chamber.

The cardiac cycle in our model lasts for a period of 1.124 s and is divided into the basic stages of diastole, systole, isovolumetric relaxation and isovolumetric contraction. In diastole, the muscles of the ventricle are relaxed and the blood is flowing from the atrium into the ventricle, and in systole the muscles are contracted and the blood is ejected from the ventricle to the aorta. In the isovolumetric relaxation, both valves are closed and the pressure drops rapidly, whereas during the isovolumetric contraction both valves are closed but the pressure rises quickly. Presently, no geometrical models of the mitral and aortic valves are included in the model. Instead, the four stages are modeled by varying the boundary conditions at marked regions on the surface of the LV model. In diastole and systole the pressure is set to the atrial respectively aortic pressure according to a given pressure curve at the opened valves. At the closed valves a no-slip boundary condition is applied which means that the blood has the same velocity as the wall. For the isovolumetric phases boundary conditions for the velocity and the pressure are set for the whole surface. The fine structure of the inner wall (trabeculation) is not taken into account. Since the flow is incompressible, and we know the volume rate and the area of the mitral opening, the magnitude of the inflow velocity can be derived. Based on clinical observations, a flat inflow profile perpendicular to the mitral opening is set as an inflow boundary condition during diastole.

The simulation starts when the ventricle has reached its maximum volume and the initial velocity is set to zero. Thus, the simulation has to be run over a few cycles to get a flow field independent of the initial conditions. Even though non-Newtonian characteristic can be observed, in large vessels the blood can be modelled as a Newtonian fluid [27] to capture the main flow features. Thus, to calculate the blood flow in the LV we use the incompressible Navier-Stokes equations.

In this paper, our geometry model of the LV is based on ultrasound measurements of the position of the inner wall at different levels in the short-axis view of the LV of one single healthy individual. Additional measurements of distances in the long-axis view complete the basic geometry of the LV. The movement of the heart wall during one heart cycle which lasts about 1 s is not uniform. Thus, 12 surface meshes describing the inner wall of the chamber at different snapshots in time are constructed by using subdivision techniques [1]. Since the computational mesh must be updated at every time step, a local ALE coordinate map is used in the space-time discretization of the Navier-Stokes equations accounting for the mesh velocity in the convective term as described in [10]. To attain a continuous movement of the wall, these surface meshes are interpolated in time by applying Hermite interpolation, and mesh smoothing algorithms are used to adjust the volume mesh to this boundary motion.

Let $\Omega^t \subset \mathbb{R}^3$ be a time-dependent domain with $t \in I := [0, \tilde{t}]$. Vectors and matrices are indicated with bold letters. Given the initial values and boundary conditions, we want to determine velocity $\mathbf{u}(\mathbf{x}, t) : \Omega^t \to \mathbb{R}^3$ and pressure $p(\mathbf{x}, t) : \Omega^t \to \mathbb{R}$ such that:

$$\rho(\dot{\mathbf{u}} + ((\mathbf{u} - \mathbf{m}) \cdot \nabla)\mathbf{u}) - \mu\Delta\mathbf{u} + \nabla p = \mathbf{0} \quad (\mathbf{x}, t) \in \Omega^t \times I, \tag{1a}$$

$$\nabla \cdot \mathbf{u} = 0 \quad (\mathbf{x}, t) \in \Omega^t \times I, \tag{1b}$$

where $\mathbf{m}$ denotes the mesh velocity in the ALE formulation. At the boundary $\mathbf{m}$ is given by the velocity of the prescribed wall motion, and within the volume $\mathbf{m}$ is equal to the velocity of the mesh smoothing algorithm which is applied for maintaining the quality of the mesh. We set the dynamic viscosity to $\mu = 0.0027\,\text{Pa s}$ and the blood density to $\rho = 1060\,\text{kg/m}^3$ [5].

2.2 Finite Element Approximation

We introduce a sequence of discrete time steps $0 := t^0 < t^1 < \cdots < t^N := \tilde{t}$ where $I^n := (t^{n-1}, t^n]$ is an time interval of length $k^n := t^n - t^{n-1}$.

$T^n = \{K\}$ specifies the spatial discretization of Ω^{t^n}, and h^n identifies the minimal diameter of the cell elements $K \in T^n$. We also introduce the space-time slab $S^n := \Omega^{t^n} \times I^n$ and the finite dimensional space of piecewise linear functions $W^n \subset H^1(\Omega^{t^n})$, where

$$H^1(\Omega^{t^n}) := \{v \in L^2(\Omega^{t^n}) | \frac{\partial v}{\partial x_k} \in L^2(\Omega^{t^n}), k = 1, 2, 3\} \tag{2}$$

$$W^n := \{v \in C(\Omega^{t^n}) | v \in P^1(K), \forall K \in T^n\}, \tag{3}$$

$$W_0^n := \{v \in W^n | v = 0 \text{ on } \partial\Omega^{t^n}\} \tag{4}$$

$$\mathbf{W_0^n} := [W_0^n]^3. \tag{5}$$

We identify the discrete solution for velocity and pressure as $\hat{\mathbf{U}} = (\mathbf{U}, P)$, the discrete mesh velocity as $\mathbf{M}$, and the test function as $\hat{\mathbf{v}} = (\mathbf{v}, q)$. In time, we choose $\mathbf{U}$ to be piecewise linear and P, $\mathbf{v}$ and q to be piecewise constant.

We now formulate the spatially and temporally discretised finite element formulation of the continuum model (1) with homogeneous Dirichlet boundary conditions for the velocity: for each space-time slab S^n, find $(\mathbf{U^n}, P^n) := (\mathbf{U(t^n)}, P(t^n))$ with $\mathbf{U^n} \in \mathbf{W_0^n}$ and $P^n \in W^n$, such that:

$$((\rho/k^n)(\mathbf{U^n} - \mathbf{U^{n-1}}) + (\rho(\bar{\mathbf{U}}^\mathbf{n} - \mathbf{M}^n) \cdot \nabla)\bar{\mathbf{U}}^\mathbf{n}, \mathbf{v}) + (\mu \nabla \bar{\mathbf{U}}^\mathbf{n}, \nabla \mathbf{v}) - (P^n, \nabla \cdot \mathbf{v}) \tag{6}$$

$$+ SD_\delta(\bar{\mathbf{U}}^\mathbf{n}, \mathbf{M}^n, P^n, \mathbf{v}, q, \rho) = 0,$$

for $\forall(\boldsymbol{v}, q) \in \mathbf{W_0^n} \times W^n$, where $\bar{\mathbf{U}}^\mathbf{n} = \frac{1}{2}(\mathbf{U^n} + \mathbf{U^{n-1}})$ and

$$(\mathbf{v}, \mathbf{w}) = \sum_{K \in T^n} \int_K \mathbf{v} \cdot \mathbf{w} \, dx, \tag{7}$$

By using the midpoint quadrature rule in time, we obtain a Crank-Nicholson time-stepping scheme. We apply a simplified Galerkin/least-square method, to stabilise the convection dominated problem, where the time derivative term is dropped since the test functions are piecewise constant in time:

$$SD_\delta(\bar{\mathbf{U}}^\mathbf{n}, \mathbf{M}^n, P^n, \mathbf{v}, q, \rho) = \tag{8}$$

$$(\delta_1 \rho(((\bar{\mathbf{U}}^\mathbf{n} - \mathbf{M^n}) \cdot \nabla)\bar{\mathbf{U}}^\mathbf{n} + \nabla P^n), \rho((\bar{\mathbf{U}}^\mathbf{n} - \mathbf{M^n}) \cdot \nabla)\mathbf{v} + \nabla q)$$

$$+ (\delta_2 \nabla \cdot \bar{\mathbf{U}}^\mathbf{n}, \nabla \cdot \mathbf{v}).$$

The stabilisation parameters are chosen as $\delta_2 = \kappa_2 \rho h^n |\mathbf{U^{n-1}}|$ and $\delta_1 = \kappa_1 \rho^{-1}(1/(k^n)^2 + |\mathbf{U^{n-1}} - \mathbf{M^{n-1}}|^2/(h^n)^2)^{-1/2}$ as defined in [9], where κ_1, κ_2 are problem independent positive constants of order $O(1)$. The timestep size k^n is set to be constant such that a CFL number of 0.1 is achieved at the initial time point. This is a conservative, but robust choice.

2.3 Computational Tools

Our Heart Solver is implemented in the HPC branch [14] of the open source Finite Element Method (FEM) library DOLFIN [22] and the adaptive flow solver Unicorn [12], which are components of FEniCS, a computing platform for automated solution of PDEs. The framework has been shown to scale well [14] and is used to efficiently solve large scale industrial problems [4, 15]. The simulations in this paper were performed on Povel and Beskow located at PDC Center for High Performance Computing at KTH Royal Institute of Technology. Povel is a cluster that offers 170 nodes and each node is equipped with four Six-Core AMD Opteron 8425HE processors and 2 GB of RAM. The nodes are connected using QDR Infiniband. Beskow is a Cray XC40 system consisting of 1676 compute nodes, where each node has two cpus (Intel E5-2698v3) with 16 cores.

3 Results

A detailed compilation of the results on convergence study and illustration of the modular capacity of the framework exemplified by the simple implementation of r-adaptivity can be found in [26]. We here only discuss the results regarding parallel efficiency and the fluid dynamics.

3.1 Parallel Efficiency

The software has proven to scale well for PDEs with fixed geometries, both strongly and weakly, for a wide range of architectures and applications, as shown in [12]. To show parallel efficiency for our LV model with a moving boundary, strong scalability tests are performed on Povel. In the case of ideal strong scaling, doubling the number of cores for the same simulation would halve the computational time. The mean time is measured to advance one step in time with the whole Heart Solver.

The measurements are conducted on two meshes of different size and the results for both meshes are presented in Fig. 1. In the first case the solver runs with a mesh containing 417′126 vertices on 1, 24, 48 and 96 cores, and in the second case a mesh with 1′578′505 vertices is used to run on 96, 192 and 384 cores. Since in the case of the small mesh the time can be measured up to the serial case, one can see that the computation time on the local mesh cannot compensate for the communication cost. This clarifies also why we achieve better scaling results when starting at 96 cores, for the bigger mesh, where the communication cost is always present.

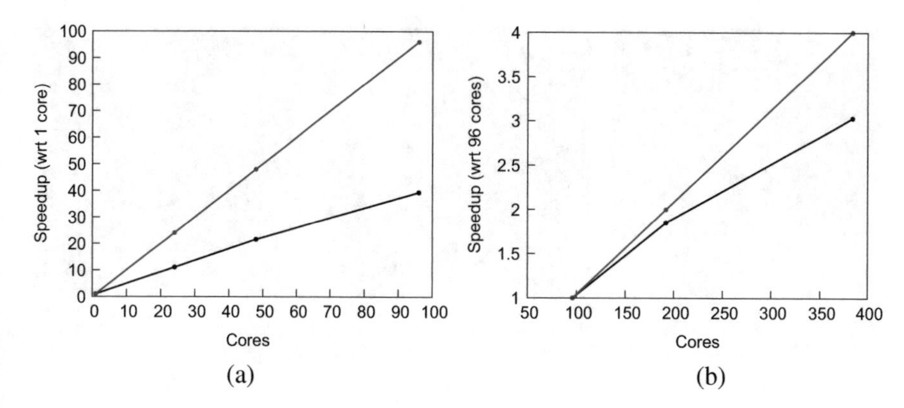

Fig. 1 Strong scalability: The mean performance time (blue) to advance one step in time with the whole Heart Solver is compared to linear speed up (black). (**a**) Heart Solver (417′126 vertices). (**b**) Heart Solver (1′578′505 vertices)

3.2 Hemodynamics

To identify cardiac diseases in an early stage by using non-invasive methods, is an aim of clinical cardiologists. Medical imaging of the heart as Doppler-echocardiography or color Doppler M-mode is a non-invasive method that can be used to gain information, but it might be difficult to correctly interpret the data due to lack of high temporal or spatial resolution [23]. CFD can therefore play an important role to complement such techniques to identify quantities indicating cardiac disease.

The blood flow in systole is an outward pushed flow and does not show significantly notable features. The diastolic flow however forms characteristic vortices as depicted in Fig. 2 which we want to describe in detail. A vortex can roughly be described as a mass of fluid flowing around a common axis. Studies are conducted to establish a link between the vortex formation and cardiac function in either normal or pathologic conditions, and different indexes to quantify the characterization of vortex rings in the LV have been proposed and are debated, see e.g. [8, 24]. To analyze the evolution of the mitral vortex ring in our simulations, we apply the λ_2-method [16], where vortex structures are identified and visualized by extreme negative isosurface values. We use the open source code Saaz to calculate λ_2 for our simulations [18]. It is difficult to determine a threshold Θ_{λ_2} for the isosurface of λ_2. Here, we manually adjust Θ_{λ_2} until we can differentiate coherent vortex structures.

After the isovolumetric relaxation period, diastole can be divided into 3 phases: the rapid filling phase, diastasis and atrial systole [7]. During the rapid filling phase a strong jet flows from the mitral opening into the LV chamber and generates a torus-shaped, well-defined vortex ring (Fig. 2a). This mitral vortex ring (MVR) is fully developed as the filling reaches its peak as shown in Fig. 2b. As the mitral vortex moves towards the apex, it impinges and slides along the LV wall (Fig. 2c). This interaction causes an elongation and rotation of the vortex ring. At the end of

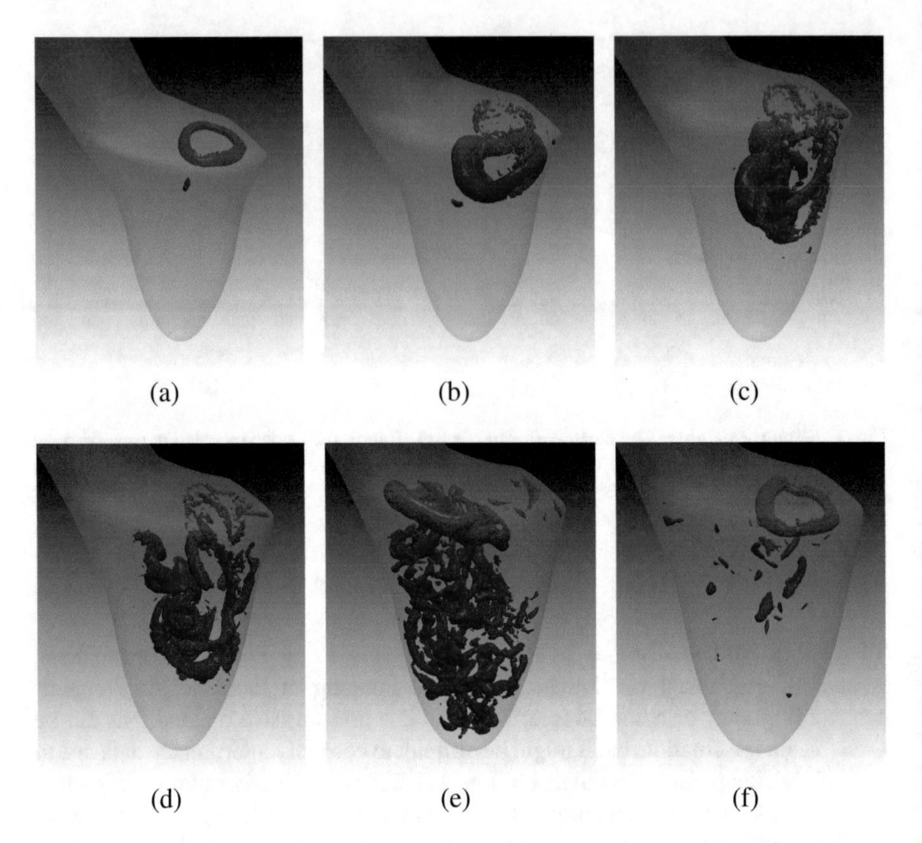

Fig. 2 Mitral vortex formation during diastole, visualized using $\Theta_{\lambda_2} = -10000$. (**a**) 2.701 s, (**b**) 2.761 s, (**c**) 2.801 s, (**d**) 2.851 s, (**e**) 3.001 s, (**f**) 3.327 s

the early filling phase the flow then slows down and the MVR is broken down into small-scale structures (Fig. 2d). A new vortex ring between the mitral and aortic valve can now be observed in the subsequent time period (diastasis), see Fig. 2e. In this region, the blood stream is split and is either redirected towards the apex or towards the aortic outflow tract. Finally, due to the atrial contraction, a second vortex ring is generated at the ventricular basal level (Fig. 2f).

The complex flow pattern clearly shows the importance of applying a three-dimensional, patient-specific geometry when studying the hemodynamics in the LV chamber.

4 Conclusion

In this paper, we have presented a numerical model to simulate the blood flow in the left ventricle based on patient-specific data using high performance computing. The flow problem is solved by a stabilized finite element method using an open source software. The results are comparable to observations done with other numerical models and imaging techniques and indicate that our approach possesses the potential to provide relevant information for diagnosis and medical treatment. The novelty of our approach is an open source patient-specific ALE Finite Element framework for simulating the blood flow in the LV of a human heart using HPC, modular both in modelling and numerical algorithms. In order to address limitations of our current model, our future work will be focused on integration of the actual movement of the cardiac valves, and automated generation of 3D geometries from experimental data with uncertainty quantification.

Acknowledgements The authors would like to thank Michael Broomé, Ulf Gustafsson, Anders Waldenström, Mats Larsson and Per Vesterlund for sharing their great knowledge and offering their support. We also would like to acknowledge the financial support from the Swedish Foundation for Strategic Research, the Swedish Research Council and the European Research Council—ERC Starting Grant 'UNICON', proposal number 202984. The research was conducted on resources provided by the Swedish National Infrastructure for Computing (SNIC) at the Center of High-Performance Computing (PDC).

References

1. Aechtner, M.: Arbitrary Lagrangian-Eulerian Finite Element Modelling of the Human Heart. Master's thesis, KTH Royal Institute of Technology, Stockholm (2009)
2. Arvidsson, P.M., Kovacs, S.J., Töger, J., Borgquist, R., Heiberg, E., Carlsson, M., Arheden, H.: Vortex ring behavior provides the epigenetic blueprint for the human heart. Sci. Rep. **6**, 22021 (2016)
3. Cooke, J., Hertzberg, J., Boardman, M., Shandas, R.: Characterizing Vortex ring behavior during ventricular filling with doppler echocardiography: an in vitro study. Ann. Biomed. Eng. **32**(2), 245–256 (2004)
4. de Abreu, R.V., Jansson, N., Hoffman, J.: Computation of aeroacoustic sources for a Gulfstream G550 nose landing gear model using adaptive FEM. Comput. Fluids **124**, 136–146 (2016)
5. Di Martino, E., Guadagni, G., Fumero, A., Ballerini, G., Spirito, R., Biglioli, P., Redaelli, A.: Fluid-structure interaction within realistic three-dimensional models of the aneurysmatic aorta as a guidance to assess the risk of rupture of the aneurysm. Med. Eng. Phys. **23**(9), 647–655 (2001)
6. Doost, S.N., Ghista, D., Su, B., Zhong, L., Morsi, Y.S.: Heart blood flow simulation: a perspective review. Biomed. Eng. Online **15**(1), 101 (2016)
7. Galderisi, M.: Diastolic dysfunction and diastolic heart failure: diagnostic, prognostic and therapeutic aspects. Cardiovasc. Ultrasound **3**(1), 9 (2005)
8. Gharib, M., Rambod, E., Kheradvar, A., Sahn, D.J., Dabiri, J.O.: Optimal vortex formation as an index of cardiac health. Proc. Natl. Acad. Sci. **103**(16), 6305–6308 (2006)
9. Hoffman, J., Johnson, C.: Computational Turbulent Incompressible Flow, Applied Mathematics, Body and Soul, vol. 4. Springer, Berlin (2007)
10. Hoffman, J., Jansson, J., Stöckli, M.: Unified continuum modeling of fluid-structure interaction. Math. Models Methods Appl. Sci. **21**(3), 491–513 (2011)
11. Hoffman, J., Jansson, J., Degirmenci, C., Jansson, N., Nazarov, M.: Unicorn: a unified continuum mechanics solver. In: Logg, A., Mardal, K.A., Wells, G. (eds.) Automated Solution of Differential Equations by the Finite Element Method. Lecture Notes in Computational Science and Engineering, vol. 84, pp. 339–361. Springer, Berlin (2012)

12. Hoffman, J., Jansson, J., Vilela de Abreu, R., Degirmenci, N.C., Jansson, N., Müller, K., Nazarov, M., Spühler, J.H.: Unicorn: parallel adaptive finite element simulation of turbulent flow and fluid-structure interaction for deforming domains and complex geometry. Comput. Fluids **80**, 310–319 (2013)
13. Hong, G.R., Pedrizzetti, G., Tonti, G., Li, P., Wei, Z., Kim, J.K., Baweja, A., Liu, S., Chung, N., Houle, H., Narula, J., Vannan, M.A.: Characterization and quantification of vortex flow in the human left ventricle by Contrast Echocardiography using vector particle image velocimetry. JACC Cardiovasc. Imaging **1**(6), 705–717 (2008)
14. Jansson, N.: High Performance Adaptive Finite Element Methods: With Applications in Aerodynamics. KTH Royal Institute of Technology, Stockholm (2013). Ph.D. thesis
15. Jansson, N., Hoffman, J., Nazarov, M.: Adaptive simulation of Turbulent flow past a full car model. In: State of the Practice Reports, SC '11, pp. 20:1–20:8. ACM, New York (2011)
16. Jeong, J., Hussain, F.: On the identification of a vortex. J. Fluid Mech. **285**, 69–94 (1995)
17. Jonson, B., Wollmer, P.: Klinisk Fysiologi: Med Nuklearmedicin och Klinisk Neurofysiologi. Liber, Stockholm (2005)
18. King, A., Arobone, E., Sarkar, S., Baden, S.B.: The Saaz framework for turbulent flow queries. In: Proceedings of the 2011 IEEE conference on e-Science. IEEE, Piscataway (2011)
19. Larsson, D., Spühler, J.H., Petersson, S., Nordenfur, T., Hoffman, J., Colarieti-Tosti, M., Winter, R., Larsson, M.: Multimodal validation of patient-specific intraventricular flow simulations from 4D echocardiography. In: 2016 IEEE International Ultrasonics Symposium (IUS), pp. 1–4 (2016)
20. Larsson, D., Spühler, J.H., Petersson, S., Nordenfur, T., Colarieti-Tosti, M., Hoffman, J., Winter, R., Larsson, M.: Patient-specific left ventricular flow simulations from transthoracic echocardiography: robustness evaluation and validation against ultrasound doppler and magnetic resonance imaging. IEEE Trans. Med. Imaging **36**(11), 2261–2275 (2017). https://doi.org/10.1109/TMI.2017.2718218
21. Lee, J., Niederer, S., Nordsletten, D., Grice, I.L., Smaill, B., Kay, D., Smith, N.: Coupling contraction, excitation, ventricular and coronary blood flow across scale and physics in the heart. Philos. Trans. R. Soc. A **367**, 2311–2331 (2009)
22. Logg, A., Wells, G.N.: DOLFIN: automated finite element computing. ACM Trans. Math. Softw. **37**(2), 20:1–20:28 (2010)
23. Nishimura, R.A., Tajik, A.J.: Evaluation of diastolic filling of left ventricle in health and disease: doppler echocardiography is the clinician's rosetta stone. J. Am. Coll. Cardiol. **30**(1), 8–18 (1997)
24. Pasipoularides, A., Vlachos, P.P., Little, W.C.: Vortex formation time is Not an index of ventricular function. J. Cardiovasc. Transl. Res. **8**(1), 54–58 (2015)
25. Spühler, J., Hoffman, J., Jansson, J., Gustafsson, U., Brommé, M., Larson, M.G., Vesterlund, P.: Simulation of the blood flow in the left ventricle using a finite element method with moving geometries based on ultrasound. In: Conference Proceedings CMBE 2011 (2011)
26. Spühler, J.H., Jansson, J., Jansson, N., Hoffman, J.: A finite element framework for high performance computer simulation of blood flow in the left ventricle of the human heart. KTH, Computational Science and Technology (CST), Stockholm (2015). Tech. Rep. 34. QC 20160212
27. Westerhof, N., Stergiopulos, N., Noble, M.I.: Snapshots of Hemodynamics an Aid for Clinical Research and Graduate Education. Springer, New York (2010)
28. WHO: cardiovascular diseases. http://www.who.int/mediacentre/factsheets/fs317/en/index.html. Accessed 12 Feb 2014

Phase Field-Based Incompressible Two-Component Liquid Flow Simulation

Babak Sayyid Hosseini and Matthias Möller

Abstract In this work, we consider a Cahn–Hilliard phase field-based computational model for immiscible and incompressible two-component liquid flows with interfacial phenomena. This diffuse-interface complex-fluid model is given by the incompressible Navier–Stokes–Cahn–Hilliard (NSCH) equations. The coupling of the flow and phase field equations is given by an extra phase induced surface tension force term in the flow equations and a fluid induced transport term in the Cahn–Hilliard (CH) equations. Galerkin-based isogeometric finite element analysis is applied for space discretization of the coupled system in velocity–pressure–phase field–chemical potential formulation. For the approximation of the velocity and pressure fields, LBB compatible non-uniform rational B-spline spaces are used which can be regarded as smooth generalizations of Taylor–Hood pairs of finite element spaces. The one-step θ-scheme is used for the discretization in time. For the validation of the two-phase flow model, we present numerical results for the challenging Rayleigh-Taylor instability flow problem in two dimensions and compare them to reference results.

Keywords Two-phase flow · Cahn–Hilliard phase field model · Navier–Stokes–Cahn–Hilliard equations · Isogeometric Analysis · Isogeometric finite elements · B-splines/NURBS · Rayleigh–Taylor instability

---

B. S. Hosseini (✉)
Institute of Applied Mathematics (LS III), TU Dortmund, Dortmund, Germany
e-mail: babak.hosseini@math.tu-dortmund.de

M. Möller
Delft Institute of Applied Mathematics, Delft University of Technology, Delft, The Netherlands
e-mail: m.moller@tudelft.nl

H. van Brummelen et al. (eds.), *Numerical Methods for Flows*,
Lecture Notes in Computational Science and Engineering 132,
https://doi.org/10.1007/978-3-030-30705-9_15

1 Introduction

Multiphase flows comprise flow of materials with different phases (i.e. gas, liquid, etc.), or materials with different chemical properties in the same phase, such as oil and water. In two-phase flows, being the most common multiphase flow configuration involving two distinct fluids, the fluids are segregated by a very thin interfacial region where surface tension effects and mass transfer due to chemical reactions may appear. Multiphase flows are ubiquitous in nature and industrial systems and are quite challenging from the point of view of mathematical modeling and simulation due to the complex physical interaction between the involved fluids including topological changes and the complexity of having to deal with unknown moving fluid-fluid interfaces. As for methodologies to address the moving interface problem, there are various methods such as volume-of-fluid, front tracking, immersed boundary, level-set and phase field methods (cf. [5, 8]).

In this work we use a phase field diffuse interface method based on the Cahn–Hilliard equation and apply Isogeometric Analysis for the discretization of the involved equations. Particularly for two-phase flows, Diffuse-interface models have gained a lot of attention due to their ability to easily handle moving contact lines and topological transitions without any need for reinitialization or advective stabilization. On a general note, diffuse interface models allow the modeling of interfacial forces as continuum forces with the effect that delta-function forces and discontinuities at the interface are smoothed by smearing them over thin yet numerically resolvable layers. The phase field method—also known as the diffuse interface model—is based on models of fluid free energy and offers a systematic physical approach by describing the interface in a physical rather than in a numerical sense. One principal advantage of diffuse interface models is their ability to describe topological transitions like droplet coalescence or break-up in a natural way. In the phase field framework, the interface is modeled by a function $\varphi(x, t)$ which represents the concentration of the fluids. The function $\varphi(x, t)$, also referred to as the order parameter, or the phase field, attains a distinct constant value in each phase and rapidly, but smoothly, changes in the interface region between the phases. For a binary fluid, a usual assumption is that φ takes values between -1 and 1, or 0 and 1. The relaxation of the order parameter is driven by local minimization of the fluid free energy subject to the phase field conservation. As a result, complex interface dynamics such as coalescence or segregation can be captured without any special procedures [1, 9].

The outline of this article is as follows: In Sect. 2 we briefly introduce the mathematical model used in this work. It serves as a basis for Sect. 3 that is dedicated to the presentation of the weak forms and discretization aspects of the mathematical model with Isogeometric Analysis. We present our numerical results in Sect. 4 and conclude the article with a short summary in Sect. 5.

2 Navier–Stokes–Cahn–Hilliard Two-Phase Flow Model

Let $\Omega = (\Omega_1 \cup \Omega_2) \subset \mathbb{R}^n$ be an arbitrary open domain, with $n = 2$ or 3 and let its boundary $\partial\Omega$ be sufficiently smooth (e.g. Lipschitz continuous). Moreover, let Γ denote the interface between the different fluids or phases occupying the subdomains Ω_1 and Ω_2 and let $\mathbf{n}$ be the outward ($\Omega_1 \to \Omega_2$) unit normal at the interface. The NSCH variable density, variable viscosity incompressible two-phase flow model (1) is obtained by the extension of the Navier–Stokes equations with a surface tension force term $\eta\nabla\varphi$, written in its potential form, and a fluid induced transport term $\mathbf{v}\cdot\nabla\varphi$ in the Cahn–Hilliard equations ((1c) + (1d)).

$$\rho(\varphi)\left(\frac{\partial \mathbf{v}}{\partial t} + (\mathbf{v}\cdot\nabla)\mathbf{v}\right) - \nabla\cdot\boldsymbol{\sigma}(\varphi) = \rho(\varphi)\mathbf{g} + \eta\nabla\varphi \text{ in } \Omega_T, \tag{1a}$$

$$\nabla\cdot\mathbf{v} = 0 \quad \text{in } \Omega_T, \tag{1b}$$

$$\frac{\partial\varphi}{\partial t} + \mathbf{v}\cdot\nabla\varphi - \nabla\cdot(m(\varphi)\nabla\eta) = 0 \quad \text{in } \Omega_T, \tag{1c}$$

$$\eta - \beta\,\frac{d\psi(\varphi)}{d\varphi} + \alpha\,\nabla^2\varphi = 0 \quad \text{in } \Omega_T, \tag{1d}$$

$$\varphi(\mathbf{x},0) = \varphi_0(\mathbf{x}), \quad \mathbf{v}(\mathbf{x},0) = \mathbf{v}_0(\mathbf{x}) \quad \text{in } \Omega, \tag{1e}$$

$$\frac{\partial\varphi}{\partial\mathbf{n}} = \frac{\partial\eta}{\partial\mathbf{n}} = 0, \quad \mathbf{v} = \mathbf{v}_D \quad \text{on } (\partial\Omega_T)_D, \tag{1f}$$

$$\left(-p\mathbf{I} + \mu(\varphi)\left(\nabla\mathbf{v} + (\nabla\mathbf{v})^T\right)\right)\cdot\mathbf{n} = \mathbf{t} \quad \text{on } (\partial\Omega_T)_N. \tag{1g}$$

Above, $\Omega_T = \Omega \times (0, T)$, $(\partial\Omega)_D$ is the Dirichlet part of the domain boundary, $\boldsymbol{\sigma}(\varphi) = -p\mathbf{I} + \mu(\varphi)\left(\nabla\mathbf{v} + (\nabla\mathbf{v})^T\right)$ denotes the (variable viscosity) fluid Cauchy stress tensor, $\mathbf{t}$ is the prescribed traction force on the Neumann boundary $(\partial\Omega)_N$, $\mathbf{g}$ is the gravitational force field and p is the pressure variable acting as a Lagrange multiplier in the course of enforcing the incompressibility condition. This basically corresponds to the model presented by Ding et al. [2] which can be seen as a generalization of "Model H" [4, 6] for the case of different densities and viscosities. In contrast to "Model H", a surface tension force term in potential form $\eta\nabla\varphi$ has replaced the divergence of the phase induced stress tensor $-\hat{\sigma}\varepsilon\,(\nabla\varphi\otimes\nabla\varphi)$. The latter, that is, $-\hat{\sigma}\varepsilon\,\mathrm{div}\,(\nabla\varphi\otimes\nabla\varphi)$, represents the phase induced force. In Eq. (1d), α and β are functions of the surface energy density $\hat{\sigma}$ and the interfacial region thickness ε.

In the Cahn–Hilliard equations ((1c) + (1d)), $\varphi \in [-1, 1]$ is a measure of phase and it holds $\varphi(x) = 1$ (respectively $\varphi(x) = -1$) if and only if fluid 1 (respectively fluid 2) is present at point x. η represents the chemical potential and the nonlinear functions $m(\varphi)$ and $\psi(\varphi)$ model the concentration dependent mobility and fluid components' immiscibility, respectively.

3 Variational Formulation and Discretization

We use Isogeometric Analysis for the approximation of the solution of the coupled equation system (1). Inspired by operator splitting techniques, it is solved in two consecutive stages in order to alleviate numerical treatment. More specifically, given a flow field $\mathbf{v}$, we first solve the phase field equations ((1c) + (1d)) in order to update the phase φ and chemical potential information η. The second step eventually uses these information to compute the surface tension force and the phase dependent values of density $\rho(\varphi(\mathbf{x}))$ and viscosity $\mu(\varphi(\mathbf{x}))$ in the course of the solution of the Navier–Stokes equations ((1a) + (1b)). As time integrator for both systems, we use the one-step θ-scheme with $\theta = 1$ or $\theta = 0.5$ yielding the first order implicit Euler or second order Crank-Nicolson scheme, respectively. For the approximation of the velocity and pressure functions in the Navier–Stokes equations, we use LBB-stable Taylor–Hood-like B-spline/NURBS[1] space pairs $\hat{\mathbf{V}}_h^{TH}/\hat{Q}_h^{TH}$ which are defined in the parametric spline domain $\hat{\Omega}$ as

$$\begin{aligned}
\hat{\mathbf{V}}_h^{TH} &\equiv \hat{\mathbf{V}}_h^{TH}(\mathbf{p}, \boldsymbol{\alpha}) = \boldsymbol{\mathcal{N}}_{\alpha_1,\alpha_2}^{p_1+1,p_2+1} = \mathcal{N}_{\alpha_1,\alpha_2}^{p_1+1,p_2+1} \times \mathcal{N}_{\alpha_1,\alpha_2}^{p_1+1,p_2+1}, \\
\hat{Q}_h^{TH} &\equiv \hat{Q}_h^{TH}(\mathbf{p}, \boldsymbol{\alpha}) = \mathcal{N}_{\alpha_1,\alpha_2}^{p_1,p_2}.
\end{aligned} \tag{2}$$

Above, $\mathcal{N}_{\alpha_1,\alpha_2}^{p_1+1,p_2+1}$ denotes a tensor product bivariate NURBS space of polynomial degrees $p_i + 1$ and continuities α_i, $i = 1, 2$, with respect to parametric spline domain directions ξ_i. We refer to Hosseini et al. [7] for a detailed description of the above spline spaces. In all performed computations we used a $\mathscr{C}^0$ $\boldsymbol{\mathcal{N}}_{0,0}^{2,2}/\mathcal{N}_{0,0}^{1,1}$ NURBS space pair for the approximation of the velocity and pressure functions. This corresponds to the Isogeometric counterpart of a Q_2Q_1 Taylor–Hood space which is well known from the finite element literature. The degree and continuity of the discrete spaces used for the approximation of the Navier–Stokes velocity and Cahn–Hilliard phase and chemical potential functions are set to be identical. In the sequel we picture the individual solution stages and outline the spatial and temporal discretization of the involved equations.

Step 1: Cahn–Hilliard Equation For the treatment of the nonlinearity in the advective Cahn–Hilliard equation, we seek for the current approximation of the solution $u^k = (\varphi^k, \eta^k)$ small perturbations $\delta u = (\delta\varphi, \delta\eta)$, such that

$$\begin{aligned}
\varphi^{k+1} &= \varphi^k + \delta\varphi, \\
\eta^{k+1} &= \eta^k + \delta\eta
\end{aligned} \tag{3}$$

[1] Non-Uniform Rational B-splines (NURBS).

satisfy the nonlinear partial differential equations ((1c) + (1d)). Under the premise that $\delta\varphi$ is sufficiently small, we linearize the nonlinear function $\psi'(\varphi)$ as:

$$\psi'(\varphi^{k+1}) = \psi'(\varphi^k) + \psi''(\varphi^k)\,\delta\varphi + \mathcal{O}((\delta\varphi)^2) \approx \psi'(\varphi^k) + \psi''(\varphi^k)\,\delta\varphi. \tag{4}$$

After the time discretization by the one-step θ-scheme, we arrive at

$$\begin{aligned} \frac{\varphi^{n+1}-\varphi^n}{\Delta t} &+ \theta\left((\mathbf{v}\cdot\nabla)\varphi^{n+1} - \nabla\cdot m\,\nabla\eta^{n+1}\right) \\ &+(1-\theta)\left((\mathbf{v}\cdot\nabla)\varphi^n - \nabla\cdot m\,\nabla\eta^n\right) = 0 \quad \text{in } \Omega_T, \\ &\eta^{n+1} - \beta\,\psi'(\varphi^{n+1}) + \alpha\nabla^2\varphi^{n+1} = 0 \quad \text{in } \Omega_T, \end{aligned} \tag{5}$$

where the boundary terms have not been displayed for the sake of lucidity. Above, in the spirit of Picard iteration, the nonlinear mobility function $m(\varphi)$ is evaluated with respect to the already available values of the phase field, that is, φ^n. This linearization allows us to treat it as a constant which simplifies its numerical treatment.

The variational form of the problem reads: Find $\varphi(\mathbf{x}, t)$ and $\eta(\mathbf{x}, t) \in \mathcal{H}^1(\Omega) \times (0, T)$, such that $\forall q, v \in \mathcal{H}_0^1(\Omega)$ it holds:

$$\begin{aligned} &\int_\Omega \frac{\varphi^{n+1}-\varphi^n}{\Delta t}\, q\,d\mathbf{x} + \theta\left(\int_\Omega (\mathbf{v}\cdot\nabla)\varphi^{n+1}\, q + m\,\nabla\eta^{n+1}\cdot\nabla q\,d\mathbf{x} - \int_{\partial\Omega} \mathbf{n}\cdot m\,\nabla\eta^{n+1}\, q\,ds\right) \\ &\qquad +(1-\theta)\left(\int_\Omega (\mathbf{v}\cdot\nabla)\varphi^n\, q + m\,\nabla\eta^n\cdot\nabla q\,d\mathbf{x} - \int_{\partial\Omega} \mathbf{n}\cdot m\,\nabla\eta^n\, q\,ds\right) = 0, \\ &\int_\Omega \eta^{n+1}\, v\,d\mathbf{x} - \int_\Omega \beta\,\frac{d\psi(\varphi^{n+1})}{d\varphi}\, v\,d\mathbf{x} - \int_\Omega \alpha\,\nabla\varphi^{n+1}\cdot\nabla v\,d\mathbf{x} + \int_{\partial\Omega} \mathbf{n}\cdot\alpha\nabla\varphi^{n+1}\,ds = 0. \end{aligned} \tag{6}$$

The application of (3) and (4) on (6) yields:

$$\begin{aligned} &\int_\Omega (\varphi^k + \delta\varphi - \varphi^n)\, q\,d\mathbf{x} + \theta\Delta t\int_\Omega (\mathbf{v}\cdot\nabla)(\varphi^k + \delta\varphi)\, q + m\,\nabla(\eta^k + \delta\eta)\cdot\nabla q\,d\mathbf{x} \\ &\qquad\qquad +(1-\theta)\Delta t\int_\Omega (\mathbf{v}\cdot\nabla)\varphi^n\, q + m\,\nabla\eta^n\cdot\nabla q\,d\mathbf{x} = 0, \\ &\int_\Omega (\eta^k + \delta\eta)\, v\,d\mathbf{x} - \int_\Omega \beta\left(\psi'(\varphi^k) + \psi''(\varphi^k)\,\delta\varphi\right) v\,d\mathbf{x} - \int_\Omega \alpha\,\nabla(\varphi^k + \delta\varphi)\cdot\nabla v\,d\mathbf{x} = 0. \end{aligned} \tag{7}$$

In Eq. (7), the indices n and k refer to the solution from the last time step and the current Newton-iterate, respectively. $(\delta\varphi, \delta\eta)$ is associated with the Newton-update. Gathering all terms with the unknowns $\delta\varphi$ and $\delta\eta$ on the left hand side, we obtain the following expressions

$$
\begin{aligned}
&\int_\Omega \underbrace{\delta\varphi\, q}_{\mathbf{M}} + \theta\Delta t \left(\underbrace{(\mathbf{v}\cdot\nabla)\delta\varphi\, q}_{\mathbf{C}} + m \underbrace{\nabla\delta\eta\cdot\nabla q}_{\mathbf{D}} \right) \, d\mathbf{x} = \\
&\int_\Omega - \underbrace{\varphi^k\, q}_{\mathbf{M}} - \theta\Delta t \left(\underbrace{(\mathbf{v}\cdot\nabla)\varphi^k\, q}_{\mathbf{C}} + m \underbrace{\nabla\eta^k\cdot\nabla q}_{\mathbf{D}} \right) \, d\mathbf{x} \\
&+ \int_\Omega \underbrace{\varphi^n\, q}_{\mathbf{M}} - (1-\theta)\Delta t \left(\underbrace{(\mathbf{v}\cdot\nabla)\varphi^n\, q}_{\mathbf{C}} + m \underbrace{\nabla\eta^n\cdot\nabla q}_{\mathbf{D}} \right) \, d\mathbf{x}, \\
&\int_\Omega \underbrace{\delta\eta\, v}_{\mathbf{M}} \, d\mathbf{x} - \int_\Omega \beta \underbrace{\psi''(\varphi^k)\,\delta\varphi\, v}_{\mathbf{N}} \, d\mathbf{x} - \int_\Omega \alpha \underbrace{\nabla\delta\varphi\cdot\nabla v}_{\mathbf{D}} \, d\mathbf{x} = \\
&- \int_\Omega \underbrace{\eta^k\, v}_{\mathbf{M}} \, d\mathbf{x} + \int_\Omega \beta \underbrace{\psi'(\varphi^k)\, v}_{\mathbf{n}} \, d\mathbf{x} + \int_\Omega \alpha \underbrace{\nabla\varphi^k\cdot\nabla v}_{\mathbf{D}} \, d\mathbf{x}.
\end{aligned}
\tag{8}
$$

The corresponding discrete system for the Newton-iteration may now be written in matrix form as

$$
\begin{aligned}
&\begin{matrix} q \\ v \end{matrix} \underbrace{\begin{pmatrix} \mathbf{M}+\theta\Delta t\mathbf{C} & \theta\Delta t m\mathbf{D} \\ -\alpha\mathbf{D}-\beta\mathbf{N} & \mathbf{M} \end{pmatrix}}_{\mathbf{J}} \begin{pmatrix} \delta\varphi \\ \delta\eta \end{pmatrix} = \\
&\underbrace{\begin{pmatrix} -\mathbf{M}-\theta\Delta t\mathbf{C} & -\theta\Delta t m\mathbf{D} \\ \alpha\mathbf{D} & -\mathbf{M} \end{pmatrix} \begin{pmatrix} \varphi^k \\ \eta^k \end{pmatrix} + \begin{pmatrix} \mathbf{0} \\ \beta\mathbf{n} \end{pmatrix} + \begin{pmatrix} \mathbf{M}-(1-\theta)\Delta t\mathbf{C} & -(1-\theta)\Delta t m\mathbf{D} \\ \mathbf{0} & \mathbf{0} \end{pmatrix} \begin{pmatrix} \varphi^n \\ \eta^n \end{pmatrix}}_{-\mathbf{F}}
\end{aligned}
\tag{9}
$$

and solved for δu in order to update the unknowns as $(\varphi^{k+1}, \eta^{k+1}) = (\varphi^k, \eta^k) + (\delta\varphi^k, \delta\eta^k)$.

Step 2: Navier–Stokes Equations This step involves the numerical approximation of the solution of the unsteady variable density and variable viscosity Navier–Stokes equations extended by a surface tension force term. The initial condition for the velocity field is required to satisfy $\nabla\cdot\mathbf{v}_0 = 0$. With $\mathbf{b}$ denoting the body force term, the variational formulation of the problems ((1a) + (1b)) reads: Find $\mathbf{v}(\mathbf{x}, t) \in \mathscr{H}_0^1(\Omega) \times (0, T)$ and $p(\mathbf{x}, t) \in \mathscr{L}_2(\Omega)/\mathbb{R} \times (0, T)$, such that for all $(\mathbf{w}, q) \in \mathscr{H}_0^1(\Omega) \times \mathscr{L}_2(\Omega)/\mathbb{R}$ it holds

$$
\begin{cases}
(\mathbf{w}, \mathbf{v}_t) + a(\mathbf{w}, \mathbf{v}) + c(\mathbf{v}; \mathbf{w}, \mathbf{v}) + b(\mathbf{w}, p) = (\mathbf{w}, \mathbf{b}) + (\mathbf{w}, \mathbf{t})_{(\partial\Omega)_N}, \\
b(q, \mathbf{v}) = 0.
\end{cases}
\tag{10}
$$

Replacement of the linear-, bilinear- and trilinear-forms with their respective definitions and application of integration by parts yields

$$
\begin{aligned}
&\underbrace{\int_{\Omega} \rho(\varphi)\mathbf{w}\cdot\mathbf{v}_t \,\mathrm{d}\Omega}_{(\mathbf{w},\mathbf{v}_t)} + \underbrace{\int_{\Omega} \mu(\varphi)\nabla\mathbf{w} : \left(\nabla\mathbf{v} + (\nabla\mathbf{v})^T\right) \mathrm{d}\Omega}_{a(\mathbf{w},\mathbf{v})} + \underbrace{\int_{\Omega} \rho(\varphi)\mathbf{w}\cdot\mathbf{v}\cdot\nabla\mathbf{v}\,\mathrm{d}\Omega}_{c(\mathbf{v};\mathbf{w},\mathbf{v})} = \\
&\underbrace{\int_{\Omega} \nabla\cdot\mathbf{w}\, p\,\mathrm{d}\Omega}_{b(\mathbf{w},p)} - \underbrace{\int_{\Omega} q\,\nabla\cdot\mathbf{v}\,\mathrm{d}\Omega}_{b(q,\mathbf{v})} + \underbrace{\int_{\Omega} \rho(\varphi)\mathbf{w}\cdot\mathbf{b} + \mathbf{w}\cdot\eta\nabla\varphi\,\mathrm{d}\Omega}_{(\mathbf{w},\mathbf{b})} + \\
&\underbrace{\int_{(\partial\Omega)_N} \mu(\varphi)\,\mathbf{w}\cdot\left(\left(\nabla\mathbf{v} + (\nabla\mathbf{v})^T\right)\cdot\mathbf{n}\right) \mathrm{d}(\partial\Omega)_N - \int_{(\partial\Omega)_N} \mathbf{w}\cdot\mathbf{n}\,p\,\mathrm{d}(\partial\Omega)_N}_{(\mathbf{w},\mathbf{t})_{(\partial\Omega)_N}} .
\end{aligned}
\tag{11}
$$

A downcast of the variational formulation (10) to the discrete level gives rise to the problem statement

$$
\begin{cases}
\text{Find } \mathbf{v}^h \in \mathscr{H}_0^1(\Omega)\cap\mathbf{V}_h^{TH}\times(0,T) \text{ and } p^h \in \mathscr{L}_2(\Omega)/\mathbb{R}\cap Q_h^{TH}\times(0,T), \text{ such that} \\
\forall(\mathbf{w}^h,q^h) \in \mathscr{H}_0^1(\Omega)\cap\mathbf{V}_h^{TH}\times\mathscr{L}_2(\Omega)/\mathbb{R}\cap Q_h^{TH} \\
(\mathbf{w}^h,\mathbf{v}_t^h) + a(\mathbf{w}^h,\mathbf{v}^h) + c(\mathbf{v}^h;\mathbf{w}^h,\mathbf{v}^h) + b(\mathbf{w}^h,p^h) = (\mathbf{w}^h,\mathbf{b}^h) + (\mathbf{w}^h,\mathbf{t}^h)_{(\partial\Omega)_N} \\
b(q^h,\mathbf{v}^h) = 0,
\end{cases}
\tag{12}
$$

with superscript h dubbing the mesh family index. Using Isogeometric Taylor-Hood finite elements and the one-step θ-scheme for the respective discretizations in space and time, we obtain the following discrete system

$$
\begin{aligned}
&\underbrace{\begin{pmatrix} \frac{1}{\Delta t}\mathbf{M}(\varphi^{n+1}) + \theta(\mathbf{D}(\varphi^{n+1}) + \mathbf{C}(v^{n+1},\varphi^{n+1})) & \mathbf{G} \\ \mathbf{G}^T & \mathbf{0} \end{pmatrix}}_{\mathbf{S}_l} \underbrace{\begin{pmatrix} \mathbf{v}^{n+1} \\ \mathbf{p}^{n+1} \end{pmatrix}}_{\mathbf{u}^{n+1}} = \\
&\underbrace{\begin{pmatrix} \frac{1}{\Delta t}\mathbf{M}(\varphi^{n}) - (1-\theta)(\mathbf{D}(\varphi^{n}) + \mathbf{C}(v^{n},\varphi^{n})) & \mathbf{0} \\ \mathbf{0} & \mathbf{0} \end{pmatrix}}_{\mathbf{S}_r} \underbrace{\begin{pmatrix} \mathbf{v}^{n} \\ \mathbf{p}^{n} \end{pmatrix}}_{\mathbf{u}^{n}} + \underbrace{\theta\,\mathbf{f}^{n+1}(\eta^{n+1},\varphi^{n+1})}_{\mathbf{b}^{n+1}} + \underbrace{(1-\theta)\mathbf{f}^{n}(\eta^{n},\varphi^{n})}_{\mathbf{b}^{n}},
\end{aligned}
\tag{13}
$$

where $\mathbf{M}, \mathbf{D}, \mathbf{C}, \mathbf{G}$, and $\mathbf{G}^{\mathbf{T}}$ denote the mass, rate of deformation, advection, gradient, and divergence matrices, respectively. The body and the surface tension force terms are discretized altogether into $\mathbf{f}$. For the treatment of the nonlinearity in the Navier–Stokes equations ((1a) + (1b)), we use the Newton-iteration

$$
\mathbf{J}(\mathbf{u}^k,\varphi^k)\,\delta\mathbf{u} = -\mathbf{F}(\mathbf{u}^k,\mathbf{u}^n,\eta^k,\eta^n,\varphi^k,\varphi^n), \tag{14a}
$$

$$
\mathbf{u}^{k+1} = \mathbf{u}^k + \delta\mathbf{u}, \tag{14b}
$$

whose right-hand side is set to be the residual of Eq. (13), that is,

$$\mathbf{F}(\mathbf{u}^{n+1}, \mathbf{u}^{n}, \eta^{n+1}, \eta^{n}, \varphi^{n+1}, \varphi^{n}) = \mathbf{S}_l\, \mathbf{u}^{n+1} - \mathbf{S}_r\, \mathbf{u}^{n} - \mathbf{b}^{n+1} - \mathbf{b}^{n}. \tag{15}$$

For a detailed description of the setup of the Jacobian $\mathbf{J}$ in Eq. (14), we refer to [7]. The linear equation systems arising from the discretization of the Cahn–Hilliard and Navier–Stokes equations are solved with direct solvers. For the solution of corresponding 3D problems involving larger systems, iterative solvers are advisable.

4 Application to the Rayleigh-Taylor Instability Problem

The Rayleigh-Taylor instability is a two-phase instability which occurs whenever two fluids of different density are accelerated against each other. Any perturbation along the interface between a heavy fluid (F_H) on top of a lighter fluid (F_L), both subject to a gravitational field, gives rise to the phenomenon of Rayleigh-Taylor instability. The initial perturbations progress from an initial linear growth phase into a non-linear one, eventually developing “mushroom head” like structures moving upwards and thinning “spikes” falling downwards. Assuming negligible viscosity and surface tension, the instability is characterized by the density disparity, measured with the Atwood number $\mathscr{A} = (\rho_H - \rho_L)/(\rho_H + \rho_L)$. For the validation of our results, we will consider the works of Tryggvason [10] and Guermond et al. [3] as reference. The former investigated the initial growth and long-time evolution of the instability for incompressible and inviscid flows with zero surface tension at $\mathscr{A} = 0.5$. Guermond et al., on the other hand, studied this instability problem at the same Atwood number, however, taking viscous effects additionally into account.

The setup of the problem is described by a rectangular computational domain $[0, d] \times [0, 4d]$, where an initial wavy interface segregates a heavier fluid in the upper domain part from a lighter fluid on the lower part. The initial interface is described by the function

$$y(x) = 2d + 0.1d\cos(2\pi x/d)$$

representing a planar interface superimposed by a perturbation of wave number $k = 1$ and amplitude $0.1d$. Note that setting the surface tension coefficient $\hat{\sigma}$ to 0, effectively downgrades the Cahn–Hilliard equations ((1c) + (1d)) to a pure transport equation well known from the level-set context. This, in turn, implies to pass on both the physical benefits inherent to phase field models and to the automatic recreation of the smooth transition of the phase field in the interface region. In order to circumvent these issues, we chose to set the surface tension coefficient to the small, yet non-zero value 0.01. As for the remaining simulation parameters we set $d = 1, \rho_H = 3, \rho_L = 1, \mu_H = \mu_L = 0.0031316$ and $g = 9.80665$, giving rise to $\mathscr{A} = 0.5$ and $Re = \rho_H d^{3/2} g^{1/2}/\mu_H = 3000$. At the top and bottom boundaries we use the no-slip boundary condition, whereas the free slip boundary condition is imposed on the vertical walls. Figure 1 depicts our results for the temporal evolution of the interface computed in the time interval [0, 1.5] with $\Delta t = 0.001, h = 2^{-7}, \varepsilon = 0.005$ and $D = 0.00004$. As anticipated, the heavier fluid on top starts to fall through the lighter fluid and gradually develops spikes which are subject to strong deformations. When it comes to the comparison of the vortex structure with the "inviscid" results of Tryggvason and the "viscous" results of Guermond et al., our viscous solution exhibits a satisfactory agreement with both, especially with the latter mentioned. Note that the data provided by the above references are computed with respect to individual scalings of the involved PDE variables in order to obtain nondimensional variables. Therefore, comparisons require the time scales of the respective simulations to be mapped to each other. Since, in contrast to the reference results, we did not perform any rescaling, our time t is mapped to Tryggvason's time $\tilde{t}$ via the relation $t = \sqrt{d/(\mathscr{A}g)}\,\tilde{t}$.

We continue the validation of our results with a quantitative analysis and conduct a comparison of the tip of the rising and falling fluids with the inviscid and viscous results provided by Tryggvason and Guermond et al., respectively. The results, depicted in Fig. 2, are in good agreement with both references whose data have individually been translated along the y-axis to facilitate comparisons. The upper curve referring to the tip of the rising fluid shows a better correlation with the data provided by Tryggvason while our curve for the falling fluid seems to perfectly match the results of Guermond. As for space and time discretization, for the results depicted in Fig. 2, Guermond uses $P_2 - P_1$ finite elements on a 49577 P_2 nodes mesh with a time step size $\Delta\tilde{t} = 5 \times 10^{-4}$. Tryggvason on the other hand uses a Lagrangian-Eulerian vortex method on a 64×128 grid. Our results depicted in Figs. 1, 2 and 3, are obtained using a fully implicit time integration ($\theta = 1$) in combination with a space discretization based on a $\mathscr{C}^0$ $\mathcal{N}^{2,2}_{0,0}/\mathcal{N}^{1,1}_{0,0}$ NURBS space pair for the approximation of the velocity and pressure functions. For mesh refinement levels $h \in \{2^{-5}, 2^{-6}, 2^{-7}\}$ the above choice of space discretization yields $\{(33410, 4257), (132354, 16705), (526850, 66177)\}$ (velocity, pressure) degrees of freedom, respectively. Note that with the above mentioned relation between t and $\tilde{t}$, our time step $\Delta t = 0.001$ is mapped to $\Delta\tilde{t} = 0.0022143$.

The analysis is finally concluded with the examination of the interface structure at a randomly selected fixed time $t = 0.79031$. As shown in Fig. 3, the results are mesh and time converged for three consecutive mesh refinement levels $h \in \{2^{-5}, 2^{-6}, 2^{-7}\}$ and time step sizes $\Delta t \in \{0.004, 0.002, 0.001\}$, respectively. The main difference between the figures is in the level of detail of the vortices. Besides, the y-coordinates of the tip of the rising and falling fluids slightly differ from one mesh refinement level to the other and are thus regarded as weakly resolution dependent. Apart from that no significant differences can be observed.

Fig. 1 The evolution of a single wavelength initial condition in the Rayleigh-Taylor instability simulation. Snapshots refer to times $t \in \{0, 0.17, 0.33, 0.5, 0.67, 0.83, 1, 1.17, 1.33, 1.5\}$ from top left to bottom right

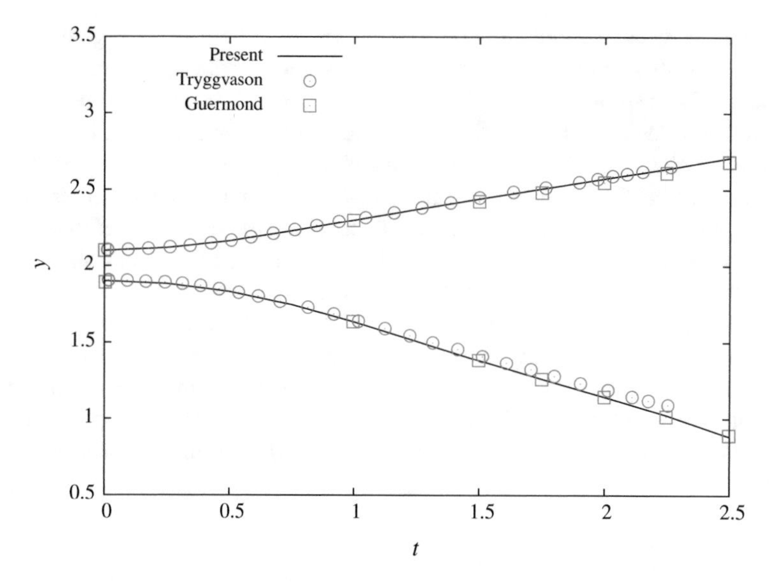

Fig. 2 The y-coordinate of the tip of the rising and falling fluid versus time

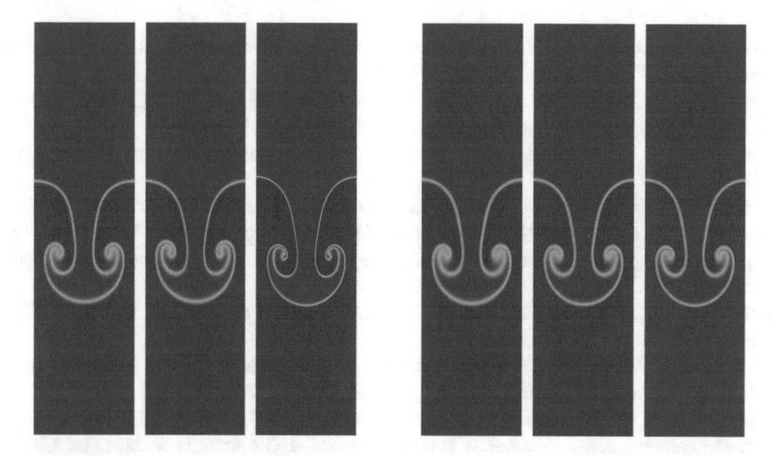

Fig. 3 Rayleigh-Taylor instability simulation at time $t = 0.79031$ with $\theta = 1$, $D = 0.00004$, $\hat{\sigma} = 0.01$. Left: Mesh converged results for $\Delta t = 0.0035$, $(h = 2^{-5}, \varepsilon = 0.02)$, $(h = 2^{-6}, \varepsilon = 0.01)$, $(h = 2^{-7}, \varepsilon = 0.005)$. Right: Time converged results for $\Delta t \in \{0.004, 0.002, 0.001\}$, $h = 2^{-5}$, $\varepsilon = 0.02$

5 Summary and Conclusions

We presented a phase field-based computational model for immiscible and incompressible two-component liquid flows based on the incompressible Navier–Stokes–Cahn–Hilliard equations. The Cahn–Hilliard equation was reformulated so as to introduce the chemical potential as an additional unknown handled by an extra equation. A weak formulation of the complex-fluid problem together with a derivation of its analytical Jacobian were presented providing enough information for hassle-free reproducibility of this work. Numerical results were presented for the Rayleigh-Taylor instability flow problem, based on Isogeometric finite element analysis of the weak formulation of the NSCH problem. The Rayleigh-Taylor instability problem has become a popular test case for numerical methods intended to study multiphase or multimaterial problems. Using the setup and reference results of Tryggvason [10] and Guermond et al. [3], we analyzed the evolution of a single wavelength interface perturbation. Qualitative comparisons of the interface shapes and quantitative analysis of the positions of the tip of the rising and falling fluid rendered our approximations to be in good correlation with the above references. Moreover, we showed our results to be mesh converged, since the produced data associated with different mesh resolutions are well comparable except for the high resolution features such as the roll-up spirals emerging at higher mesh refinement levels.

References

1. Badalassi, V.E., Ceniceros, H.D., Banerjee, S.: Computation of multiphase systems with phase field models. J. Comput. Phys. **190**(2), 371–397 (2003)
2. Ding, H., Spelt, P.D.M., Shu, C.: Diffuse interface model for incompressible two-phase flows with large density ratios. J. Comput. Phys. **226**(2), 2078–2095 (2007)
3. Guermond, J.-L., Quartapelle, L.: A projection FEM for variable density incompressible flows. J. Comput. Phys. **165**(1), 167–188 (2000)
4. Gurtin, M.E., Polignone, D., Vinals, J.: Two-phase binary fluids and immiscible fluids described by an order parameter. Math. Models Methods Appl. Sci. **6**(6), 815–831 (1996)
5. Hirt, C.W., Nichols, B.D.: Volume of fluid (VOF) method for the dynamics of free boundaries. J. Comput. Phys. **39**(1), 201–225 (1981)
6. Hohenberg, P.C., Halperin, B.I.: Theory of dynamic critical phenomena. Rev. Mod. Phys. **49**, 435–479 (1977)
7. Hosseini, B.S., Möller, M., Turek, S.: Isogeometric analysis of the Navier-Stokes equations with Taylor-Hood B-spline elements. Appl. Math. Comput. **267**, 264–281 (2015)
8. Sethian, J.A.: Level set methods and fast marching methods: evolving interfaces in computational geometry, fluid mechanics, computer vision, and materials science. In: Cambridge Monographs on Applied and Computational Mathematics. Cambridge University Press, Cambridge (1999)
9. Sun, Y., Beckermann, C.: Sharp interface tracking using the phase-field equation. J. Comput. Phys. **220**(2), 626–653 (2007)
10. Tryggvason, G.: Numerical simulations of the Rayleigh-Taylor instability. J. Comput. Phys. **75**(2), 253–282 (1988)

A Study on the Performance Portability of the Finite Element Assembly Process Within the Albany Land Ice Solver

Jerry Watkins, Irina Tezaur, and Irina Demeshko

Abstract This paper presents a performance analysis of the finite element assembly process of the Albany Land Ice solver. The analysis shows that a speedup over traditional MPI-only simulations is achieved on multiple architectures including Intel Haswell CPUs, Intel Xeon Phi Knights Landing and IBM POWER8/NVIDIA P100 platforms. A scalability study also shows that performance remains reasonably close among all architectures. These results are obtained on a single codebase without architecture-dependent code optimizations by utilizing abstractions in shared memory parallelism from the Kokkos library and is part of an ongoing process of achieving performance portability for the Albany Land Ice code.

Keywords Land ice · Performance portability · Kokkos · Finite element assembly · GPUs · KNLs

1 Introduction

High resolution simulations of the evolution of the polar ice sheets play a crucial role in the ongoing effort to develop more accurate and reliable regional and global integrated Earth-system models (ESMs) for probabilistic sea-level projections. These types of simulations often require a massive amount of memory and computation from large supercomputing clusters to provide sufficient accuracy and resolution. The current list of fastest supercomputers [34] shows a diverse set of computing platforms typically including multicore/manycore processors and GPU accelerators. This heterogeneity in supercomputing architectures has become the

---

J. Watkins (✉) · I. Tezaur
Sandia National Laboratories, Livermore, CA, USA
e-mail: jwatkin@sandia.gov; ikalash@sandia.gov

I. Demeshko
Los Alamos National Laboratory, Los Alamos, NM, USA
e-mail: irina@lanl.gov

H. van Brummelen et al. (eds.), *Numerical Methods for Flows*,
Lecture Notes in Computational Science and Engineering 132,
https://doi.org/10.1007/978-3-030-30705-9_16

norm and will continue to be a challenge for software developers as the high performance computing (HPC) community moves towards exascale. In order to take full advantage of available resources, performance portability has become an increasingly important subject for the simulation of ice sheets.

The term "performance portability" is difficult to define and there is no consensus on a clear definition [19]. In general, performance portability for an application means that a reasonable level of performance is achieved across a wide variety of computing architectures with the same source code. Here, "performance" and "variety" are admittedly subjective. In [22, 23], Pennycook et al. quantified performance portability through performance efficiencies based on hardware and application performance for a given set of hardware platforms. For the purposes of this paper, performance portability will be characterized by execution time and scalability efficiencies for multicore/manycore processors and GPUs.

There have been a number of approaches to performance portability for applications including directives such as OpenMP and OpenACC, and frameworks such as Kokkos [9], RAJA [14] and OCCA [17]. Performance portability for finite element assembly has also been executed on a variety of different software packages including Hiflow3 [1] and Firedrake/FEniCS/PyOP2 [16, 26, 27]. These packages are discussed briefly in [7] where the authors present the performance portability strategy used within the Albany multiphysics, finite element code [29].

Current generation ice sheet modeling [6, 11, 15, 28, 36] has seen a dramatic increase in fidelity partly due to the inability of previous generation ice sheet models to accurately predict dynamic behavior [31]. While the models have improved, a majority of codes rely solely on Message Passing Interface (MPI) libraries to achieve performance gains which limits the achievable performance on the latest supercomputing architectures, for example, supercomputers with GPUs. GPUs could provide a substantial amount of performance for ice sheet modeling if properly utilized [4]. Though it may be tempting to construct highly optimized implementations for each emerging architecture, this type of software development will become increasingly harder to maintain as future HPC architectures become increasingly more complex. This motivates the need for fundamental abstractions to be present at the application level during code development.

Towards the development of a fully performance portable implementation of a land ice model, this paper provides an analysis of the performance portability of the finite element assembly process within the Albany Land Ice solver. We show that this process can be executed on many different architectures including Intel Haswell CPUs, Intel Xeon Phi Knights Landing (KNL) and NVIDIA P100 GPUs and a speedup is achieved over traditional MPI-only simulations through the use of abstractions in shared memory parallelism from Kokkos. A scalability study is also presented, showing that the process is able to achieve a high level of performance when scaling to multiple devices on the latest supercomputing clusters, the National Energy Research Scientific Computing Center's (NERSC) Cori and Sandia National Laboratories' (SNL's) internal testbed called Ride.

The remainder of this paper is organized as follows. In Sect. 2, a description of the Albany Land Ice code is given along with a brief discussion on the finite element assembly process. Section 3 provides a performance analysis via device comparisons and scalability studies to verify the performance portability of the code. Conclusions are drawn from these results in Sect. 4.

2 Albany Land Ice

Ice sheets and glaciers are typically modeled as an incompressible fluid in a low Reynolds number regime with a power-law viscous rheology. More specifically, it is widely accepted that the momentum balance of the ice is governed by a nonlinear Stokes flow system of partial differential equations[1] (PDEs). Here, we utilize a simplified first-order approximation [8, 30] to this so-called full Stokes model, referred to as the "Blatter-Pattyn" model [3, 20], or simply the "first-order (FO) Stokes model". For a detailed description of the model, the reader is referred to [32].

2.1 Numerical Implementation Within the Albany Code

The FO Stokes model is discretized in an open-source,[2] C++, multi-physics, unstructured grid, implicit, parallel, scalable and robust finite element code base known as Albany [29]. Albany makes heavy use of numerous computational libraries from the open-source[3] Trilinos suite [12], which are "glued" together within this code through the use of Template-Based Generic Programming (TBGP) [21].

The FO Stokes-based momentum balance land-ice solver in Albany is referred to as Albany Land Ice. For a detailed description of Albany Land Ice (previously known as Albany/FELIX), the reader is referred to [32]. The key methods implemented in Albany Land Ice are summarized below.

- ***Classical Galerkin finite element method (FEM) discretization***: In Albany Land Ice the classical Galerkin FEM was selected to discretize the FO Stokes model for its flexibility in using unstructured meshes (e.g., grids with increased resolution in areas of large velocity gradients such as in the vicinity of outlet glaciers) and straightforward implementation of the basal sliding boundary condition. The STK package of Trilinos was used for mesh database structures and mesh I/O. The Intrepid2 package of Trilinos was used as a finite element shape function library with general integration kernels.

[1]The interested reader is referred to Appendix A of [32] for the nonlinear Stokes flow equations.

[2]Available on github: https://github.com/SNLComputation/Albany.

[3]Available on github: https://github.com/trilinos/Trilinos.

- ***Newton's method with automatic differentiation (AD) Jacobians***: Once the large, sparse system of nonlinear algebraic equations for the ice velocities is created following discretization by the FEM, the fully-coupled nonlinear system is solved using damped Newton's method. An analytic Jacobian matrix is computed at each iteration of Newton's method using AD, available through the Sacado package of Trilinos.
- ***Iterative linear solver with ILU or AMG preconditioning***: Within each Newton iteration detailed above, a number of linear systems arise. These systems are solved using a preconditioned iterative method (CG or GMRES, both available through the Belos package of Trilinos). Although the model is symmetric, and hence amenable to the CG iterative linear solver, it was found that faster convergence can be obtained with the GMRES iterative linear solver for some problems (e.g., simulations of the AIS). Two options for the preconditioner are considered: an ILU additive Schwarz preconditioner with 0 overlap and 0 level-of-fill, and a recently proposed (introduced in [32, 33]; detailed in [35]) AMG preconditioner, constructed based on the idea of semi-coarsening (i.e., coarsening only in the structured dimension, in this case z–dimension). These preconditioners are available through the Ifpack and ML packages of Trilinos, respectively.
- ***Adjoint-based optimization for ice sheet initialization:*** To calculate the ice sheet initial conditions, namely the basal sliding and basal topography fields, we formulate and solve a PDE-constrained optimization problem that minimizes the mismatch between model output and observations (detailed in [24]). The optimization is performed using the Limited–memory Broyden–Fletcher–Goldfarb–Shanno (LBFGS) method [37], as implemented in the ROL Rapid Optimization Library of Trilinos. The cost function gradients with respect to the parameter fields are computed using adjoints.
- ***Performance portability:*** Albany follows an "MPI+X" programming model where MPI is used for distributed memory parallelism and the Kokkos library [9] of Trilinos is used for shared memory parallelism. This enables performance portability across a variety of different multicore/manycore processors and GPU accelerators under a single codebase. Abstractions are used to obtain optimal data layouts and hardware features, reducing the complexity of application code. Further implementation details can found in [7].

The accuracy of the Albany Land Ice FO Stokes solver has been verified extensively using the method of manufactured solutions and canonical land ice benchmarks, as well as by performing a mesh convergence study using a realistic Greenland Ice Sheet (GIS) geometry [32]. In order to enable dynamic simulations of ice sheet evolution as well as coupling to Earth System Models (ESM), Albany Land Ice has been coupled to two existing codes which solve the thickness and temperature evolution equations (conservation of mass and energy, respectively), namely the Community Ice Sheet Model (CISM) [5] and the Model for Prediction Across Scales (MPAS) [18]. The resulting models, termed CALI and MALI, respectively, have been used to perform large-scale realistic simulations of the

Greenland and Antarctic Ice Sheets [13, 25], and, in the case of CALI, validated using real data from two NASA satellites [10, 25].

2.2 Finite Element Assembly

A dynamic land-ice simulation is comprised of numerous steady-state stress-velocity diagnostic solves. Each diagnostic solve can be split into two major parts: the linear solver and the finite element assembly. Performance portability for linear solvers is a widely researched topic and Albany utilizes Trilinos to obtain the latest solvers suitable for solving ice sheet problems. This paper focuses on the performance portability of the finite element assembly process for which there have been relatively fewer studies. Figure 1 shows the main work flow for a single nonlinear iteration of the Albany Land Ice solver. The finite element assembly process can be further split into the following components where the goal is use the global solution to obtain a global residual and Jacobian for the linear solver:

- ***Import***: Imports the global solution from a nonoverlapping data structure where each MPI rank owns a unique part of the solution to an overlapping data structure where some data exists on multiple ranks. This gives each rank access to relevant solution data without any further communication. In this case, the solution is the ice velocity vector.
- ***Gather***: Gathers solution values from an overlapping data structure to an element local data structure where data is indexed according to element and local node. This process also includes the gathering of geometry and field data.
- ***Interpolate***: Interpolates the solution and solution gradient from nodal points to quadrature points. Other field variables may also require interpolation.

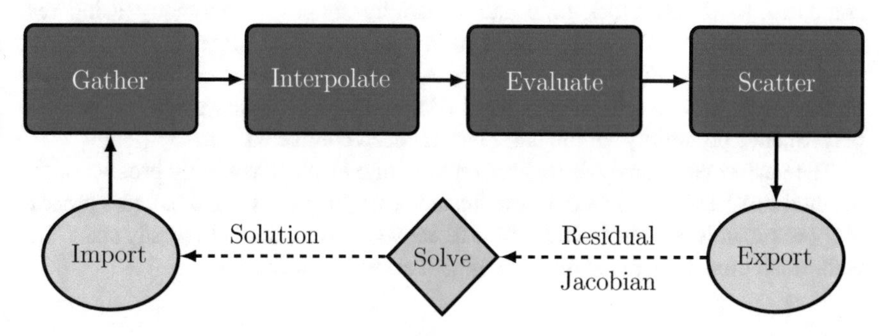

Fig. 1 A flow chart showing the main work flow for a single nonlinear iteration of the Albany Land Ice solver. Finite element assembly begins and ends with distributed memory assembly (DMAssembly) (in yellow) which constructs data structures which run more efficiently during the linear solver and finite element assembly phases. This is also where MPI communication occurs. The remaining shared memory assembly (SMAssembly) processes (in blue) perform global and local assembly and are parallelized over elements using Kokkos `parallel_for`

- ***Evaluate***: Evaluates the residual, Jacobian and source terms. These operators are templated in order to take advantage of automatic differentiation for analytical Jacobians. In this case, the FO Stokes model is evaluated.
- ***Scatter***: Scatters residual and Jacobian values from an element local data structure to an overlapping data structure. Kokkos `atomic_fetch_add` is used to avoid race conditions during global assembly.
- ***Export***: Exports the residual and Jacobian from an overlapping data structure to a nonoverlapping data structure where information is updated across MPI ranks. This global structure allows for efficient use of linear solvers.

Import and *Export* are handled by the Tpetra package [2] of Trilinos and uses both MPI and Kokkos to construct the overlapping and nonoverlapping data structures.

Kokkos utilizes execution and memory spaces to determine where routines are executed and where memory is stored. For this work, the Kokkos `Serial` execution space is used for MPI-only simulations, `OpenMP` is used for MPI+OpenMP and `Serial` and `Cuda` is used for MPI+GPU. The memory in element local data structures is contiguous along elements for GPUs (`LayoutLeft`) and strided along elements for CPUs (`LayoutRight`). CUDA UVM is used to avoid manual memory transfers between host and device and CUDA-Aware MPI is used to provide direct GPU to GPU communication. Both of these features our abstracted in Kokkos and Tpetra, respectively. These configurations are described in more detail in [7] where a similar configuration was used with the exception of CUDA-Aware MPI.

3 Performance Analysis

The performance of Albany Land Ice is quantified by performing architecture comparisons of wall-clock time and scalability studies on geometry using two tetrahedral meshes. These are described in Table 1 and an example of the coarser mesh is shown in Fig. 2. In this study, a no-slip boundary condition is chosen on the lower surface instead of a basal friction (slip) boundary condition because performance portability for the latter has not been completed yet.

The performance analysis focuses on the finite element assembly process of the residual and Jacobian needed for the linear solver. In Albany Land Ice, this process is approximately 50% of the total time to converge the solution to steady state. The wall-clock time is averaged over 100 global assembly evaluations.

Table 1 Greenland ice sheet mesh resolutions

Mesh	Resolution	Number of elements
GIS4k-20k	4–20 km	1,505,790
GIS1k-7k	1–7 km	14,397,900

Fig. 2 Surface of the 1,505,790 element tetrahedral mesh of our GIS geometry (referred to as GIS4k-20k). The mesh is constructed from 10 layers of the bottom surface

3.1 Architectures

Albany Land Ice is used to simulate the flow of ice on two supercomputing clusters: Cori and Ride. A summary of the hardware, software and compiler information for each configuration can be found in Table 2.

The code is compiled using Kokkos `Serial` and `OpenMP` on Cori and `Serial` and `Cuda` on Ride. In this study, MPI ranks are mapped to cores and OpenMP threads are mapped to hardware threads. These were found to be the optimal

Table 2 Cluster and build configurations

Name	Cori (HW)	Cori (KNL)	Ride (P100)
CPU	Intel Xeon E5-2698 v3 ("Haswell")	Intel Xeon Phi 7250 ("Knights Landing")	IBM POWER8 (8-core)
# of cores	16	68	8
Threads/core	2	4	8
GPU	None	None	NVIDIA Tesla P100
Node arch	2 CPUs	1 CPU	2 CPUs + 4 GPUs
Memory/node	128 GB DDR4	96 GB DDR4 + 16 GB MCDRAM	512 GB DDR4
Interconnect	Cray aries	Cray aries	Mellanox ConnectX-4 IB
Compiler	Intel 18.0.1.163	Intel 18.0.1.163	gcc 5.4.0
MPI	cray-mpich 7.6.2	cray-mpich 7.6.2	openmpi 1.10.4
NVCC	None	None	8.0.44

configurations when utilizing MPI and OpenMP. Mapping MPI ranks to hardware threads reduces performance due to scheduling issues, as multiple ranks compete for the resources on a single core. Mapping OpenMP threads to cores also reduces performance since those cores could have been utilized for coarser grain parallelism (i.e. MPI). For the simulations using GPUs, each MPI rank is assigned a single core and GPU; therefore, there are CPU cores which are not being utilized within a node.

The following naming convention is used to describe the different mappings for each simulation,

$$r\,(\mathrm{MPI} + j\mathrm{X})\,, \qquad \mathrm{X} \in \{\mathrm{OMP}, \mathrm{GPU}\}, \tag{1}$$

where r is the number of MPI ranks, j is the number of OpenMP threads or GPUs per rank and X is the architecture used for shared memory parallelism (e.g. OMP for OpenMP).

3.2 Architecture Comparison

An architecture comparison is performed in order to quantify the performance of the finite element assembly process among different computing platforms. Figure 3 plots the wall-clock time against different devices for the GIS4k-20k mesh described in Table 1 where the devices considered are the Haswell, KNL and P100 GPU.

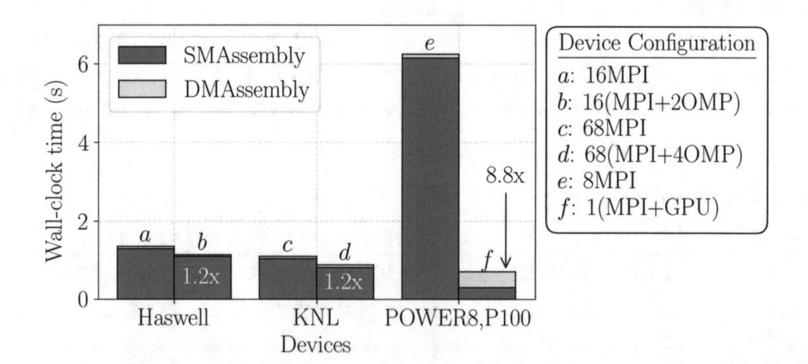

Fig. 3 A device comparison of finite element assembly wall-clock times (averaged over 100 evaluations) for Kokkos `Serial`, `OpenMP` and `Cuda` execution spaces on the GIS4k-20k mesh. A speedup is achieved across all execution spaces. In this case, the P100 (case f) performs the best

3.3 Scalability Study

A scalability study is performed in order to quantify the performance among different processors when scaling to multiple devices or larger problems. Strong scaling is used to determine how well the code utilizes additional resources to achieve a faster simulation of the same problem. Strong scaling efficiency is given by $((t_1/t_d)/d)100\%$ where t is the wall-clock time, d is the number of devices and (t_1/t_d) is the speedup which, ideally, is proportional to the number devices. Figure 4 shows the wall-clock time on up to 32 devices for the GIS4k-20k mesh while Table 3 reports the efficiency on up to 32 devices for Haswell, KNL and P100.

Weak scaling is used to determine how well the code is able to maintain the same wall-clock time when simulating larger problems with a proportionally larger amount of resources. Weak scaling efficiency is given by $(((t_1/N_1)/(t_d/N_d))/d)100\%$ where N is the number of elements. Here, N is used to normalize the time with respect to the number of elements in the problem. Table 3 summarizes the resulting efficiencies for each case where the finer GIS1k-7k mesh is used as the larger problem in the weak scalability study.

The strong and weak scaling efficiencies in Table 3 shows that all cases are able to maintain a reasonable amount of performance across multiple devices. In this study, the Haswell performs best for strong scaling while the KNL performs best

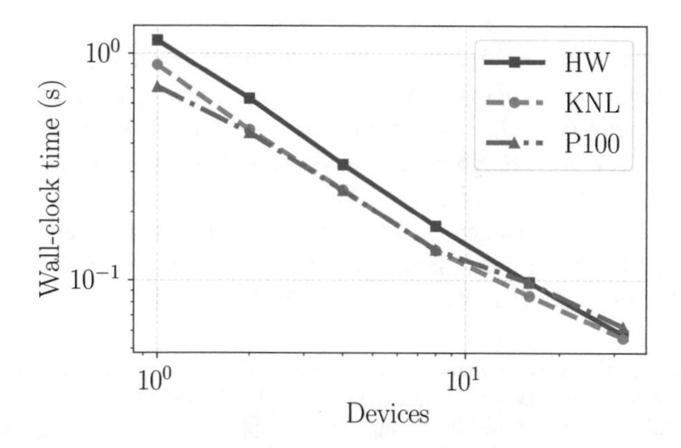

Fig. 4 Strong scalability wall-clock time on up to 32 devices (Haswell, KNL, P100) on the GIS4k-20k mesh. The wall-clock time among architectures remains similar on up to 32 devices

Table 3 A comparison of scalability efficiencies where one device is used as a reference

Device	Device configuration	Strong scaling[a]	Weak scaling[b]
Haswell	16(MPI+2OMP)	62.1%	88.8%
KNL	68(MPI+4OMP)	49.9%	94.5%
P100	1(MPI+GPU)	35.7%	69.2%

[a]Strong scaling performed on 32 devices (512 HW cores, 2176 KNL cores, 32 GPUs)
[b]Weak scaling performed on 10 devices (160 HW cores, 680 KNL cores, 10 GPUs)

for weak scaling. In both scalability studies, the P100 performs the worst due to the higher demand in distributed memory assembly. Despite the loss in strong scaling efficiency in the KNL and P100, Fig. 4 shows that the wall-clock time remains similar among all architectures when scaling on up to 32 devices.

It's important to note that the results only show the performance of the finite element assembly process. A performance portable linear solver is needed to obtain a more accurate performance analysis of Albany Land Ice.

4 Conclusions

In this paper, the finite element assembly process within an Albany Land Ice solver is presented and analyzed in order to study how well the code performs on a variety of difference computing architectures. A performance analysis showed that the assembly process performs reasonable well on Intel Haswell CPUs, Intel Xeon Phi KNLs and NVIDIA P100 GPUs on a single code base without architecture-dependent code optimizations. By utilizing GPUs and hardware threads for shared memory parallelism, a speedup over traditional MPI-only simulations is shown across all architectures. A scalability study showed that performance remained reasonably close among all architectures. The study also identified deficiencies in distributed memory assembly on GPU architectures which is a subject of future optimization within the Tpetra package in Trilinos.

Since performance portability is subjective, obtaining a performance portable implementation of a land ice solver remains an ongoing process. Future work will focus on optimizing the current implementation to obtain better performance across all architectures through code refactoring and hierarchical parallelism. Other possible optimizations which remain to be studied include explicit vectorization on CPUs, multiple CUDA instances on GPUs for better node utilization and explicit data management to minimize memory transfers. Performance portability for linear solvers is also an ongoing research topic within Trilinos.

Acknowledgements This work was supported under the Biological and Environmental Research (BER) Scientific Discovery through Advanced Computing (SciDAC) Partnership: a collaboration between the Advanced Scientific Computing Research (ASCR) and BER programs funded by the U.S. Department of Energy's Office of Science. The refactoring of Albany to a Kokkos programming model was also supported by Frameworks, Algorithms and Scalable Technologies for Mathematics (FASTMath) SciDAC Institute.

This research used resources of the National Energy Research Scientific Computing Center, which is supported by the Office of Science of the U.S. Department of Energy under Contract No. DE-AC02-05CH11231.

Sandia National Laboratories is a multimission laboratory managed and operated by National Technology and Engineering Solutions of Sandia LLC, a wholly owned subsidiary of Honeywell International Inc. for the U.S. Department of Energy's National Nuclear Security Administration under contract DE-NA0003525.

References

1. Anzt, H., Augustin, W., Baumann, M., Bockelmann, H., Gengenbach, T., Hahn, T., Heuveline, V., Ketelaer, E., Lukarski, D., Otzen, A., et al.: Hiflow3–a flexible and hardware-aware parallel finite element package. Preprint Series of the Engineering Mathematics and Computing Lab, 6 (2010)
2. Baker, C.G., Heroux, M.A.: Tpetra, and the use of generic programming in scientific computing. Sci. Program. **20**(2), 115–128 (2012)
3. Blatter, H.: Velocity and stress fields in grounded glaciers: a simple algorithm for including deviatoric stress gradients. J. Glaciol. **41**(138), 333–344 (1995)
4. Brædstrup, C.F., Damsgaard, A., Egholm, D.L.: Ice-sheet modelling accelerated by graphics cards. Comput. Geosci. **72**, 210–220 (2014)
5. CISM/The Community Ice Sheet Model. https://cism.github.io/index.html. Accessed 30 April 2018
6. Cornford, S.L., Martin, D.F., Graves, D.T., Ranken, D.F., A.M., Brocq, L., Gladstone, R.M., Payne, A.J., Ng, E.G., Lipscomb, W.H.: Adaptive mesh, finite volume modeling of marine ice sheets. J. Comput. Phys. **232**(1), 529–549 (2013)
7. Demeshko, I., Watkins, J., Tezaur, I.K., Guba, O., Spotz, W.F., Salinger, A.G., Pawlowski, R.P., Heroux, M.A.: Toward performance portability of the Albany finite element analysis code using the Kokkos library. Int. J. High Perform. Comput. Appl. **33**(2), 332–352 (2019)
8. Dukowicz, J.K., Price, S.F., Lipscomb, W.H.: Consistent approximations and boundary conditions for ice-sheet dynamics from a principle of least action. J. Glaciol. **56**(197), 480–496 (2010)
9. Edwards, H.C., Trott, C.R., Sunderland, D.: Kokkos: enabling manycore performance portability through polymorphic memory access patterns. J. Parallel Distrib. Comput. **74**(12), 3202–3216 (2014)
10. Evans, K.J., Kennedy, J.H., Lu, D., Forrester, M.M., Price, S., Fyke, J., Bennett, A.R., Hoffman, M.J., Tezaur, I., Zender, C.S., Vizcaíno, M.: LIVVkit 2.1: automated and extensible ice sheet model validation. Geosci. Model Dev. **12**(3), 1067–1086 (2019)
11. Gagliardini, O., Zwinger, T., Gillet-Chaulet, F., Durand, G., Favier, L., de Fleurian, B., Greve, R., Malinen, M., Martín, C., Råback, P., et al.: Capabilities and performance of Elmer/Ice, a new-generation ice sheet model. Geosci. Model Dev. **6**(4), 1299–1318 (2013)
12. Heroux, M.A., Bartlett, R.A., Howle, V.E., Hoekstra, R.J., Hu, J.J., Kolda, T.G., Lehoucq, R.B., Long, K.R., Pawlowski, R.P., Phipps, E.T., et al.: An overview of the Trilinos project. ACM Trans. Math. Softw. **31**(3), 397–423 (2005)
13. Hoffman, M.J., Perego, M., Price, S.F., Lipscomb, W.H., Zhang, T., Jacobsen, D., Tezaur, I., Salinger, A.G., Tuminaro, R. and Bertagna, L.: MPAS-Albany Land Ice (MALI): a variable-resolution ice sheet model for Earth system modeling using Voronoi grids. Geosci. Model Dev. **11**(9), 3747-3780 (2018)
14. Hornung, R.D., Keasler, J.A.: The RAJA portability layer: overview and status. Technical report, Lawrence Livermore National Lab. (LLNL), Livermore (2014)
15. Larour, E., Seroussi, H., Morlighem, M., Rignot, E.: Continental scale, high order, high spatial resolution, ice sheet modeling using the Ice Sheet System Model (ISSM). J. Geophys. Res. Earth Surf. **117**(F1) (2012)
16. Markall, G.R., Slemmer, A., Ham, D.A., Kelly, P.H.J., Cantwell, C.D., Sherwin, S.J.: Finite element assembly strategies on multi-core and many-core architectures. Int. J. Numer. Methods Fluids **71**(1), 80–97 (2013)
17. Medina, D.S., St-Cyr, A., Warburton, T.: OCCA: a unified approach to multi-threading languages. arXiv preprint arXiv:1403.0968 (2014)
18. MPAS-Albany Land Ice. https://mpas-dev.github.io/land_ice/land_ice.html. Accessed 30 April 2018
19. Neely, J.R.: DOE centers of excellence performance portability meeting. Technical report, Lawrence Livermore National Lab. (LLNL), Livermore (2016)

20. Pattyn, F.: A new three-dimensional higher-order thermomechanical ice sheet model: basic sensitivity, ice stream development, and ice flow across subglacial lakes. J. Geophys. Res. Solid Earth **108**(B8) (2003)
21. Pawlowski, R.P., Phipps, E.T., Salinger, A.G.: Automating embedded analysis capabilities and managing software complexity in multiphysics simulation, part I: template-based generic programming. Sci. Program. **20**(2), 197–219 (2012)
22. Pennycook, S.J., Sewall, J.D., Lee, V.W.: A metric for performance portability. arXiv preprint arXiv:1611.07409 (2016)
23. Pennycook, S.J., Sewall, J.D., Lee, V.W.: Implications of a metric for performance portability. Futur. Gener. Comput. Syst. **92**, 947–958 (2017)
24. Perego, M., Price, S., Stadler, G.: Optimal initial conditions for coupling ice sheet models to Earth system models. J. Geophys. Res. Earth Surf. **119**(9), 1894–1917 (2014)
25. Price, S.F., Hoffman, M.J., Bonin, J.A., Howat, I.M., Neumann, T., Saba, J., Tezaur, I., Guerber, J., Chambers, D.P., Evans, K.J., et al.: An ice sheet model validation framework for the Greenland ice sheet. Geosci. Model Dev. **10**(1), 255–270 (2017)
26. Rathgeber, F., Markall, G.R., Mitchell, L., Loriant, N., Ham, D.A., Bertolli, C., Kelly, P.H.J.: PyOP2: a high-level framework for performance-portable simulations on unstructured meshes. In: 2012 SC Companion High Performance Computing, Networking, Storage and Analysis (SCC), pp. 1116–1123. IEEE, Piscataway (2012)
27. Rathgeber, F., Ham, D.A., Mitchell, L., Lange, M., Luporini, Fabio, A., McRae, T.T., Bercea, G.-T., Markall, G.R., Kelly, P.H.J.: Firedrake: automating the finite element method by composing abstractions. ACM Trans. Math. Softw. **43**(3), 24 (2017)
28. Rutt, I.C., Hagdorn, M., Hulton, N.R.J., Payne, A.J.: The Glimmer community ice sheet model. J. Geophys. Res. Earth Surf. **114**(F2) (2009)
29. Salinger, A.G., Bartlett, R.A., Bradley, A.M., Chen, Q., Demeshko, I.P., Gao, X., Hansen, G.A., Mota, A., Muller, R.P., Nielsen, E., et al.: Albany: using component-based design to develop a flexible, generic multiphysics analysis code. Int. J. Multiscale Comput. Eng. **14**(4), 415–438 (2016)
30. Schoof, C., Hindmarsh, R.C.A.: Thin-film flows with wall slip: an asymptotic analysis of higher order glacier flow models. Q. J. Mech. Appl. Math. **63**(1), 73–114 (2010)
31. Solomon, S.: Climate Change 2007-the Physical Science Basis: Working Group I Contribution to the Fourth Assessment Report of the IPCC, vol. 4. Cambridge University Press, Cambridge (2007)
32. Tezaur, I.K., Perego, M., Salinger, A.G., Tuminaro, R.S., Price, S.F.: Albany/FELIX: a parallel, scalable and robust, finite element, first-order Stokes approximation ice sheet solver built for advanced analysis. Geosci. Model Dev. **8**(4), 1197 (2015)
33. Tezaur, I.K., Tuminaro, R.S., Perego, M., Salinger, A.G., Price, S.F.: On the scalability of the Albany/FELIX first-order Stokes approximation ice sheet solver for large-scale simulations of the Greenland and Antarctic ice sheets. Proc. Comput. Sci. **51**, 2026–2035 (2015)
34. TOP500 Project: November 2017 TOP500 list. https://www.top500.org/lists/2017/11/. Accessed 5 April 2018
35. Tuminaro, R., Perego, M., Tezaur, I., Salinger, A., Price, S.: A matrix dependent/algebraic multigrid approach for extruded meshes with applications to ice sheet modeling. SIAM J. Sci. Comput. **38**(5), C504–C532 (2016)
36. Winkelmann, R., Martin, M.A., Haseloff, M., Albrecht, T., Bueler, E., Khroulev, C., Levermann, A.: The Potsdam parallel ice sheet model (PISM-PIK)-part 1: model description. Cryosphere **5**(3), 715 (2011)
37. Wright, S., Nocedal, J.: Numerical optimization. Springer Science, vol. 35, pp. 67–68. Springer, Berlin (1999)

A Multimesh Finite Element Method for the Stokes Problem

August Johansson, Mats G. Larson, and Anders Logg

Abstract The multimesh finite element method enables the solution of partial differential equations on a computational mesh composed by multiple arbitrarily overlapping meshes. The discretization is based on a continuous–discontinuous function space with interface conditions enforced by means of Nitsche's method. In this contribution, we consider the Stokes problem as a first step towards flow applications. The multimesh formulation leads to so called cut elements in the underlying meshes close to overlaps. These demand stabilization to ensure coercivity and stability of the stiffness matrix. We employ a consistent least-squares term on the overlap to ensure that the inf-sup condition holds. We here present the method for the Stokes problem, discuss the implementation, and verify that we have optimal convergence.

Keywords FEM · Unfitted mesh · Non-matching mesh · Multimesh · CutFEM · Nitsche

---

A. Johansson (✉)
Simula Research Laboratory, Lysaker, Norway

Mathematics and Cybernetics, SINTEF Digital, Oslo, Norway
e-mail: august.johansson@sintef.no

M. G. Larson
Department of Mathematics, Umeå University, Umeå, Sweden
e-mail: mats.larson@math.umu.se

A. Logg
Department of Mathematical Sciences, Chalmers University of Technology and University of Gothenburg, Göteborg, Sweden
e-mail: logg@chalmers.se

H. van Brummelen et al. (eds.), *Numerical Methods for Flows*,
Lecture Notes in Computational Science and Engineering 132,
https://doi.org/10.1007/978-3-030-30705-9_17

1 Introduction

Consider the Stokes problem

$$-\Delta \boldsymbol{u} + \nabla p = \boldsymbol{f} \quad \text{in } \Omega, \tag{1}$$

$$\operatorname{div} \boldsymbol{u} = 0 \quad \text{in } \Omega, \tag{2}$$

$$\boldsymbol{u} = \boldsymbol{0} \quad \text{on } \partial\Omega, \tag{3}$$

for the velocity $\boldsymbol{u} : \Omega \to \mathbb{R}^d$ and pressure $\boldsymbol{p} : \Omega \to \mathbb{R}$ in a polytopic domain $\Omega \subset \mathbb{R}^d$, $d = 2, 3$.

The Stokes problem is considered here as a first step towards a multimesh formulation for multi-body flow problems, and ultimately fluid–structure interaction problems, in which each body is discretized by an individual boundary-fitted mesh and the boundary-fitted meshes move freely on top of a fixed background mesh. The applications for such a formulation are many, e.g., the simulation of blood platelets in a blood stream, the optimization of the configuration of an array of wind turbines, or the investigation of the effect of building locations in a simulation of urban wind conditions and pollution. Common to these applications is that the multimesh method removes the need for costly mesh (re)generation and allows the platelets, wind turbines or buildings to be moved around freely in the domain, either in each timestep as a part of a dynamic simulation, or in each iteration as part of an optimization problem.

The multimesh formulation presented here is a generalization of the formulation presented and analyzed in [5] for two domains. For comparison, the multimesh discretization of the Poisson problem for arbitrarily many intersecting meshes is presented in [6] and analyzed in [8].

2 Notation

We first review the notation for domains, interfaces, meshes and overlaps used to formulate the multimesh finite element method. For a more detailed exposition, we refer to [6].

Notation for domains

Let $\Omega = \widehat{\Omega}_0 \subset \mathbb{R}^d$, $d = 2, 3$, be a domain with polytopic boundary (the background domain).
Let $\widehat{\Omega}_i \subset \widehat{\Omega}_0$, $i = 1, \dots, N$ be the so-called *predomains* with polytopic boundaries (see Fig. 1).
Let $\Omega_i = \widehat{\Omega}_i \setminus \bigcup_{j=i+1}^{N} \widehat{\Omega}_j$, $i = 0, \dots, N$ be a partition of Ω (see Fig. 2).

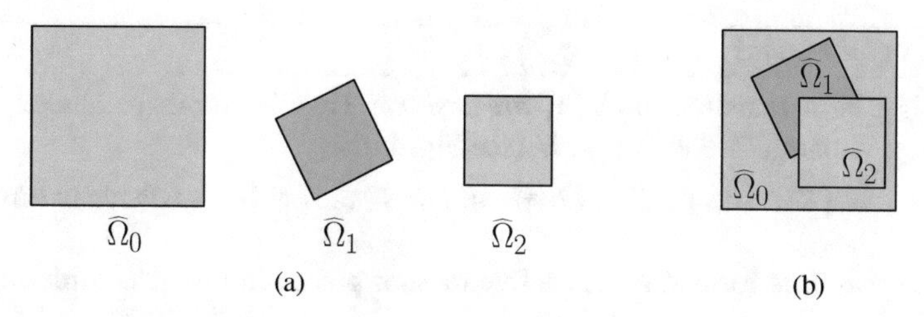

Fig. 1 (**a**) Three polygonal predomains. (**b**) The predomains are placed on top of each other in an ordering such that $\widehat{\Omega}_0$ is placed lowest, $\widehat{\Omega}_1$ is in the middle and $\widehat{\Omega}_2$ is on top

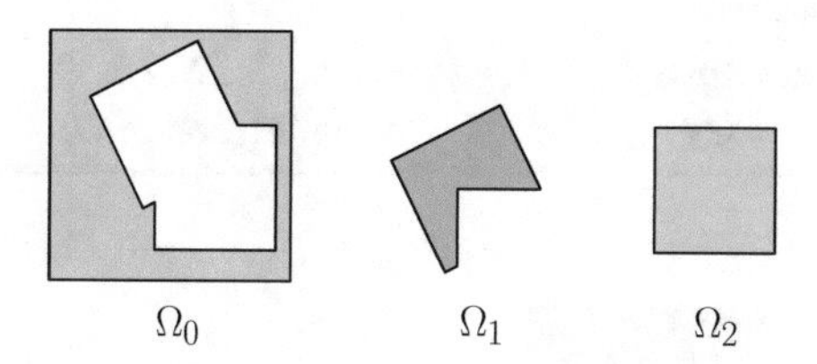

Fig. 2 Partition of $\Omega = \Omega_0 \cup \Omega_1 \cup \Omega_2$. Note that $\Omega_2 = \widehat{\Omega}_2$

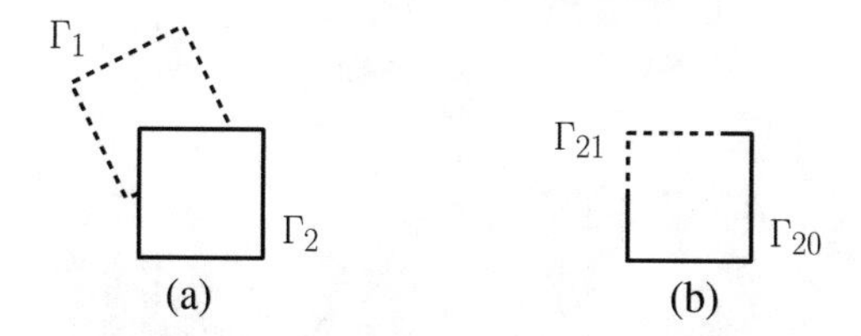

Fig. 3 (**a**) The two interfaces of the domains in Fig. 1: $\Gamma_1 = \partial\widehat{\Omega}_1 \setminus \widehat{\Omega}_2$ (dashed line) and $\Gamma_2 = \partial\widehat{\Omega}_2$ (filled line). Note that Γ_1 is not a closed curve. (**b**) Partition of $\Gamma_2 = \Gamma_{20} \cup \Gamma_{21}$

Remark 1 To simplify the presentation, the domains $\Omega_1, \ldots, \Omega_N$ are not allowed to intersect the boundary of Ω.

Notation for interfaces

Let the *interface* Γ_i be defined by $\Gamma_i = \partial\widehat{\Omega}_i \setminus \bigcup_{j=i+1}^{N} \widehat{\Omega}_j$, $i = 1, \ldots, N-1$ (see Fig. 3a).
Let $\Gamma_{ij} = \Gamma_i \cap \Omega_j$, $i > j$ be a partition of Γ_i (see Fig. 3b).

Notation for meshes

Let $\widehat{\mathcal{K}}_{h,i}$ be a quasi-uniform [3] *premesh* on $\widehat{\Omega}_i$ with mesh parameter $h_i = \max_{K\in\widehat{\mathcal{K}}_{h,i}} \operatorname{diam}(K)$, $i = 0, \dots, N$ (see Fig. 4a).
Let $\mathcal{K}_{h,i} = \{K \in \widehat{\mathcal{K}}_{h,i} : K \cap \Omega_i \neq \emptyset\}$, $i = 0, \dots, N$ be the *active meshes* (see Fig. 4b).
The *multimesh* is formed by the active meshes placed in the given ordering (see Fig. 5b).
Let $\Omega_{h,i} = \bigcup_{K\in\mathcal{K}_{h,i}} K$, $i = 0, \dots, N$ be the *active domains*.

Notation for overlaps

Let $\mathcal{O}_i$ denote the *overlap* defined by $\mathcal{O}_i = \Omega_{h,i} \setminus \Omega_i$, $i = 0, \dots, N-1$.
Let $\mathcal{O}_{ij} = \mathcal{O}_i \cap \Omega_j = \Omega_{h,i} \cap \Omega_j$, $i < j$ be a partition of $\mathcal{O}_i$.

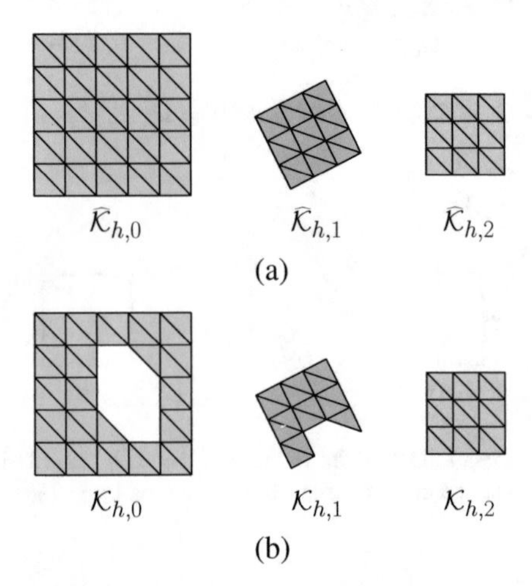

Fig. 4 (**a**) The three premeshes. (**b**) The corresponding active meshes (cf. Fig. 1)

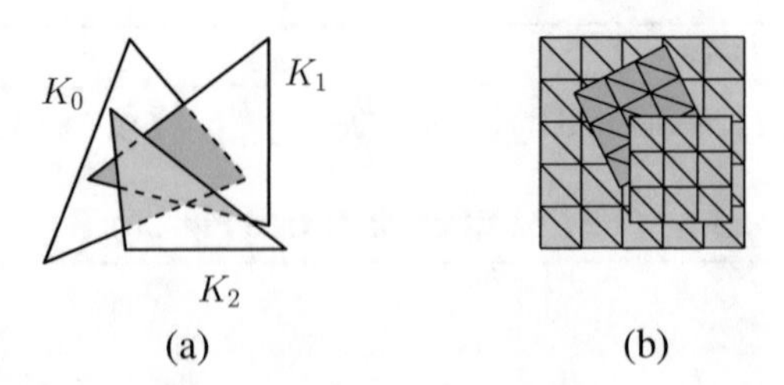

Fig. 5 (**a**) Given three ordered triangles K_0, K_1 and K_2, the overlaps are $\mathcal{O}_{01}$ in green, $\mathcal{O}_{02}$ in red and $\mathcal{O}_{12}$ in blue. (**b**) The multimesh of the domains in Fig. 1b consists of the active meshes in Fig. 4b

3 Multimesh Finite Element Method

To formulate the multimesh finite element for the Stokes problem (1) and (2), we assume for each (active) mesh $\mathcal{K}_{h,i}$ the existence of a pair of inf-sub stable spaces $\boldsymbol{V}_{h,i} \times Q_{h,i}$, $i = 0, 1, \ldots, N$, away from the interface. To be precise, we assume inf-sup stability in $\omega_{h,i} \subset \Omega_{h,i}$ in the sense of (5) below, where $\omega_{h,i}$ is close to Ω_i in the sense that $\Omega_{h,i} \setminus \omega_{h,i} \subset U_\delta(\Gamma_i)$, where

$$U_\delta(\Gamma_i) = \bigcup_{\boldsymbol{x}\in\Gamma_i} B_\delta(\boldsymbol{x}) \tag{4}$$

and $B_\delta(\boldsymbol{x})$ is a ball of radius δ centered at $\boldsymbol{x}$. In other words, $U_\delta(\Gamma_i)$ is the tubular neighborhood of Γ with thickness δ. In the numerical examples, we let $\omega_{h,i}$ be the union of elements in $\mathcal{K}_{h,i}$ with empty intersection with Γ_{ij}, $j > i$.

The inf-sup condition may expressed on each submesh $\omega_{h,i}$ by

$$\|p_i - \lambda_{\omega_{h,i}}(p)\|_{\omega_{h,i}} \lesssim \sup_{\boldsymbol{v}\in \boldsymbol{W}_{h,i}} \frac{(\operatorname{div} \boldsymbol{v}, p)_{\omega_{h,i}}}{\|D\boldsymbol{v}\|_{\omega_{h,i}}}, \tag{5}$$

where $p_i = p|_{\Omega_{h,i}}$, $\lambda_{\omega_{h,i}}(p)$ is the average of p over $\omega_{h,i}$ and $\boldsymbol{W}_{h,i}$ is the subspace of $\boldsymbol{V}_{h,i}$ defined by

$$\boldsymbol{W}_{h,i} = \{\boldsymbol{v} \in \boldsymbol{V}_{h,i} : \boldsymbol{v} = \boldsymbol{0} \text{ on } \overline{\Omega_{h,i} \setminus \omega_{h,i}}\}. \tag{6}$$

We now define the multimesh finite element space as the direct sum

$$\boldsymbol{V}_h \times Q_h = \bigoplus_{i=0}^{N} \boldsymbol{V}_{h,i} \times Q_{h,i}, \tag{7}$$

where $\boldsymbol{V}_h$ and Q_h consist of piecewise polynomial of degree k and l, respectively. This means that an element $\boldsymbol{v} \in \boldsymbol{V}_h$ is a tuple $(\boldsymbol{v}_0, \ldots, \boldsymbol{v}_N)$, and the inclusion $\boldsymbol{V}_h \hookrightarrow L^2(\Omega)$ is defined by $\boldsymbol{v}(\boldsymbol{x}) = \boldsymbol{v}_i(\boldsymbol{x})$ for $\boldsymbol{x} \in \Omega_i$. A similar interpretation is done for $q \in Q_h$. We consider here Taylor-Hood elements [3] with $k \geq 2$, $l = k - 1$, for which the condition (5) is fulfilled, but non-conforming elements are also possible.

We now consider the following asymmetric finite element method: Find $(\boldsymbol{u}_h, p_h) \in \boldsymbol{V}_h \times Q_h$ such that $A_h((\boldsymbol{u}_h, p_h), (\boldsymbol{v}, q)) = l_h(\boldsymbol{v})$ for all $(\boldsymbol{v}, q) \in \boldsymbol{V}_h \times Q_h$, where

$$A_h((\boldsymbol{u}, p), (\boldsymbol{v}, q)) = a_h(\boldsymbol{u}, \boldsymbol{v}) + s_h(\boldsymbol{u}, \boldsymbol{v}) + b_h(\boldsymbol{u}, q) + b_h(\boldsymbol{v}, p) + d_h((\boldsymbol{u}, p), (\boldsymbol{v}, q)), \tag{8}$$

$$a_h(\boldsymbol{u}, \boldsymbol{v}) = \sum_{i=0}^{N} (D\boldsymbol{u}_i, D\boldsymbol{v}_i)_{\Omega_i} \tag{9}$$

$$- \sum_{i=1}^{N} \sum_{j=0}^{i-1} \big((\langle (D\boldsymbol{u}) \cdot \boldsymbol{n}_i \rangle, [\boldsymbol{v}])_{\Gamma_{ij}} + ([\boldsymbol{u}], \langle (D\boldsymbol{v}) \cdot \boldsymbol{n}_i \rangle)_{\Gamma_{ij}} \big)$$

$$+ \sum_{i=1}^{N} \sum_{j=0}^{i-1} \beta_0 h^{-1} ([\boldsymbol{u}], [\boldsymbol{v}])_{\Gamma_{ij}},$$

$$s_h(\boldsymbol{u}, \boldsymbol{v}) = \sum_{i=0}^{N-1} \sum_{j=i+1}^{N} \beta_1 ([D\boldsymbol{u}_i], [D\boldsymbol{v}_i])_{\mathcal{O}_{ij}}, \tag{10}$$

$$b_h(\boldsymbol{u}, q) = - \sum_{i=0}^{N} (\operatorname{div} \boldsymbol{u}_i, q_i)_{\Omega_i} + \sum_{i=1}^{N} \sum_{j=0}^{i-1} ([\boldsymbol{n}_i \cdot \boldsymbol{u}], \langle q \rangle)_{\Gamma_{ij}}, \tag{11}$$

$$d_h((\boldsymbol{u}, p), (\boldsymbol{v}, q)) = \sum_{i=0}^{N} \delta h^2 (\Delta \boldsymbol{u}_i - \nabla p_i, \Delta \boldsymbol{v}_i + \nabla q_i)_{\Omega_{h,i} \setminus \omega_{h,i}}, \tag{12}$$

$$l_h(\boldsymbol{v}) = \sum_{i=0}^{N} (\boldsymbol{f}, \boldsymbol{v}_i)_{\Omega_i} - \sum_{i=0}^{N} \delta h^2 (\boldsymbol{f}, \Delta \boldsymbol{v}_i + \nabla q_i)_{\Omega_{h,i} \setminus \omega_{h,i}}. \tag{13}$$

Here, β_0 and β_1 are stabilization parameters that must be sufficiently large to ensure that the bilinear form A_h is coercive; cf. [5] for an analysis of the two-domain case.

For simplicity, we use the global mesh size h here and throughout the presentation. If the meshes are of substantially different sizes, it may be beneficial to introduce the individual mesh sizes h_i in (12) and the average $h_i^{-1} + h_j^{-1}$ in (9).

Note that since Γ_i is partitioned into interfaces Γ_{ij} relative to underlying meshes, the sums of the interface terms are over $0 \leq j < i \leq N$. In contrast, the sums of the overlap terms are over $0 \leq i < j \leq N$ since the overlap $\mathcal{O}_i$ is partitioned into overlaps $\mathcal{O}_{ij}$ relative to overlapping meshes.

The jump terms on $\mathcal{O}_{ij}$ and Γ_{ij} are defined by $[\boldsymbol{v}] = \boldsymbol{v}_i - \boldsymbol{v}_j$, where $\boldsymbol{v}_i$ and $\boldsymbol{v}_j$ are the finite element solutions (components) on the active meshes $\mathcal{K}_{h,i}$ and $\mathcal{K}_{h,j}$. The average normal flux is defined on Γ_{ij} by

$$\langle \boldsymbol{n}_i \cdot \nabla \boldsymbol{v} \rangle = (\boldsymbol{n}_i \cdot \nabla \boldsymbol{v}_i + \boldsymbol{n}_i \cdot \nabla \boldsymbol{v}_j)/2. \tag{14}$$

Here, any convex combination is valid [4].

The proposed formulation (8) is identical to the one proposed in [5] with sums over all domains and interfaces. Also note the similarity with the multimesh formulation for the Poisson problem presented in [6], the difference being the additional least-squares term d_h (and the corresponding term in l_h) since we only assume inf-sup stability in $\omega_{h,i}$, see (5) and the discussion above. If we do not assume inf-sup stability anywhere (e.g. if we would use a velocity-pressure element of equal order), the least-squares term should be applied over the whole domain as in [10]. Please cf. [10] for the use of a symmetric d_h.

Other stabilization terms may be considered. By norm equivalence, the stabilization term $s_h(\boldsymbol{u}, \boldsymbol{v})$ may alternatively be formulated as

$$s_h(\boldsymbol{u}, \boldsymbol{v}) = \sum_{i=0}^{N-1} \sum_{j=i+1}^{N} \beta_2 h^{-2}([\boldsymbol{u}], [\boldsymbol{v}])_{\mathcal{O}_{ij}}. \tag{15}$$

where β_2 is a stabilization parameter; see [8].

Note that the finite element method weakly approximates continuity in the sense that $[\boldsymbol{u}_h] = \boldsymbol{0}$ and $[\boldsymbol{n}_i \cdot \nabla \boldsymbol{u}_h] = 0$ on all interfaces.

4 Implementation

We have implemented the multimesh finite element method as part of the software framework FEniCS [2, 9]. One of the main features of FEniCS is the form language UFL [1] which allows variational forms to be expressed in near-mathematical notation. However, to express the multimesh finite element method (8), a number of custom measures must be introduced. In particular, new measures must be introduced for integrals over cut cells, interfaces and overlaps. These measures are then mapped to quadrature rules that are computed at runtime. An overview of these algorithms and the implementation is given in [7].

To express the multimesh finite element method, we let `dX` denote the integration over domains $\Omega_i, i = 0, \ldots, N$, including cut cells. Integration over Γ_{ij} and $\mathcal{O}_{ij}$ are expressed using the measures `dI` and `dO`, respectively. We let `dC` denote integration over $\Omega_{h,i} \setminus \omega_{h,i}$. Now the multimesh finite element method for the Stokes problem may be expressed as

```
a_h = inner(grad(u), grad(v))*dX \
    - inner(avg(grad(u)), tensor_jump(v, n))*dI \
    - inner(avg(grad(v)), tensor_jump(u, n))*dI \
    + beta_0/h * inner(jump(u), jump(v))*dI
s_h = beta_1 * inner(jump(grad(u)), jump(grad(v)))*dO
b_h = lambda v, q: inner(-div(v), q)*dX \
              + inner(jump(v, n), avg(q))*dI
d_h = delta*h**2 * inner(-div(grad(u)) + grad(p), \
                   -div(grad(v)) - grad(q))*dC
```

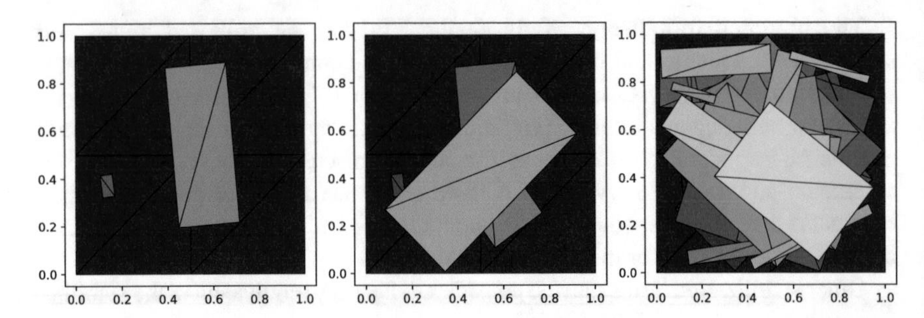

Fig. 6 A sequence of N meshes are randomly placed on top of a fixed background mesh of the unit square shown here for $N = 2, 4$ and 32 using the coarsest refinement level

This makes it easy to implement the somewhat lengthy form (8), as well as investigate the effect of different stabilization terms.

5 Numerical Results

To investigate the convergence of the multimesh finite element method, we solve the Stokes problem in the unit square with the following exact solution

$$\boldsymbol{u}(x, y) = 2\pi \sin(\pi x) \sin(\pi y) \cdot (\cos(\pi y) \sin(\pi x), -\cos(\pi x) \sin(\pi y)), \tag{16}$$

$$p(x, y) = \sin(2\pi x) \sin(2\pi y), \tag{17}$$

and corresponding right hand side. We use $P_k P_{k-1}$ Taylor–Hood elements with $k \in \{2, 3, 4\}$ and we use $N \in \{1, 2, 4, 8, 16, 32\}$ randomly placed domains as in [6] (see Fig. 6). Due to the random placement of domains, some domains are completely hidden and will not contribute to the solution. For $N = 8$, this is the case for one domain, for $N = 16$, three domains and for $N = 32$, four domains are completely hidden. This is automatically handled by the computational geometry routines. Convergence results are presented in Fig. 7 as well as in Table 1.

6 Discussion

The results presented in Table 1 and Fig. 7 show the expected order of convergence for the velocity in the $L^2(\Omega)$ norm $(k + 1)$, for the velocity in the $H_0^1(\Omega)$ norm (k), and for the pressure in the $L^2(\Omega)$ norm (k).

A detailed inspection of Fig. 7 reveals that, as expected, the multimesh discretization yields larger errors than the single mesh discretization (standard Taylor–Hood on one single mesh). The errors introduced by the multimesh discretization are

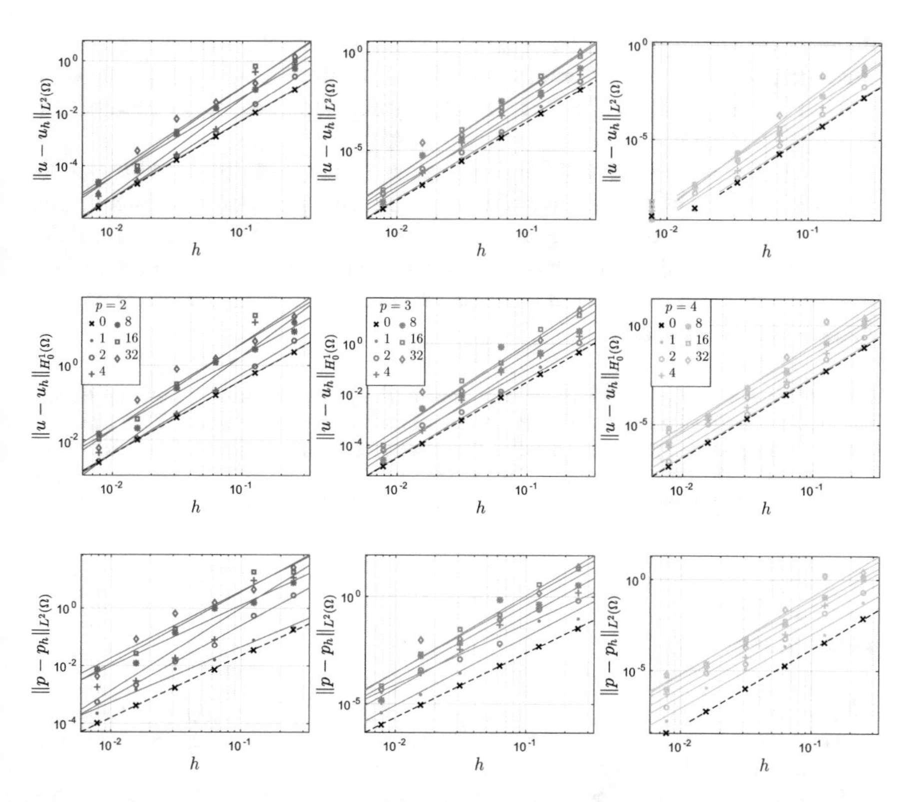

Fig. 7 Convergence results for $k = 2$, 3 and 4 (left to right) using up to 32 meshes (single mesh results are $N = 0$). From top to bottom we have the velocity error in the $L^2(\Omega)$ norm, the velocity error in the $H_0^1(\Omega)$ norm, and the pressure error in the $L^2(\Omega)$ norm. Results less than 10^{-8} are not included in the convergence lines due to limits in floating point precision

Table 1 Error rates for $\boldsymbol{e}_{L^2} = \|\boldsymbol{u} - \boldsymbol{u}_h\|_{L^2(\Omega)}$, $\boldsymbol{e}_{H_0^1} = \|\boldsymbol{u} - \boldsymbol{u}_h\|_{H_0^1(\Omega)}$ and $e_{L^2} = \|p - p_h\|_{L^2(\Omega)}$

	$k = 2$			$k = 3$			$k = 4$		
N	$\boldsymbol{e}_{L^2}$	$\boldsymbol{e}_{H_0^1}$	e_{L^2}	$\boldsymbol{e}_{L^2}$	$\boldsymbol{e}_{H_0^1}$	e_{L^2}	$\boldsymbol{e}_{L^2}$	$\boldsymbol{e}_{H_0^1}$	e_{L^2}
0	2.9952	1.9709	2.1466	4.0289	2.9966	3.0028	4.9508	3.9844	4.2286
1	2.9750	1.9658	1.9291	4.1153	3.0912	3.1932	4.8861	4.0006	4.0587
2	3.2764	2.1472	2.5036	3.9087	2.9021	2.8489	4.8677	4.0416	4.0832
4	3.6666	2.5971	2.8597	4.3996	3.3125	3.3957	5.2741	4.0966	4.1609
8	3.0359	1.9697	2.1163	4.3412	3.2258	3.4213	4.8840	3.9409	4.0169
16	3.4131	2.3298	2.4794	4.5033	3.3907	3.5910	5.4702	3.9664	4.0729
32	3.2832	2.1505	2.3255	4.4196	3.2922	3.4362	5.7538	4.3191	4.2848

one to two orders of magnitude larger than the single mesh error. However, the convergence rate is optimal and it should be noted that the results presented here are for an extreme scenario where a large number of meshes are simultaneously overlapping; see Fig. 6. For a normal application, such as the simulation of flow around a collection of objects, each object would be embedded in a boundary-fitted mesh and only a small number of meshes would simultaneously overlap (in addition to each mesh overlapping the fixed background mesh), corresponding to the situation when two or more objects are close.

The presented method and implementation demonstrate the viability of the multimesh method as an attractive alternative to existing methods for discretization of PDEs on domains undergoing large deformations. In particular, the discretization and the implementation are robust to thin intersections and rounding errors, both of which are bound to appear in a simulation involving a large number of meshes, timesteps or configurations.

Acknowledgements August Johansson was supported by the Research Council of Norway through the FRIPRO Program at Simula Research Laboratory, project number 25123. Mats G. Larson was supported in part by the Swedish Foundation for Strategic Research Grant No. AM13-0029, the Swedish Research Council Grants Nos. 2013-4708, 2017-03911, and the Swedish Research Programme Essence. Anders Logg was supported by the Swedish Research Council Grant No. 2014-6093.

References

1. Alnæs, M.S., Logg, A., Oelgaard, K.B., Rognes, M.E., Wells, G.N.: Unified form language: a domain-specific language for weak formulations of partial differential equations. ACM Trans. Math. Softw. **40** (2014). https://doi.org/10.1145/2566630
2. Alnæs, M.S., Blechta, J., Hake, J., Johansson, A., Kehlet, B., Logg, A., Richardson, C., Ring, J., Rognes, M.E., Wells, G.N.: The FEniCS project version 1.5. Arch. Numer. Softw. **3**(100) (2015). https://doi.org/10.11588/ans.2015.100.20553
3. Brenner, S.C., Scott, L.R.: The Mathematical Theory of Finite Element Methods. Springer, New York (2008). https://doi.org/10.1007/978-0-387-75934-0
4. Hansbo, A., Hansbo, P., Larson, M.G.: A finite element method on composite grids based on Nitsche's method. ESAIM-Math. Model. Num. **37**(3), 495–514 (2003). https://doi.org/10.1051/m2an:2003039
5. Johansson, A., Larson, M.G., Logg, A.: High order cut finite element methods for the Stokes problem. Adv. Model. Simul. Eng. Sci. **2**(1), 1–23 (2015). https://doi.org/10.1186/s40323-015-0043-7
6. Johansson, A., Kehlet, B., Larson, M.G., Logg, A.: MultiMesh finite element methods: solving PDEs on multiple intersecting meshes. Comput. Methods Appl. Mech. Eng. **343**, 672–689 (2019). https://doi.org/10.1016/j.cma.2018.09.009
7. Johansson, A., Kehlet, B., Logg, A.: Construction of quadrature rules on general polygonal and polyhedral domains in cut finite element methods (in preparation, 2018)
8. Johansson, A., Larson, M.G., Logg, A.: MultiMesh finite elements with flexible mesh sizes. Submitted to CMAME (2019). arXiv preprint. https://arxiv.org/abs/1804.06455
9. Logg, A., Mardal, K.A., Wells, G.N., et al.: Automated Solution of Differential Equations by the Finite Element Method. Springer, Berlin (2012). https://doi.org/10.1007/978-3-642-23099-8
10. Massing, A., Larson, M.G., Logg, A., Rognes, M.E.: A stabilized Nitsche overlapping mesh method for the Stokes problem. Numer. Math. 1–29 (2014). https://doi.org/10.1007/s00211-013-0603-z

A Variational Multi-Scale Anisotropic Mesh Adaptation Scheme for Aerothermal Problems

Youssef Mesri, Alban Bazile, and Elie Hachem

Abstract We propose a new mesh adaptation technique to solve the thermal problem of the impingement jet cooling. It relies on a subscales error estimator computed with bubble functions to locate and evaluate the PDE-dependent approximation error. Then, a new metric tensor $\mathscr{H}^{new}_{aniso}$ based on the subscales error estimator is suggested for anisotropic mesh adaptation. We combine the coarse scales anisotropic interpolation error indicator with the subscales error estimator allowing us to take into account the anisotropic variations of the solution but also the sub-grid information. The results show that the resulting meshes of this parallel adaptive framework allow to capture the turbulently generated flow specificities of the impingement jet cooling and in particular, the secondary vortexes.

Keywords Mesh adaption technique · Impingement jet cooling · VMS error estimator · Navier–Stokes equations · Heat transfer equations

1 Introduction

The simulation of aerothermal problems requires a significant amount mesh elements to ensure highly accurate solutions. Consequently, engineers are continuously looking for a trade-off between high precision level and prohibitive computational costs. A practical way to reconcile high precision with feasible computational costs consists in distributing the computational loads over the domain in space. Indeed, mesh adaptation allows to improve efficiency of numerical methods by local modifications of the mesh. This technique requires an error analysis on the mesh to locate the areas and the directions that need a large number of elements. In the past, a significant work was therefore made on error estimates. In particular, for

---

Y. Mesri (✉) · A. Bazile · E. Hachem
MINES ParisTech, PSL - Research University, CEMEF - Centre for Material Forming, CNRS UMR 7635, Sophia-Antipolis Cedex, France
e-mail: youssef.mesri@mines-paristech.fr

H. van Brummelen et al. (eds.), *Numerical Methods for Flows*,
Lecture Notes in Computational Science and Engineering 132,
https://doi.org/10.1007/978-3-030-30705-9_18

anisotropic mesh adaptation, the interpolation based error indicator has been well developed. From the interpolation error analysis, several recent results in [1] and references therein have brought renewed focus on metric-based mesh adaptation where the underling metric is derived from a recovered Hessian.

Despite the practical advantages of the Hessian based anisotropic mesh adaptation technique, the proposed interpolation based error indicators are usually not sharp enough to capture the fine scale features related to the dynamic flow solution [2]. To take them into account, we need to consider the PDE-dependent approximation error. As illustrated in [3], multiscale error estimators are usually well suited to estimate variational finite element errors. However, the development of such error estimators requires a mathematical framework for the error analysis. This framework can be provided by the Variational Multiscale method (VMS) that we use to stabilize our continuous finite element scheme. The VMS approach introduced in [4] consists in the splitting of the solution into a resolved part (i.e. coarse scales) and an unresolved part (i.e. subscales). The resolution of the unresolved part gives a direct access to the sub-mesh scale information of the solution and allows us to compute an approximation error estimator without solving any additional equation.

In this paper, we propose a subscales error estimator for the convective-diffusive thermal problem of the impingement jet cooling. This subscale error estimator is computed using a linear combination of bubble functions to establish a *pointwise* computation of the error. It has been developed by Irisarri et al. for 2D transport equation in [5]. Results on the estimated localization of the subscales error for the impingement jet cooling are presented in the last section.

Furthermore, we propose a new metric tensor $\mathscr{H}_{aniso}^{new}$ for *anisotropic* mesh adaptation. This metric is built on the combination of both the coarse scales interpolation error indicator and the subscale error estimator. Doing so, the new anisotropic metric tensor $\mathscr{H}_{aniso}^{new}$ allows to take into account the anisotropic variations of the solution on the mesh and also relies upon the sub-grid information of the solution.

2 Numerical Methods for the Thermal Resolution of the Impinging Jet Cooling

The aerothermal impinging jet cooling problem is solved by coupling incompressible Navier–Stokes equations and the convection-diffusion heat transfer equation. The solutions $(\mathbf{v}, p)$ of the unsteady incompressible Navier–Stokes equations are obtained thanks to a Variational MultiScale (VMS) finite element solver [6] and the temperature u solution of the heat transfer problem is obtained by using the Streamline Upwind Petrov-Galerkin (SUPG) finite element scheme [7]. In the following, the focus is on the Variational Multiscale error analysis of the thermal problem to derive an a posteriori error estimator.

2.1 Variational Multi-Scale Scheme for the Convection-Diffusion Equation

We describe the convection-diffusion problem as the following:

$$\begin{cases} \mathscr{L}u = -a\Delta u + \mathbf{v}\cdot\nabla u = f \quad in\ \Omega \\ u = g \quad on\ \Gamma_g \end{cases} \tag{1}$$

where $\mathscr{L}$ is a linear differential operator. f is the source term of the equation; g is the value of the Dirichlet boundary condition; a is the diffusion coefficient and $\mathbf{v}$ the velocity field.
The variational formulation of Eq. (1) is given by:

$$\begin{cases} \text{Find } u \in \mathscr{S} = \{u \in H^1(\Omega) \mid u = g \ on\ \Gamma_g\} \text{ such that:} \\ a(w,u) = (w,f), \ \forall w \in \mathscr{V} = \{w \in H^1(\Omega) \mid w = 0 \ on\ \Gamma_g\} \\ \text{where } a(.,.) \text{ is a bilinear form, and } (.,.) \text{ the } L^2(\Omega) \text{ inner product.} \end{cases} \tag{2}$$

Applying the Finite Element Method (FEM), we mesh the domain into n_{el} non-overlapping elements Ω_e. We write $\tilde{\Omega}$ and $\tilde{\Gamma}$ as:

$$\tilde{\Omega} = \bigcup_{e=1}^{n_{el}} \Omega_e \quad \tilde{\Gamma} = \bigcup_{e=1}^{n_{el}} \Gamma_e \setminus \Gamma \tag{3}$$

With these definitions, we apply the standard Galerkin method:

$$\begin{cases} \text{Find } u_h \in \mathscr{S}^h = \{u_h \in \mathscr{S} \mid u_{h|\Omega_e} \in \mathbb{P}_1, \ \forall \Omega_e \in \tilde{\Omega}\} \text{ such that,} \\ a(w_h, u_h) = (w_h, f), \ \forall w_h \in \mathscr{V}^h = \{u_h \in \mathscr{V} \mid u_{h|\Omega_e} \in \mathbb{P}_1, \ \forall \Omega_e \in \tilde{\Omega}\} \end{cases} \tag{4}$$

It is well known that this formulation is unstable and leads to spurious oscillations when the convective term of the equation is dominant. For this reason, we stabilize the formulation using the SUPG scheme. To do so, for all terms in Eq. (4), we replace the weighting function w_h by a new weighting function $w_h + \tau_e \mathbf{v}\cdot\nabla w_h$. This modification of the formulation is usually interpreted as adding more weight upstream and reducing it downstream. It adds an artificial weighted diffusion along the streamline direction. Concerning the choice of the stabilizing parameter τ_e, we refer for example to Codina in [6].

Remark 1 In order to study the error of this numerical scheme, we use the Variational Multiscale analysis. In fact, it has to be noted that the SUPG scheme can be considered as a particular form of the generalized VMS formulation. Indeed, the stabilizing term can also be seen as the effect of the subscales on the coarse scales. Hughes gives more details about this concordance in [4].

In the VMS formalism, the solution and test functions spaces are decomposed into two sub-spaces: a mesh scale subspace (or coarse scales) $(\mathscr{S}^h, \mathscr{V}^h)$ and an under-mesh scale subspace (or subscales) $(\mathscr{S}', \mathscr{V}')$ such that $\mathscr{S} = \mathscr{S}^h \oplus \mathscr{S}'$ and $\mathscr{V} = \mathscr{V}^h \oplus \mathscr{V}'$. Therefore, we can decompose the solution and test functions as follows:

$$\begin{aligned} u &= u_h + u', \quad u_h \in \mathscr{S}_h, \quad u' \in \mathscr{S}' \\ w &= w_h + w', \quad w_h \in \mathscr{W}_h, \quad w' \in \mathscr{S}' \end{aligned} \tag{5}$$

Thanks to the orthogonality between the coarse scales subspace and the subscales subspace, the variational form can be split into a coarse scales sub-problem and a subscales sub-problem [6]:

$$\begin{aligned} a(w_h, u_h) + a(w_h, u') &= (w_h, f) \;\; \forall w_h \in \mathscr{V}^h \\ a(w', u_h) + a(w', u') &= (w', f) \;\; \forall w' \in \mathscr{V}' \end{aligned} \tag{6}$$

We start by solving the subscales sub-problem (second equation). For smooth functions on the element interior but rough across the inter-element boundaries, the integration by parts leads to the following equation:

$$\begin{aligned} a(w', u') &= -a(w', u_h) + (w', f) \\ a(w', u') &= -(w', \mathscr{L}u_h - f) - (w', [\mathbb{B}u_h])_{\tilde{\Gamma}} \;\; \forall w' \in \mathscr{V}' \end{aligned} \tag{7}$$

where the jump term [.] represents the difference of the fluxes on both sides of the element boundaries and $\mathbb{B}$ the trace operator of $\mathscr{L}$ on $\tilde{\Gamma}$. An analytic solution of problem (7) can be found in [5].

2.2 A Posteriori Error Estimation on Solution's Subscales

The subscale u' is computed thanks to a *pointwise* error estimation as in [5]. It uses a set of bubble functions as a substitution of the subscales Green's functions. This computation method consists in decomposing the error into two components according to the nature of the residuals:

$$u'(\mathbf{x}) = u'_{bub}(\mathbf{x}) + u'_{poll}(\mathbf{x}). \tag{8}$$

where $\mathbf{x}$ is the space coordinates vector. The first term u'_{bub} is the internal residual error and it is related to the local internal residual, $f - \mathscr{L}u_h$, inside the elements. As we will see later, this part of the error is modeled locally thanks to a set of bubble functions. The second term u'_{poll} is the inter-element error. It represents the pollution error due to sources of errors outside the element. It is negligible when considering convection-dominated regime [5]. Consequently, in this paper we will consider only the internal residual error. This error component is expressed with a combination of bubble functions:

$$u'(\mathbf{x}) \approx u'_{bub}(\mathbf{x}) = \sum_{i=1}^{n_{bub}} c_i^b b_i(\mathbf{x}) \tag{9}$$

Considering bubbles functions of order 3, we have:

$$u'_{bub}(\mathbf{x}) = c_1^b b_1(\mathbf{x}) + c_2^b b_2(\mathbf{x}) + c_3^b b_3(\mathbf{x}) \tag{10}$$

with c_i^b unknown coefficients to be determined.

The definition of the first bubble function $b_1(\mathbf{x})$ is the following:

$$b_1(\mathbf{x}) = (d+1)^{d+1} \prod_{i=1}^{d+1} \hat{\lambda}_i \tag{11}$$

where d is the dimension of the problem and $\hat{\lambda}_i$ are the barycentric coordinates in the reference element. The next bubble functions $b_2(\mathbf{x})$ and $b_3(\mathbf{x})$ are built by adding the monomials of the Pascal triangle with center in the barycenter $c_e = (\xi_e, \eta_e)$ of the element. For example, in 2D, in the reference element: $\Omega_{ref} = \{(\xi, \eta) : 0 \leq \xi \leq 1; 0 \leq \eta \leq 1 - \xi\}$, as in [2], we choose the following bubble functions:

$$\begin{aligned} b_1(\xi, \eta) &= 27 \times \xi\eta(1 - \xi - \eta) \\ b_2(\xi, \eta) &= 27 \times \xi\eta(1 - \xi - \eta)(\xi - \xi_b) \\ b_3(\xi, \eta) &= 27 \times \xi\eta(1 - \xi - \eta)(\eta - \eta_b) \end{aligned} \tag{12}$$

with $c_e = (\xi_b, \eta_b)$, $\xi_b = 1/3$ and $\eta_b = 1/3$.

Then, approximating $u'_{bub}(\mathbf{x})$ by a Taylor series and neglecting the second order terms, we have an expression of $u'_{bub}(\mathbf{x})$ close to the centroid $\mathbf{c}_i$ of the element. It allows us to compute the coefficients c_i^b of Eq. (10) by identification of each term with the terms of the Taylor development. Making the assumption that the residual $f - \mathscr{L}u_h \in \mathbb{P}_0$ (constant by element), we only keep the first term of Eq. (10). Therefore, the internal residual can be expressed as follows:

$$u'_{bub}(\mathbf{x}) = b_1(\mathbf{x})(f - \mathscr{L}u_h)(\mathbf{c}_i) \tag{13}$$

Developing the residual with the convection-diffusion operator $\mathscr{L}$, we have:

$$u'_{bub}(\mathbf{x}) = b_1(\mathbf{x})(f(\mathbf{c}_i) + a\Delta u_h(\mathbf{c}_i) - \mathbf{v} \cdot \nabla u_h(\mathbf{c}_i)) \tag{14}$$

Then, $u_h \in \mathbb{P}_1$, therefore, $\nabla u_h(\mathbf{c}_i)$ is a constant inside the element and $\Delta u_h(\mathbf{c}_i) = 0$. For this reason, we finally get:

$$u'_{bub}(\mathbf{x}) = b_1(\mathbf{x})(f(\mathbf{c}_i) - \mathbf{v} \cdot \nabla u_h(\mathbf{c}_i)) \tag{15}$$

The above computation of the error estimator is *pointwise*. In fact, the error estimator is given at each point $\mathbf{x}$ of the domain. However, to include the error information in the mesh adaptation, the error is evaluated at the integration points of the bubble function inside each element.

3 Mesh Adaptation

To describe the new anisotropic mesh adaptation techniques developed in this paper, we start by a reminder on the Hessian based anisotropic mesh adaptation. Then, a combination between the interpolation error indicator and the previous subscales error estimator is proposed. From it, we derive a new metric tensor $\mathcal{H}^{new}_{aniso}$ that allows to drive the remeshing mechanics.

3.1 Principles of Hessian Based Anisotropic Mesh Adaptation

To start, let us consider a certain triangulation Ω_h. We can derive an upper bound of the approximation error using an interpolation error analysis in the L^p norm. Referring to [1] and references therein, this upper bound is expressed thanks to the recovered Hessian of the approximated solution u_h. In fact, using $\mathbb{P}_1$ linear elements, we usually cannot compute directly the Hessian of the solution. Instead, we compute an approximation called the *recovered* Hessian matrix $H_R(u_h(x))$. However, the *recovered* Hessian matrix is not a metric because it is not positive definite. Therefore, we define the following metric tensor:

$$\mathcal{H}_{aniso} = \mathcal{R}\Lambda\mathcal{R}^{\mathrm{T}} \tag{16}$$

where $\mathcal{R}$ is the orthogonal matrix built with the eigenvectors $(e_i)_{i=1,d}$ of $H_R(u_h(x))$ and $\Lambda = diag(|\lambda_1|, \ldots, |\lambda_d|)$ is the diagonal matrix of absolute value of the eigenvalues of $H_R(u_h(x))$. The eigenvalues are then classified in ascending order and we have:

$$|\lambda_d| > \ldots > |\lambda_1| \tag{17}$$

Following the work of Mesri et al. in [1], we can introduce the following local error indicator in the L^p norm:

$$\eta_{\Omega_e} = d|\Omega_e|^{\frac{1}{p}}|\lambda_d(x_0)|h_d^2 \tag{18}$$

where $|\lambda_d(x_0)|$ is the maximum eigenvalue of $H_R(u_h(x))$ at the center x_0 of the element Ω_e, $|\Omega_e|$ is the volume of the element and h_d is the length of the element in direction d.

3.2 A New Anisotropic Mesh Adaptation Technique Based on the Subscales Error Estimator

In this section, we propose a new anisotropic mesh adaptation technique. It uses a new anisotropic local error indicator $\eta_{\Omega_e,new}$ that takes into account (1) the interpolation error indicator and (2) the subscales error estimator. To do so, we derive a new anisotropic metric tensor $\mathcal{H}_{aniso}^{new}$ that allows to take into account the anisotropic variations of the solution on the mesh but also relies on the subscales error estimator.

We consider the previous anisotropic local error indicator defined in Eq. (18). Now, the unknown of the re-meshing problem is $h_{d,new}$. In fact, we want the new mesh size to take into account the subscale error estimates. Thus, we propose a new anisotropic local error indicator:

$$\eta_{\Omega_e,new} = d|\Omega_e|^{\frac{1}{p}} \times |\lambda_d(x_0)| \times \frac{||u'||_{L^\infty(\Omega_e)}}{u'_{TOL}} \times h_{d,new}^2 \tag{19}$$

From here, we can define the new anisotropic metric tensor as:

$$\boxed{\mathcal{H}_{aniso}^{new} = \mathcal{R}\Lambda\mathcal{R}^T = \frac{||u'||_{L^\infty(\Omega_e)}}{u'_{TOL}}|\lambda_1|e_1 \otimes e_1 + \ldots + \frac{||u'||_{L^\infty(\Omega_e)}}{u'_{TOL}}|\lambda_d|e_d \otimes e_d} \tag{20}$$

Doing so, we keep the anisotropic effects from the solution variations but we isotropically scale this effect by the subscales error estimator. With this new error indicator and following the same proof than in Mesri et al. [1], we define a new optimization problem under the constraint of a fixed number of elements as follows:

$$\begin{cases} \underset{h_{1,\Omega_e,new},\ldots,h_{d,\Omega_e,new}}{\text{minimizes}} & F(h_{1,\Omega_e,new},\ldots,h_{d,\Omega_e,new}) = \sum_{\Omega_e \in \Omega_h} (\eta_{\Omega_e,new})^p \\ \text{subject to} & \mathcal{N}_{\Omega'_h} = C_0^{-1} \sum_{\Omega_e \in \Omega_h} \int_{\Omega_e} \prod_{i=1}^{d} \frac{1}{h_{i,\Omega_e,new}} d\Omega_e \end{cases}$$

where C_0 is the volume of a regular tetrahedron. The solution of this optimization problem is the vector of the new mesh sizes $h_{1,\Omega_e,new}, \ldots, h_{d,\Omega_e,new}$ in the eigen vectors directions e_1, e_2, e_3.

4 Results and Discussions

4.1 Test Case Description

We consider in this paper, the study of an unconfined three dimensional turbulent isothermal round jet, impinging normally on a hot plate. The configuration of the impingement jet is given on Fig. 1. It represents a 2D slice view of the 3D computation. The geometrical and physical parameters used in this paper are given on Table 1. Referring to previous works on the subject [8], the nozzle to plate distance H is 2 times the jet diameter $D = 26$ mm and the injection Reynolds number is $Re_{inj} = 23000$.

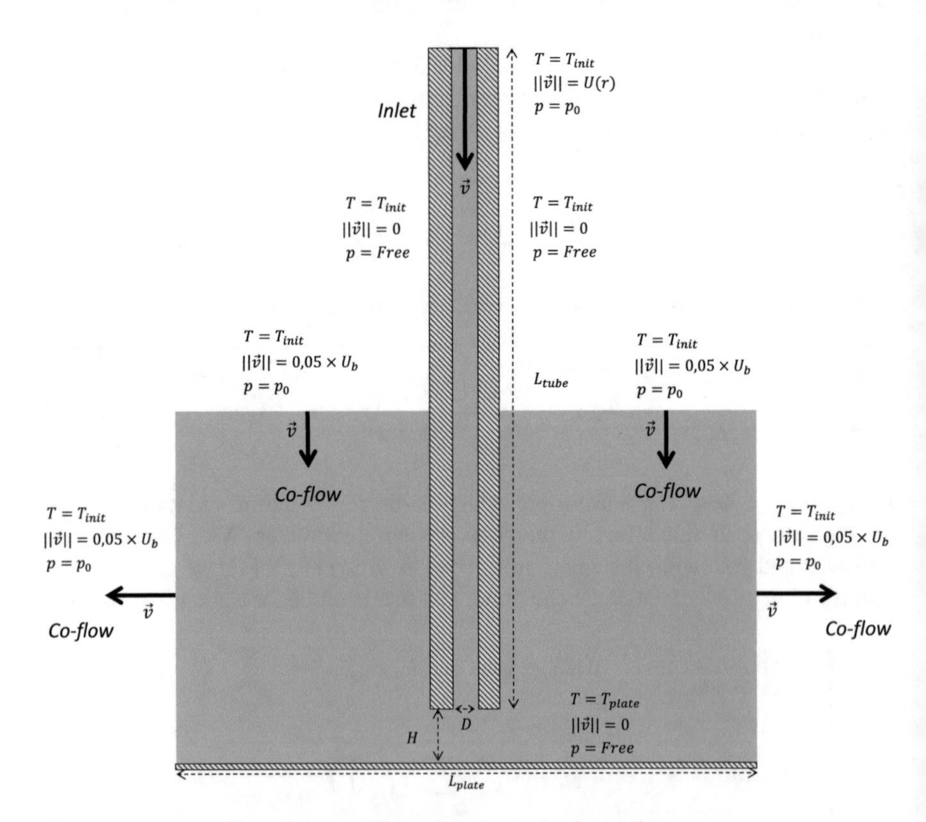

Fig. 1 Geometry and boundary conditions of the impinging jet cooling

Table 1 Problem parameters

Parameter	Numerical value
U_c	19.3 m/s
U_b	15.9 m/s
Re_{inj}	23,000
η_{fluid}	$18e^{-6}$ m^2/s
D	0.026 m
H	0.052 m
L_{plate}	0.520 m
L_{tube}	0.572 m
T_{init}	300 K
T_{plate}	330 K
p_0	0 Pa

At initialization, the velocity is null everywhere in the domain except on the boundary conditions. A mean velocity profile is imposed at the inlet and the bulk velocity is given by:

$$U_b = \frac{Re_{inj} \times \eta_{fluid}}{D} \tag{21}$$

with η_{fluid} the air kinematic viscosity (see Table 1). Following the work of Cooper et al. in [8], the axial velocity is expressed using a specific power law profile for turbulent pipe flows. In the direction z, we have:

$$U(r) = \left(1 - \frac{2r}{D}\right)^{1/7.23} \times \frac{U_b}{0.811 + 0.038(log(Re) - 4)} \tag{22}$$

where r represents the distance to jet center axis. As in experimental studies [8], the initial, ambient and jet temperatures are equals and taken at $T_{init} = 300$ K. The plate is treated as an isothermal wall with a temperature of $T_{plate} = 330$ K.

In this section, we evaluate the efficiency of our new mesh adaptation techniques in the resolution of the thermal problem of the impingement jet cooling.

4.2 *Flow Dynamics, Localization of the Subscales Error and Resulting Meshes*

The fundamental phenomenon that occurs during the impingement jet cooling concerns the rebound of the primary structures when the jet impacts the hot plate. This rebound takes place during the transitional flow and therefore, requires an unsteady resolution of the problem (see Fig. 2). In particular, it generates secondary vortexes in the opposite direction near the wall that have an important impact on the convective heat transfers. In fact, the injection of cold air induced by the secondary

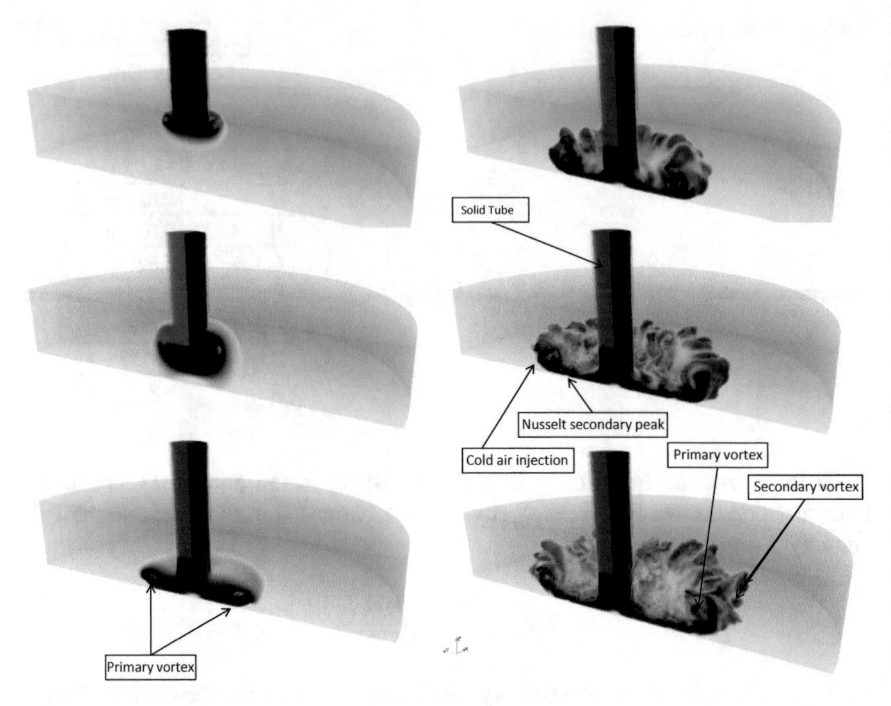

Fig. 2 Unsteady computation of the primary vortex rebound

vortex enhances the cooling and generates a secondary peak in the Nusselt number distribution. Therefore, it is necessary that mesh adaptation captures accurately the thermal exchanges that take place in the secondary vortexes.

On Fig. 3, we observe that the proposed subscales error estimator $\frac{||u'||_{L^\infty(\Omega_e)}}{u'_{TOL}}$ allows us to locate the unresolved part of the thermal solution. Then, thanks to our new anisotropic mesh adaptation technique, the mesh is adapted anisotropically according to this sub-scale information. Doing so, it allows us to capture accurately and dynamically the structure of the secondary vortexes during the unsteady resolution. The different adaptive simulations have been performed in parallel on our Lab's cluster. It consists of 2000 Intel Xeon cores (E5-2670 and E5-2680 chips) interconnected with infiniband network.

5 Conclusions

We proposed a multiscale error estimator to drive anisotropic mesh adaptation procedure for finite element flow problems. This new error estimator relies on the combination of both an interpolation based anisotropic error indicator and a subscales error estimator. The results show that this combination allows to

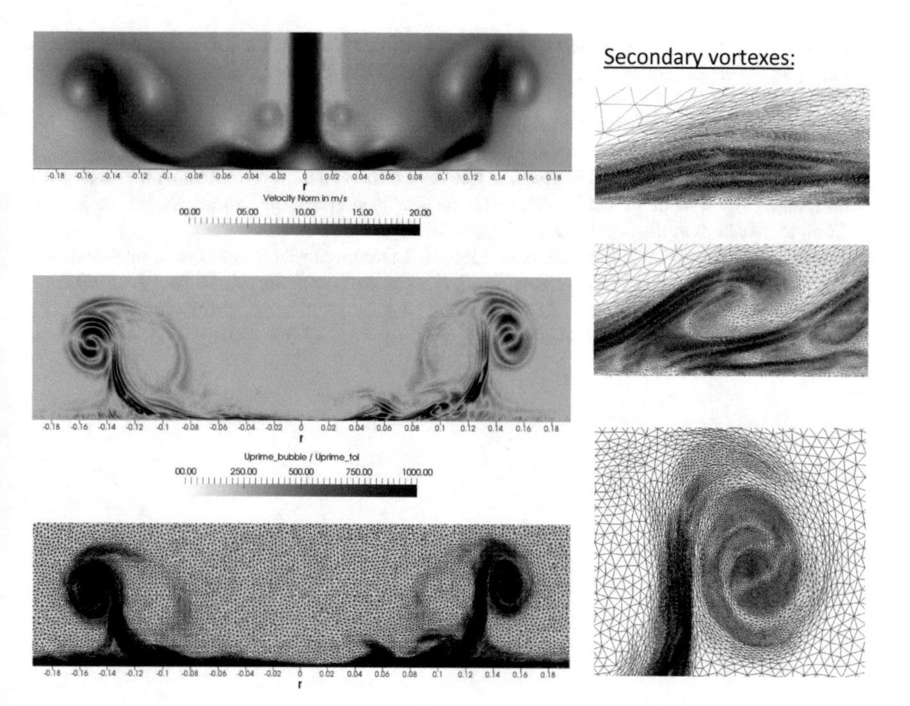

Fig. 3 Localization of the subscales error and new anisotropic mesh adaptation

get relevant anisotropically adapted meshes that capture complex turbulent flow structures which are the secondary vortexes generated by the single impingement jet simulation. Furthermore, the re-definition of the re-meshing optimization problem allows us to include the sub-grid information in mesh adaptation with respect to the constraint of a fixed number of elements. The proposed anisotropic error estimator has been integrated in a parallel adaptive finite element framework to deal with realistic 2D and 3D aerothermal flow [9].

References

1. Mesri, Y., Khalloufi, M., Hachem, E.: On optimal simplicial 3d meshes for minimizing the Hessian-based errors. Appl. Numer. Math. **106**, 235–249 (2016)
2. Bazile, A., Hachem, E., Larroya-Huguet, J.C., Mesri, Y.: Variational Multiscale error estimator for anisotropic adaptive fluid mechanic simulations: application to convection–diffusion problems. Comput. Methods Appl. Mech. Eng. **331**, 94–115 (2018)
3. Granzow, B.N., Shephard, M.S., Oberai, A.A.: Output-based error estimation and mesh adaptation for variational multiscale methods. Comput. Methods Appl. Mech. Eng. **322**, 441–459 (2017)
4. Hughes, T.J.R.: Multiscale phenomena: Green's functions, the Dirichlet-to-Neumann formulation, subgrid scale models, bubbles and the origins of stabilized methods. Comput. Methods Appl. Mech. Eng. **127**(1), 387–401 (1995)

5. Irisarri, D., Hauke, G.: A posteriori pointwise error computation for 2-D transport equations based on the variational multiscale method. Comput. Methods Appl. Mech. Eng. **311**, 648–670 (2016)
6. Codina, R.: Stabilized finite element approximation of transient incompressible flows using orthogonal subscales. Comput. Methods Appl. Mech. Eng. **191**(39–40), 4295–4321 (2002)
7. Brooks, A.N., Hughes, T.J.R., Streamline upwind/Petrov-Galerkin formulations for convection dominated flows with particular emphasis on the incompressible Navier-Stokes equations. Comput. Methods Appl. Mech. Eng. **32**(1), 199–259 (1982)
8. Cooper, D., Jackson, D.C., Launder, B.E., Liao, G.X.: Impinging jet studies for turbulence model assessment—I. Flow-field experiments. Int. J. Heat Mass Transf. **36**(10), 2675–2684 (1993)
9. Mesri, Y., Digonnet, H., Coupez, T.: Hierarchical adaptive multi-mesh partitioning algorithm on heterogeneous systems. Lect. Notes Comput. Sci. Eng. **74**, 299–306 (2010)

Density-Based Inverse Homogenization with Anisotropically Adapted Elements

Nicola Ferro, Stefano Micheletti, and Simona Perotto

Abstract The optimization of manufacturable extremal elastic materials can be carried out via topology optimization using the homogenization method. We combine here a standard density-based inverse homogenization technique with an anisotropic mesh adaptation procedure in the context of a finite element discretization. In this way, the optimized layouts are intrinsically smooth and ready to be manufactured.

Keywords Topology optimization · Inverse homogenization · Metamaterials · SIMP method · Anisotropic mesh adaptation · Finite elements

1 Introduction

The design of performant and light structures has been gaining popularity for the last years thanks to the rise and development of Additive Manufacturing (AM) techniques. Differently from subtractive methods, AM enjoys great versatility in the achievable shapes and presents very few limitations.

In this framework, topology optimization (TO) has proved to be the reference mathematical method suitable for designing innovative and performant structures of engineering interest. Essentially, it consists in the allocation of material in the so-called design domain, ensuring the optimization of a given functional and, at the same time, the satisfaction of design requirements. The final result of TO is an optimized structure, where areas of full material and void alternate so that the new topology guarantees the desired production specifications.

With a particular focus on the linear elastic problem, it is observed that the stiffness of an optimal designed structure, subject to given loads and constraints, is increased by inserting small substructures [2]. Consequently, different authors have investigated the possibility of employing topology optimization at a microscale

N. Ferro (✉) · S. Micheletti · S. Perotto
MOX, Dipartimento di Matematica, Politecnico di Milano, Milano, Italy
e-mail: nicola.ferro@polimi.it; stefano.micheletti@polimi.it; simona.perotto@polimi.it

H. van Brummelen et al. (eds.), *Numerical Methods for Flows*,
Lecture Notes in Computational Science and Engineering 132,
https://doi.org/10.1007/978-3-030-30705-9_19

as well, aiming at yielding optimized microstructures (metamaterials) [24, 26]. The ultimate goal is to combine the microscopic optimized structures with a standard TO performed at the macroscale. This link is made possible by employing homogenization techniques, which are widely used to incorporate the information provided by the microscale into macroscale models [1, 3, 20].

In this work, we enrich such an approach by resorting to a numerical discretization of the linear elastic problem based on a standard finite element solver combined with a mesh adaptation procedure. In particular, in Sect. 2, we briefly present a density-based approach for a generic topology optimization problem. In Sect. 3, the homogenization procedure is presented. We distinguish between a direct and an inverse method, consisting in prescribing the desired macroscopic effective values in order to retrieve the optimal microstructure. Section 4 is devoted to the numerical approximation and to the anisotropic setting used for the finite element discretization. In particular, we examine the mathematical tool employed to anisotropically adapt a two-dimensional mesh to the problem at hand, coupling such a procedure with the inverse homogenization technique. In Sect. 5, some numerical results are provided in order to assess the proposed algorithm, and finally some conclusions are drawn in Sect. 6.

2 A Density-Based Method for Topology Optimization

We consider the SIMP formulation for topology optimization to address the structural optimization problem [2]. In this context, the optimal layout of a material is determined in terms of an auxiliary scalar field, say ρ, defined over the domain Ω. In particular, ρ is a relative density belonging to $L^\infty(\Omega, [0, 1])$, determining presence of full material ($\rho = 1$) or void ($\rho = 0$). The optimization problem is set once the objective function $\mathscr{C}$ and the design requirements are defined, while a balance equation $\mathscr{S}$ constrains the optimization. Then, in order to account for changes in the topology, the state equation $\mathscr{S}$ is properly modified to include the density variable in the formulation. The final optimization problem thus reads

$$\min_{\rho\in L^\infty(\Omega)} \mathscr{C}(\rho) \; : \; \begin{cases} \text{State equation } \mathscr{S}(\rho) \text{ is satisfied} \\ \text{Boundary conditions} \\ \int_\Omega \rho \, d\Omega \; \leq \alpha|\Omega| \\ \rho_{\min} \leq \rho \leq 1, \end{cases} \tag{1}$$

where α is the maximum volume fraction we wish to ensure in the final configuration, and $\rho_{\min}$ is a lower bound for the density, to avoid the possible ill-posedness of $\mathscr{S}$.

In particular, $\mathscr{S}$ is chosen accordingly to the physical phenomenon under investigation, i.e., to the application at hand. For instance, for the optimization of elastic structures, the state equation can be represented by the linear elastic equation, whereas, when considering the optimization of the energy dissipation of a steady flow, one can identify $\mathscr{S}$ with the Stokes equations. In the specific case of the present work, we deal with the optimization of the design of elastic microstructures. A homogenized version of the elastic equations will represent the reference state equation as detailed in the following section. Concerning the inclusion of the density variable in the state equation, a suitable power law of ρ is usually employed to weigh the main physical constants in $\mathscr{S}$, such as the standard Lamé constants, λ and μ, for the elastic problem or the inverse permeability of the fluid for the Stokes equations.

3 The Homogenization Procedure

The homogenization method is an asymptotic technique whose goal is to assign macroscopic effective properties to microscopic entities, which are arranged periodically. This approach plays a crucial role in multiscale simulations since it allows one to deal with the macroscale only, the effects of the microscale being inherited through homogenization. The technique has been widely investigated both theoretically [3, 20] and numerically [1], and it is a well-established practice.

In this section, we analyze also the converse technique, known as inverse homogenization [14, 19, 21, 22]. This can be formulated as a control problem or, specifically, as a topology optimization problem. The aim is to find the optimal arrangement of material at the microscale so that desired effective properties are guaranteed at the macroscale. Notice that the flow of information is opposite with respect to the classical homogenization. The macroscale is fixed or prescribed, whereas the microscale is modified to match the desired requirements.

3.1 The Direct Method

Direct homogenization has been employed in different fields of application to modify the macroscale model according to the microscale layout [6, 12, 20]. This technique relies on the periodic arrangement of a microstructure which constitutes the base cell, Y. Such elementary entity represents the domain of interest and it is analyzed in order to retrieve its effect on the macroscale.

Let us consider the linear elasticity equation [11]

$$-\nabla \cdot \sigma(\mathbf{u}) = \mathbf{f} \quad \text{in } \Omega, \tag{2}$$

where $\Omega \subset \mathbb{R}^2$ is the domain under investigation at the macroscale, $\mathbf{f}$ is the volumetric forcing term, $\mathbf{u} = [u_1, u_2]^T$ is the displacement field, and σ is the stress tensor. For the sake of generality, we stick to the convention of denoting by E_{ijkl}

the fourth-order stiffness tensor, so that the stress tensor has components

$$\sigma_{ij} = E_{ijkl}\varepsilon_{kl} \text{ with } \varepsilon_{kl} = \frac{1}{2}\left(\frac{\partial u_k}{\partial x_l} + \frac{\partial u_l}{\partial x_k}\right),$$

where x_l, with $l = 1, 2$, are the spatial coordinates, ε_{kl} are the components of the strain tensor ε, and we have adopted the Einstein notation to manage index summation.

The homogenization technique relies on the repetition of the base cell Y. In order to preserve this physical feature, we impose periodic boundary conditions. In this way, we enforce that the displacement field $\mathbf{u}$ is equal in correspondence with opposite boundaries [5].

Then, the actual objective becomes to compute the homogenized (or effective) stiffness tensor, E^H, representing a macroscopic mean value of the tensor E, after neglecting the microscale fluctuations E^*. To this end, we resort to an asymptotic expansion of the displacement field $\mathbf{u}$ with respect to the base cell size, considering only the first two terms. Then, following [3, 21], it can be shown that the homogenized tensor E^H is given by

$$E^H_{ijkl} = \frac{1}{|Y|}\int_Y E_{ijpq}(\varepsilon^{0,kl}_{pq} - \varepsilon^{*,kl}_{pq})\, dY, \tag{3}$$

where $|Y|$ is the measure of the cell Y, $\varepsilon^{0,kl}$ identifies a fixed strain field, chosen among the four linearly independent possible fields (k, l being equal to $1, 2$), while $\varepsilon^{*,kl} \in V$ is the Y-periodic fluctuation strain, i.e., the weak solution to the equation

$$\int_Y E_{ijpq}\varepsilon^{*,kl}_{pq}\varepsilon_{ij}(v)\, dY = \int_Y E_{ijpq}\varepsilon^{0,kl}_{pq}\varepsilon_{ij}(v)\, dY, \quad \forall v \in V, \tag{4}$$

$V \subset [H^1(Y)]^4$ being a periodic Sobolev function space. Thus, by combining (3) and (4), we obtain the final form of the effective stiffness tensor [1, 21]

$$E^H_{ijkl} = \frac{1}{|Y|}\int_Y E_{pqrs}(\varepsilon^{0,kl}_{pq} - \varepsilon^{*,kl}_{pq})(\varepsilon^{0,ij}_{rs} - \varepsilon^{*,ij}_{rs})\, dY. \tag{5}$$

Equations (4) and (5) constitute the state equations to be employed in the inverse homogenization technique, as detailed in the following section.

3.2 The Inverse Method

We refer to inverse homogenization as to the procedure concerning the design of a base cell, Y, whose contribution to the macroscale, according to the direct homogenization process in the previous section, is prescribed [19, 22]. In order to modify the formulation of the direct method, we have to account for variations in the initial distribution of material in the base cell. This goal can be pursued via topology optimization, yielding optimized structures according to specific, user-defined, constraints and objectives.

The same paradigm as in Sect. 2 is now exploited to incorporate the cell design in the homogenization problem. Let us fix the objective function, $\mathscr{J}$, as a control over the quadratic deviation between the computed value of the homogenized stiffness tensor, E^H, and the requested one, E^W, i.e.,

$$\mathscr{J} = \sum_{ijkl} (E^H_{ijkl}(\rho) - E^W_{ijkl})^2.$$

Hence, the minimization of $\mathscr{J}$ should lead to a micro-design, whose macro-features are the ones desired by the user [22]. Thus, the final system for the micro-optimization is obtained by solving the following problem

$$\min_{\rho \in L^\infty(Y)} \mathscr{J}(\rho) \ : \begin{cases} (4)_\rho - (5)_\rho \text{ are satisfied} \\ + \text{Periodicity conditions} \\ \int_Y \rho \, dY \ \leq \alpha |Y| \\ \rho_{\min} \leq \rho \leq 1, \end{cases} \tag{6}$$

where $(4)_\rho - (5)_\rho$ represent Eqs. (4) and (5) after replacing E_{ijkl} with $\rho^p E_{ijkl}$, in order to include the design variable ρ in the formulation, p being a penalization exponent.

4 The Numerical Discretization

Problem (6) can be numerically solved via a finite element discretization [7]. After introducing a conforming tessellation $\mathscr{T}_h = \{K\}$ of Y, with K the generic triangle, we denote by V_h^r the associated finite element spaces of piecewise polynomials of degree $r > 0$, with h the maximum diameter of the mesh elements.

The topology optimization problem discretized via a finite element scheme is known to suffer from several numerical issues [15, 23]. Some of these can be tackled with a suitable choice of the spaces employed to discretize displacement and density or via filtering techniques. Here, we propose to contain any post-processing phase by exploiting the intrinsic smoothness of the optimized density field yielded using ad-hoc meshes. In particular, we choose to discretize problem (6) on a sequence of anisotropically adapted grids and, consequently, we modify the optimization algorithm to deliver smooth and, essentially, directly manufacturable structures.

4.1 The Anisotropic Setting

We resort to an anisotropic adaptive procedure driven by the density field ρ, which is expected to sharply change from 0 to 1 in correspondence with the boundaries of the structure. The expected strong gradients across the material-void interface justify the employment of anisotropic meshes as an ideal tool to sharply describe the directional features of the density field.

We follow a metric-based procedure in order to generate the optimal mesh to discretize the problem [10]. Essentially, the adaptation procedure relies on an a posteriori error estimator, merging the error information with the geometric properties of the grid. In particular, we employ an anisotropic variant of the Zienkiewicz-Zhu estimator [25], to evaluate the H^1-seminorm of the discretization error, which is expected to be the most effective measure for detecting the material-void interface. Following [16], the elementwise contribution to the anisotropic error estimator is

$$\eta_K^2 = \frac{1}{\lambda_{1,K}\lambda_{2,K}} \sum_{i=1}^{2} \lambda_{i,K}^2 \left(\mathbf{r}_{i,K}^T \, G_{\Delta_K}(E_\nabla) \mathbf{r}_{i,K} \right), \tag{7}$$

where $\lambda_{1,K}$ and $\lambda_{2,K}$ are the lengths of the semi-axes of the ellipse circumscribed to element K, while $\mathbf{r}_{1,K}$ and $\mathbf{r}_{2,K}$ represent the directions of such axes. The quantity $E_\nabla = \left[P(\nabla\rho_h) - \nabla\rho_h\right]_{\Delta_K}$ is the recovered error associated with the density ρ, where $P(\nabla\rho_h)|_{\Delta_K} = |\Delta_K|^{-1} \sum_{T\in\Delta_K} |T| \nabla\rho_h|_T$ denotes the recovered gradient computed on the patch Δ_K of the elements sharing at least a vertex with K, $|\cdot|$ being the measure operator, and $\nabla\rho_h$ is the gradient of the discrete density [9, 17]. Finally, $G_{\Delta_K}(\cdot) \in \mathbb{R}^{2\times 2}$ is the symmetric positive semidefinite matrix with entries

$$[G_{\Delta_K}(\mathbf{w})]_{i,j} = \sum_{T\in\Delta_K} \int_T w_i \, w_j \, dT \quad \text{with } i, j = 1, 2, \tag{8}$$

for any vector-valued function $\mathbf{w} = (w_1, w_2)^T \in [L^2(\Omega)]^2$. Then, the global error estimator is given by $\eta^2 = \sum_{K\in\mathscr{T}_h} \eta_K^2$.

The mesh adaptation is carried out by minimizing the number of elements of the adapted mesh, while requiring an upper bound TOLAD to the global error estimator η together with an error equidistribution criterion. This gives rise to an elementwise constrained optimization problem which admits a unique analytic solution. Specifically, by introducing the aspect ratio $s_K = \lambda_{1,K}/\lambda_{2,K} \geq 1$ measuring the deformation of element K, the adapted grid is characterized by the following quantities

$$s_K^{adapt} = \sqrt{g_1/g_2}, \quad \mathbf{r}_{1,K}^{adapt} = \mathbf{g}_2, \quad \mathbf{r}_{2,K}^{adapt} = \mathbf{g}_1,$$

where $\{g_i, \mathbf{g}_i\}_{i=1,2}$ are the eigen-pairs associated with the scaled matrix $\widehat{G}_{\Delta_K}(E_\nabla) = G_{\Delta_K}(E_\nabla)/|\Delta_K|$, with $g_1 \geq g_2 > 0$, $\{\mathbf{g}_i\}_{i=1,2}$ orthonormal vectors.

Finally, imposing the equidistribution, i.e., $\eta_K^2 = \texttt{TOLAD}^2/\#\mathcal{T}_h$, with $\#\mathcal{T}_h$ the mesh cardinality, we obtain the geometric information identifying the new adapted mesh, i.e.,

$$\lambda_{1,K}^{adapt} = g_2^{-1/2}\left(\frac{\texttt{TOLAD}^2}{2\#\mathcal{T}_h\,|\widehat{\Delta}_K|}\right)^{1/2}, \quad \lambda_{2,K}^{adapt} = g_1^{-1/2}\left(\frac{\texttt{TOLAD}^2}{2\#\mathcal{T}_h\,|\widehat{\Delta}_K|}\right)^{1/2},$$
$$\mathbf{r}_{1,K}^{adapt} = \mathbf{g}_2, \quad \mathbf{r}_{2,K}^{adapt} = \mathbf{g}_1, \tag{9}$$

with $|\widehat{\Delta}_K| = |\Delta_K|/(\lambda_{1,K}\lambda_{2,K})$.

4.2 The Adaptive Algorithm

The algorithm employed to merge the topology optimization of the base cell Y with the mesh adaptation procedure described above is here presented. We name it microSIMPATY algorithm since it is inspired by the algorithm SIMPATY in [18].

In Algorithm 1, `optimize` is a numerical routine for the inverse topology optimization, which stops whenever the maximum number of iterations, `Mit`, is exceeded, or the prescribed tolerance, `TOPT`, is satisfied. Beside the objective function $\mathcal{J}(\rho)$, the corresponding derivative with respect to ρ is required by the `optimize` algorithm, as well as other possible constraints to be imposed, with the associated derivatives. Such sensitivities are analytically computed following a Lagrangian approach [4]. Function `adapt` is a routine performing the mesh adaptation starting from the metric derived in (9). The algorithm is terminated by two stopping criteria, one based on the number of iterations, the other on the stagnation of the number of elements between two consecutive mesh adaptations to within `CTOL`.

Algorithm 1 microSIMPATY

Input : `CTOL`, `TOLAD`, `TOPT`, `kmax`, $\rho_{\min}$, $\mathcal{T}_h^{(0)}$

1: Set: ρ_h^0, k = 0, errC = 1 + `CTOL`
2: **while** errC > `CTOL` & k < `kmax` **do**
3: ρ_h^{k+1} = `optimize`(ρ_h^k, `Mit`, `TOPT`, $\rho_{\min}$, $\mathcal{J}(\rho)$, $\nabla_\rho\mathcal{J}(\rho)$, ...);
4: $\mathcal{T}_h^{(k+1)}$ = `adapt`($\mathcal{T}_h^{(k)}$, ρ_h^{k+1}, `TOLAD`);
5: errC = $|\#\mathcal{T}_h^{(k+1)} - \#\mathcal{T}_h^{(k)}|/\#\mathcal{T}_h^{(k)}$;

5 Numerical Results

The following numerical verification has been carried out with `FreeFem++` [13], which provides the users with built-in functions for both optimization [8] and metric-based mesh adaptation. In both the considered test cases, we deal with the design of a 1[m] $\times$1[m] base cell with negative Poisson ratio $\nu = \lambda/[2(\lambda + \mu)]$, corresponding to E_{1122}. We choose $p = 4$ for the penalization exponent in (6). The material employed has Young modulus equal to 0.91[Pa] and Poisson ratio $\nu = 0.3$. Finally, ρ_h^0 is set to $|\sin(2\pi x_1)\sin(2\pi x_2)|$.

Case 1 In Fig. 1, the results for $E_{1122}^W = -1$ are shown. We require a volume fraction $\alpha = 0.3$, we start with an initial structured mesh consisting of 1800 elements, and we pick `TOLAD` $= 10^{-5}$, `CTOL` $= 10^{-4}$, `TOPT` $= 10^{-3}$, $\rho_{\min} = 10^{-4}$, `kmax` $= 20$, while the maximum number of iterations, `Mit`, is set to 35 for the first three iterations and to 10 for the next ones. The algorithm stops after 20 iterations, delivering a structure with $E_{1122}^H = -0.65$. The final design thus obtained is comparable with the one in [14, Figure 3], while the quality of the solution is increased when resorting to the microSIMPATY algorithm, no filtering techniques being required. In Fig. 1, bottom-right, we show the last adapted grid. Notice that the elements are highly stretched and concentrated in correspondence with the void-solid interface.

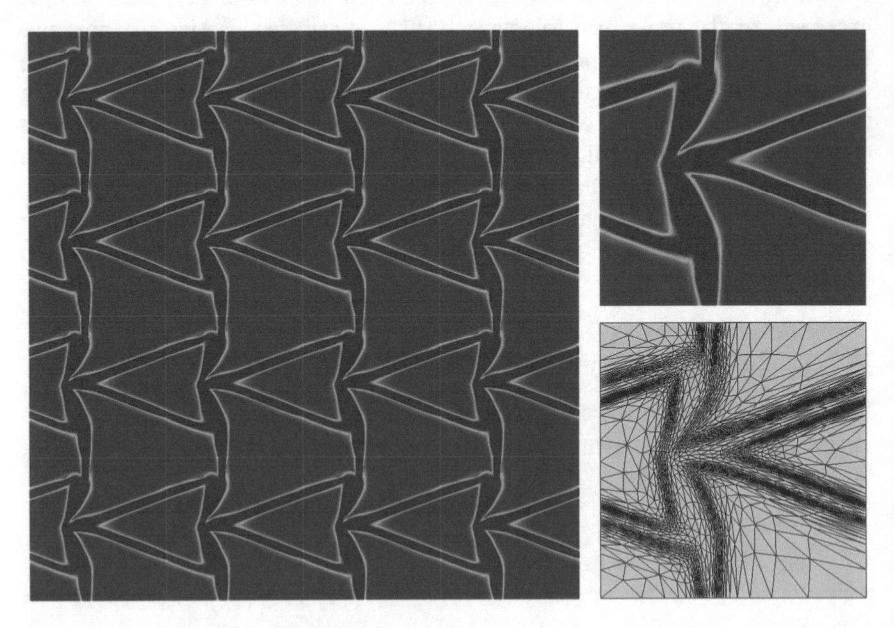

Fig. 1 Optimized microstructure for $E_{1122}^W = -1$: 4×4 periodic arrangement of the base cell (left), base cell (top-right) and corresponding adapted mesh (bottom-right)

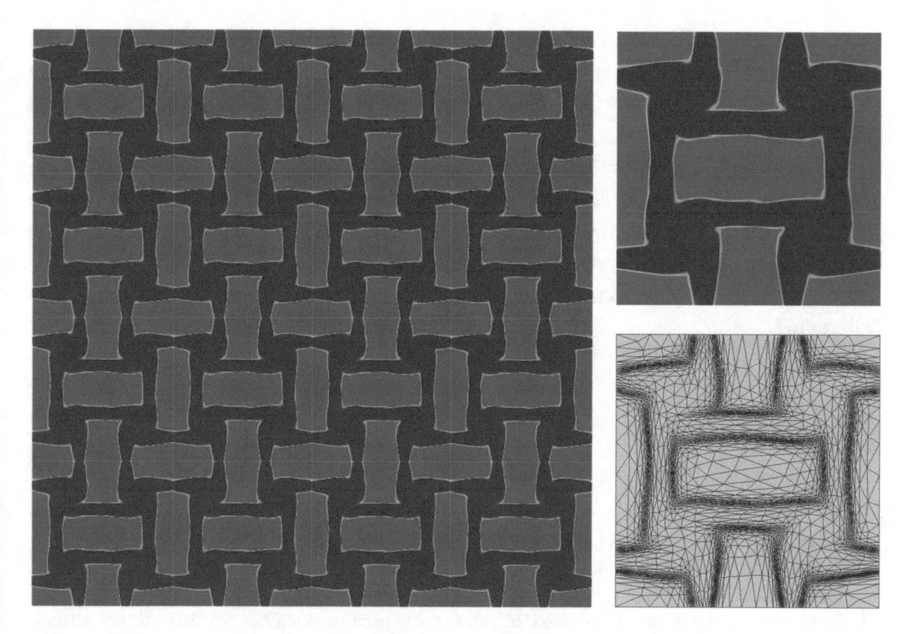

Fig. 2 Optimized microstructure for $E^W_{1122} = -0.7$: 4×4 periodic arrangement of the base cell (left), base cell (top-right) and corresponding adapted mesh (bottom-right)

The cardinality of such final mesh is 2620 and its maximum aspect ratio is 97.76.

Case 2 The second case concerns the optimization of a micro-design with $E^W_{1122} = -0.7$ and $\alpha = 0.5$ (see [21, Figure 2.17]). As for the previous test, we perform 20 iterations, starting from a structured mesh of 1800 elements and picking the same parameters as in the previous case, except for `Mit`, which is now set to 35 at the first iteration, to 25 until the fifth one, and to 15 for the later iterations. The results in Fig. 2 show a very smooth solution, where intermediate densities are very limited to a thin boundary layer, whose quality is enhanced by the adapted grid. In the final mesh, the directionalities of the density field are properly detected, making 4266 elements enough for a sharply-defined solution, with a maximum value for the aspect ratio equal to 85.58. The final structure delivers an effective Poisson ratio equal to -0.54.

6 Conclusions

In this work, we presented an algorithm to optimize microstructures according to user-defined requirements, based on the inverse homogenization method, properly merged with an anisotropic mesh adaptation procedure.

The structures derived in Sect. 5 are consistent with the ones available in the literature and exhibit a remarkable smoothness along structure boundaries, the thin material/void layers being sharply detected by the adapted mesh. This feature confirms the benefits due to microSIMPATY algorithm.

Nevertheless, the optimization process depends on several parameters to be accurately tuned in order to meet user requirements. For this reason, we plan to perform a more rigorous investigation in such a direction, especially to make the homogenized stiffness tensor closer to the requested one.

Finally, with a view to real applications, we are extending the algorithm to a 3D framework.

References

1. Andreassen, E., Andreasen, C.S.: How to determine composite material properties using numerical homogenization. Comput. Mater. Sci. **83**, 488–495 (2014)
2. Bendsøe, M.P., Sigmund, O.: Topology Optimization – Theory, Methods and Applications. Springer, Berlin (2003)
3. Bensoussan, A., Lions, J.L., Papanicolaou, G.: Asymptotic Analysis for Periodic Structures. AMS Chelsea Publishing, Providence (2011)
4. Bertsekas, D.P.: Constrained Optimization and Lagrange Multiplier Methods. Computer Science and Applied Mathematics. Academic, New York (1982)
5. Brezis, H.: Functional Analysis, Sobolev Spaces and Partial Differential Equations. Universitext. Springer, New York (2011)
6. Chung, P.W., Tamma, K.K., Namburu, R.R.: Homogenization of temperature-dependent thermal conductivity in composite materials. J. Thermophys. Heat Transf. **15**(1), 10–17 (2001)
7. Ciarlet, P.G.: The Finite Element Method for Elliptic Problems. North-Holland, Amsterdam (1978)
8. Curtis, F.E., Schenk, O., Wächter, A.: An interior-point algorithm for large-scale nonlinear optimization with inexact step computations. SIAM J. Sci. Comput. **32**(6), 3447–3475 (2010)
9. Farrell, P.E., Micheletti, S., Perotto, S.: An anisotropic Zienkiewicz-Zhu-type error estimator for 3D applications. Int. J. Numer. Methods Eng. **85**(6), 671–692 (2011)
10. Frey, P.J., George, P.L.: Mesh Generation: Application to Finite Elements. Wiley, Hoboken (2008)
11. Gould, P.L.: Introduction to Linear Elasticity. Springer, Paris (1994)
12. Hashin, Z., Shtrikman, S.: A variational approach to the theory of the effective magnetic permeability of multiphase materials. J. Appl. Phys. **33**(10), 3125–3131 (1962)
13. Hecht, F.: New development in freefem++. J. Numer. Math. **20**(3–4), 251–266 (2012)
14. Larsen, U.D., Signund, O., Bouwsta, S.: Design and fabrication of compliant micromechanisms and structures with negative Poisson's ratio. J. Microelectron. Syst. **6**(2), 99–106 (1997)
15. Lazarov, B.S., Sigmund, O.: Filters in topology optimization based on Helmholtz-type differential equations. Int. J. Numer. Methods Eng. **86**(6), 765–781 (2011)
16. Micheletti, S., Perotto, S.: Anisotropic adaptation via a Zienkiewicz–Zhu error estimator for 2D elliptic problems. In: Kreiss, G., Lötstedt, P., Målqvist, A., Neytcheva, M. (eds.), Numerical Mathematics and Advanced Applications, pp. 645–653. Springer, Berlin (2010)
17. Micheletti, S., Perotto, S., Farrell, P.E.: A recovery-based error estimator for anisotropic mesh adaptation in CFD. Bol. Soc. Esp. Mat. Apl. SeMA **50**, 115–137 (2010)

18. Micheletti, S., Perotto, S., Soli, L.: Ottimizzazione topologica adattativa per la fabbricazione stratificata additiva (2017). Italian patent application No. 102016000118131, filed on November 22, 2016 (extended as Adaptive topology optimization for additive layer manufacturing, International patent application PCT No. PCT/IB2017/057323)
19. Neves, M.M., Rodrigues, H., Guedes, J.M.: Optimal design of periodic linear elastic microstructures. Comput. Struct. **76**(1–3), 421–429 (2000)
20. Sánchez-Palencia, E.: Homogenization Method for the Study of Composite Media. In: Asymptotic Analysis, II. Lecture Notes in Mathematics, vol. 985, pp. 192–214. Springer, Berlin (1983)
21. Sigmund, O.: Design of Material Structures Using Topology Optimization. Technical University of Denmark, Lyngby (1994)
22. Sigmund, O.: Materials with prescribed constitutive parameters: an inverse homogenization problem. Int. J. Solids Struct. **31**(17), 2313–2329 (1994)
23. Sigmund, O., Petersson, J.: Numerical instabilities in topology optimization: a survey on procedures dealing with checkerboards, mesh-dependencies and local minima. Struct. Multidiscip. Optim. **16**(1), 68–75 (1998)
24. Yin, L., Ananthasuresh, G.: Topology optimization of compliant mechanisms with multiple materials using a peak function material interpolation scheme. Struct. Multidiscip. Optim. **23**(1), 49–62 (2001)
25. Zienkiewicz, O.C., Zhu, J.Z.: A simple error estimator and adaptive procedure for practical engineering analysis. Int. J. Numer. Methods Eng. **24**(2), 337–357 (1987)
26. Zuo, W., Saitou, K.: Multi-material topology optimization using ordered SIMP interpolation. Struct. Multidiscip. Optim. **55**(2), 477–491 (2017)

Bathymetry Reconstruction Using Inverse Shallow Water Models: Finite Element Discretization and Regularization

Hennes Hajduk, Dmitri Kuzmin, and Vadym Aizinger

Abstract In the present paper, we use modified shallow water equations (SWE) to reconstruct the bottom topography (also called bathymetry) of a flow domain without resorting to traditional inverse modeling techniques such as adjoint methods. The discretization in space is performed using a piecewise linear discontinuous Galerkin (DG) approximation of the free surface elevation and (linear) continuous finite elements for the bathymetry. Our approach guarantees compatibility of the discrete forward and inverse problems: for a given DG solution of the forward SWE problem, the underlying continuous bathymetry can be recovered exactly. To ensure well-posedness of the modified SWE and reduce sensitivity of the results to noisy data, a regularization term is added to the equation for the water height. A numerical study is performed to demonstrate the ability of the proposed method to recover bathymetry in a robust and accurate manner.

Keywords Bathymetry reconstruction · Shallow water equations · Continuous/discontinuous Galerkin method · Inverse problem

1 Introduction

The shallow water equations are among the most popular mathematical models for applications in environmental fluid mechanics. The geometry of a computational domain for SWE simulations of coastal, riverine, and estuarine flow problems can be determined using inexpensive and highly accurate measurement techniques for boundaries corresponding to coastlines and the free surface. However, the

H. Hajduk · D. Kuzmin (✉)
TU Dortmund University, Institute of Applied Mathematics (LS III), Dortmund, Germany
e-mail: hennes.hajduk@tu-dortmund.de; kuzmin@math.uni-dortmund.de

V. Aizinger
Chair of Scientific Computing, University of Bayreuth, Bayreuth, Germany
e-mail: vadym.aizinger@awi.de

H. van Brummelen et al. (eds.), *Numerical Methods for Flows*,
Lecture Notes in Computational Science and Engineering 132,
https://doi.org/10.1007/978-3-030-30705-9_20

resolution and accuracy of experimental data for the bottom topography (also called bathymetry) of many regions are very poor. As an alternative to direct measurements, the missing bathymetry data can be reconstructed by solving a (modified) SWE system as originally proposed in [7, 8]. The bathymetry enters the momentum equations as a source term which has a strong influence on the accuracy of simulations. In many applications such as tsunami predictions, numerical results are highly sensitive to errors in bathymetry data. The most common measurement techniques for bathymetry and their respective limitations (see [14, 18]) are as follows:

- Surveys by ships are suitable only for local measurements in small regions;
- LiDAR/LaDAR (**Li**ght/**La**ser **D**etection **A**nd **R**anging) using equipment installed on ships or aircraft is expensive and has limited coverage;
- Multi-spectral satellite imaging is only practical for shallow and clear water.

Discussions of other issues associated with direct bathymetry measurements can be found, e.g., in [14, 16, 20]. In the present paper, we explore the possibility of using SWE-based models for bathymetry reconstruction from the water surface elevation which is much easier to measure remotely (e.g. by satellite altimetry). The proposed approach involves solving a degenerate hyperbolic inverse problem, in which the roles of the free surface elevation and the bottom topography are interchanged [12]. The first proof of concept for bathymetry reconstructions by this technique was proposed by Gessese et al. in [8] using a finite difference discretization of a one-dimensional SWE system for stationary sub- and transcritical configurations. A generalization to the 2D case was presented in [7] and further developed in [9, 10]. The main objective of the present work is the design of a special finite element discretization that ensures compatibility of the forward and the inverse problems. We also address the ill-posedness issue by adding a regularization term which also improves the reconstruction quality in the presence of noise.

2 Formulation of the Forward and Inverse Problems

The shallow water equations are derived from the incompressible Navier-Stokes equations using the hydrostatic pressure assumption and averaging in the vertical direction [5, 19]. The result is the system of conservation laws

$$\frac{\partial H}{\partial t} + \nabla \cdot (H\boldsymbol{u}) = 0, \tag{1}$$

$$\frac{\partial (Hu)}{\partial t} + \nabla \cdot (Hu\boldsymbol{u}) + \frac{g}{2}\frac{\partial H^2}{\partial x} + g\,H\,\frac{\partial b}{\partial x} + \tau_{bf} Hu - f_c\,Hv = 0, \tag{2}$$

$$\frac{\partial (Hv)}{\partial t} + \nabla \cdot (Hv\boldsymbol{u}) + \frac{g}{2}\frac{\partial H^2}{\partial y} + g\,H\,\frac{\partial b}{\partial y} + \tau_{bf} Hv + f_c\,Hu = 0, \tag{3}$$

where $\boldsymbol{u} = [u, v]^T$ is the depth-averaged velocity, and $H = \xi - b$ is the total water height, that is, the difference between the free surface elevation ξ and the bathymetry b both measured with respect to the same level. The terms depending on τ_{bf} and f_c are due to the bottom friction and the Coriolis force, respectively. In a compact form, the SWE system can be written as

$$\frac{\partial \mathbf{U}}{\partial t} + \nabla \cdot \mathbf{F}(\mathbf{U}) = \mathbf{S}(\mathbf{U}, \nabla b), \tag{4}$$

where

$$\mathbf{U} = \begin{bmatrix} H \\ H\boldsymbol{u} \end{bmatrix}, \ \mathbf{F}(\mathbf{U}) = \begin{bmatrix} H\boldsymbol{u} \\ H\boldsymbol{u} \otimes \boldsymbol{u} + \frac{gH^2}{2}\mathbb{1} \end{bmatrix}, \ \mathbf{S}(\mathbf{U}, \nabla b) = \begin{bmatrix} 0 \\ f_c\, Hv - \tau_{bf} Hu - gH\frac{\partial b}{\partial x} \\ -f_c\, Hu - \tau_{bf} Hv - gH\frac{\partial b}{\partial y} \end{bmatrix}.$$

In the context of bathymetry reconstruction, the forward and inverse problems require numerical solution of system (1)–(3). In the forward problem, the free surface elevation is given by $\xi = H + b$, where b is a known bathymetry. The bathymetry gradient which appears in (2), (3) is known as well, so the source term $g\,H\,\nabla b$ depends linearly on H. In the inverse problem, the bathymetry is unknown and defined by $b = \xi - H$, where ξ is given. Since the gravitational force term of the inverse momentum equation contains the unknown bathymetry gradient, the inverse problem exhibits entirely different mathematical behavior. As a consequence of swapping the roles of ξ and b, system (1)–(3) becomes degenerate hyperbolic and more difficult to solve (see [12] for an in-depth analysis of the potentially ill-posed inverse problem).

The possible lack of uniqueness can be cured by adding a regularization term of the form $\epsilon \Delta b$ to (1). This regularization resembles the Brezzi-Pitkäranta stabilization method [3] for equal-order numerical approximations to the velocity and pressure in the incompressible Navier-Stokes equations; however, in the setting of the shallow water equations, such a term would induce artificial currents in the presence of bathymetry gradients. The Laplacian can be replaced by a total variation regularization term [4] or another anisotropic diffusion operator. In this work, we define the regularization term at the discrete level in terms of finite element matrices (see below).

3 Discretization of the SWE System

Let $\Omega \subset \mathbb{R}^2$ be a bounded Lipschitz domain with a polygonal boundary. Given a conforming triangulation $\mathcal{T}_h := \{T_1, \ldots, T_K\}$ of Ω, we define the DG space $\mathbb{V}_h := \{v_h \in L^2(\Omega) : v_h|_T \in \mathbb{P}_1(T)\ \forall T \in \mathcal{T}_h\}$ and the corresponding continuous Galerkin (CG) space $\mathbb{W}_h := \mathbb{V}_h \cap C^0(\overline{\Omega})$. The space $\mathbb{V}_h$ is spanned by $3K$ piecewise-linear basis functions ψ_{kj}, $k = 1, \ldots, K,\ j = 1, 2, 3$. The dimension of

$\mathbb{W}_h$ equals the number of vertices $\mathbf{x}_1, \ldots, \mathbf{x}_L$ of $\mathcal{T}_h$. The Lagrange basis functions $\varphi_1, \ldots, \varphi_L$ have the property that $\varphi_i(\boldsymbol{x}_j) = \delta_{ij},\ i, j = 1, \ldots, L$.

System (1)–(3) is discretized using the space $\mathbb{V}_h$ for H, $H\boldsymbol{u}$, ξ and the space $\mathbb{W}_h$ for b. Since the number of equations must be equal to the number of unknowns, the variational forms of the discrete forward and inverse problems differ in the choice of the test function space for the continuity equation. Time integration is performed using an explicit second-order SSP Runge-Kutta scheme [11], i.e., Heun's method.

Space Discretization of the Forward Problem

In the semi-discrete forward problem, we seek the coefficients of surface elevation $\xi_h \in \mathbb{V}_h$ and momentum $(H\boldsymbol{u})_h \in (\mathbb{V}_h)^2$. In practice, it is more convenient to formulate the semi-discrete forward problem in terms of $\mathbf{U}_h = [H_h, (H\boldsymbol{u})_h]^T \in (\mathbb{V}_h)^3$ and calculate the surface elevation $\xi_h = H_h + b_h \in \mathbb{V}_h$ by adding the known continuous bathymetry $b_h \in \mathbb{W}_h$ to the discontinuous water height $H_h \in \mathbb{V}_h$. For any element $T^- \in \mathcal{T}_h$ and any test function $\mathbf{v}_h \in (\mathbb{V}_h)^3$, the (element-local) DG form of system (4) is given by [1, 12]

$$\int_{T^-} \mathbf{v}_h \cdot \partial_t \mathbf{U}_h \,\mathrm{d}\boldsymbol{x} - \int_{T^-} \nabla \mathbf{v}_h : \mathbf{F}(\mathbf{U}_h) \,\mathrm{d}\boldsymbol{x} + \int_{\partial T^-} \mathbf{v}_h \cdot \widehat{\mathbf{F}}(\mathbf{U}_h^-, \mathbf{U}_h^+; \boldsymbol{\nu}_{T^-}) \,\mathrm{d}s = \int_{T^-} \mathbf{v}_h \cdot \mathbf{S}(\mathbf{U}_h, \nabla b_h) \,\mathrm{d}\boldsymbol{x}\ , \tag{5}$$

where $\boldsymbol{\nu}_{T^-}$ is the unit outward normal and $\widehat{\mathbf{F}}(\mathbf{U}_h^-, \mathbf{U}_h^+; \boldsymbol{\nu}_{T^-})$ is a numerical flux defined in terms of the one-sided limits $\mathbf{U}_h^{\pm}$ (see [12] for details). In the numerical study below, we use the Roe flux or the Lax-Friedrichs flux.

Summing over all elements, we obtain a semi-discrete problem of the form

$$(v_h, \partial_t H_h) + a_H(v_h, \mathbf{U}_h) = f_H(v_h) \quad \forall v_h \in \mathbb{V}_h, \tag{6}$$

$$(\mathbf{v}_h, \partial_t (H\boldsymbol{u})_h) + a_{\boldsymbol{u}}(\mathbf{v}_h, \mathbf{U}_h) = (\mathbf{v}_h, \mathbf{S}_h(\mathbf{U}_h, \nabla b_h)) + \mathbf{f}_{\boldsymbol{u}}(\mathbf{v}_h) \quad \forall \mathbf{v}_h \in (\mathbb{V}_h)^2, \tag{7}$$

where $(\cdot, \cdot)$ is the L^2 scalar product on Ω. The forms $a_H(\cdot, \cdot)$ and $a_{\boldsymbol{u}}(\cdot, \cdot)$ consist of volume integrals depending on $\nabla \mathbf{v}_h : \mathbf{F}(\mathbf{U}_h)$ and jump terms depending on $(\mathbf{v}_h^+ - \mathbf{v}_h^-) \cdot \widehat{\mathbf{F}}(\mathbf{U}_h^-, \mathbf{U}_h^+; \boldsymbol{\nu}_{T^-})$. The linear forms $f_H(\cdot)$ and $\mathbf{f}_{\boldsymbol{u}}(\cdot)$ contain the contribution of weakly imposed boundary conditions.

Space Discretization of the Inverse Problem

In the inverse problem, the water height $H_h \in \mathbb{V}_h$ is uniquely determined by the $3K$ known coefficients of the surface elevation $\xi_h \in \mathbb{V}_h$ and L unknown coefficients of the bathymetry $b_h \in \mathbb{W}_h$. Hence, the dimension of the test function space for system (6), (7) exceeds the number of unknowns. Substituting $\xi_h - b_h$ for H_h, we replace the continuity equation (6) by

$$(w_h, \partial_t b_h) - a_H(w_h, \mathbf{U}_h) = (w_h, \partial_t \xi_h) - f_H(w_h) \qquad \forall w_h \in \mathbb{W}_h, \tag{8}$$

while keeping the momentum equation (7) unchanged. This yields a system of $L + 6K$ equations for $L + 6K$ unknowns. Since $\mathbb{W}_h$ is a subspace of $\mathbb{V}_h$, a DG approximation $\mathbf{U}_h = [H_h, (H\boldsymbol{u})_h]^T$ satisfying (6), (7) will satisfy (7), (8) as well. If the given surface elevation ξ_h corresponds to a solution of the discrete forward problem, the underlying bathymetry b_h must be a (possibly non-unique) solution of the discrete inverse problem.

Pseudo-Time Stepping and Regularization

Any explicit SSP Runge-Kutta time discretization of system (8), (7) can be expressed as a convex combination of forward Euler updates. Let $\mathbf{M} = (m_{ij})$ denote the consistent mass matrix with entries $m_{ij} = (\varphi_i, \varphi_j),\ i, j = 1, \ldots, L$, where φ_i are the continuous Lagrange basis functions spanning the space $\mathbb{W}_h$. Row-sum mass lumping yields the diagonal approximation $\mathbf{M}_L = (m_i\delta_{ij})$, where $m_i = (\varphi_i, 1)$. To deal with the issue of ill-posedness, we march the bathymetry $b_h \in \mathbb{W}_h$ to a steady state using the regularized matrix form

$$\mathbf{M}_L \frac{b^{n+1} - b^n}{\Delta t} = R(\mathbf{U}_h^n) + \epsilon(\mathbf{M} - \mathbf{M}_L)b^n \tag{9}$$

of a generic forward Euler step for pseudo-time integration of (8). The first term on the right-hand side of the above linear system is defined by

$$R_i(\mathbf{U}_h^n) = \left(\varphi_i, \frac{\xi_h^{n+1} - \xi_h^n}{\Delta t}\right) + a_H\left(\varphi_i, \mathbf{U}_h^n\right) - f_H(\varphi_i), \tag{10}$$

where Δt is the pseudo-time step. The regularization term $\epsilon(\mathbf{M} - \mathbf{M}_L)b^n$ has the same form as the pressure stabilization term proposed by Becker and Hansbo [2] for a finite element discretization of the Stokes system. In our experience, the use of a discrete Laplace operator in place of $\mathbf{M} - \mathbf{M}_L$ produces similar results. We envisage that the use of anisotropic diffusion operators such as the one employed in [4] for total variation-based image denoising purposes can lead to more accurate reconstructions of small-scale features.

4 Numerical Results for the Inverse Problem

Our numerical discretization utilizes the FESTUNG toolbox [6, 15, 17] and is described in detail in [13]. We consider the domain $\Omega = (0, 1\,\text{km}) \times (0, 1\,\text{km})$ and utilize a triangular unstructured grid with $\Delta x = 40\,\text{m}$ to solve the forward and inverse problems on a time interval of 3 h with $\Delta t = 0.1$ s. The employed parameter settings are $g = 9.81\,\text{m/s}^2$, $f_c = 3 \cdot 10^{-5}\,\text{s}^{-1}$, $c_f = 10^{-3}\,\text{s}^{-1}$. Bathymetry for solving the forward problem is specified as a rather complex yet smooth function (see [12] and Fig. 1). The boundary conditions are as follows: in both problems, the normal fluxes are set to zero on the upper and lower boundary, and the flux

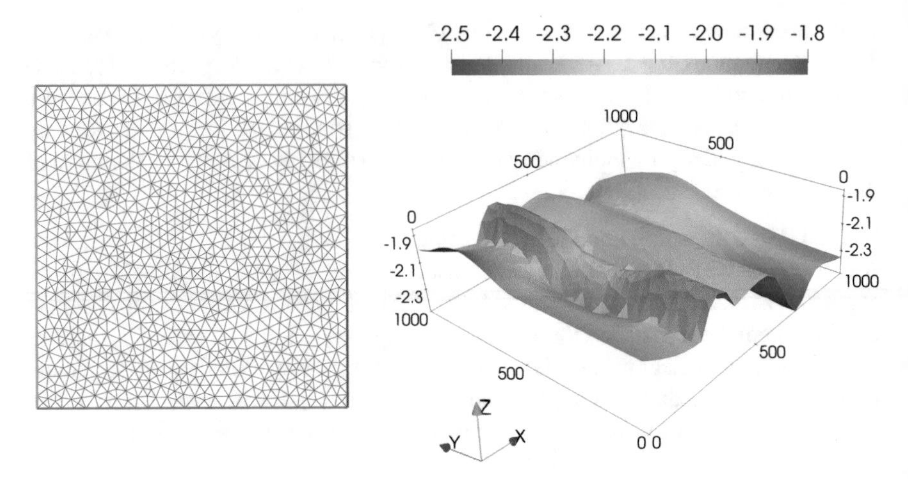

Fig. 1 Computational domain and mesh (left), exact bathymetry (right)

$\boldsymbol{Hu} = [4\ 0]^T$ is prescribed on the left (or inflow) boundary. In the forward problem, $\xi \equiv 0$ is used at the outlet, whereas in the inverse problem, the bathymetry is prescribed at the inlet.

First, we solve the forward problem with the initial condition $\xi \equiv 0$ and $\boldsymbol{Hu} \equiv [4\ 0]^T$. The steady-state result (as presented in [12]) is subsequently used as input for the inverse problem, where the initial bathymetry is set to $b \equiv -2$, and initial momentum is as in the forward problem. Running the code until the change in bathymetry between pseudo-time steps becomes sufficiently small, we obtain a very accurate reconstruction (the L^∞ error is $7.85 \cdot 10^{-6}$ m). Excellent results are also obtained if a non-stationary free surface elevation is used as input for the inverse problem, similarly to the example considered in [12].

To study the effect of noisy input data on the free surface elevation, we add random perturbations ranging in $(-10^{-4}\,\text{m}, 10^{-4}\,\text{m})$ to the free surface values in each grid vertex. Figure 2 (left) shows a typical reconstruction result for such a case. The amplification of data errors in the reconstruction indicates the ill-posedness of the inverse problem, and further study shows that the reconstruction error is even worse on refined grids. However, interesting effects can be observed if one substitutes the 'noisy' result shown in Fig. 2 (left) as the bathymetry for the forward problem: a surface elevation field differing from the original perturbed steady state is produced. Remarkably, using this new steady state as input for the inverse problem results in the exact same oscillatory steady-state bathymetry as in Fig. 2 (left). We attribute this phenomenon to the space relation $\mathbb{W}_h \subset \mathbb{V}_h$: proper reconstruction of free surface elevation $\xi_h \in \mathbb{V}_h$ from a solution $b_h \in \mathbb{W}_h$ of the inverse problem may be impossible due to the larger DG space $\mathbb{V}_h$. On the other hand, the continuous bathymetry seems to be uniquely determined by the free surface and the inflow boundary condition as long as the velocities are non-zero—which is an encouraging result.

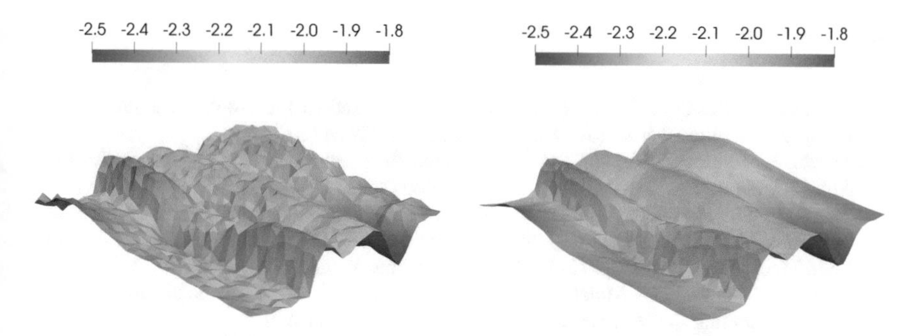

Fig. 2 Bathymetry reconstruction from noisy data without (left) and with (right) artificial diffusion

Finally, we demonstrate how to improve the oscillatory reconstruction via inclusion of diffusive terms: so far the parameter ε was set to zero. From heuristic testing we found the best possible reconstruction is possible with $\varepsilon = 0.08\,\mathrm{m}^2/\mathrm{s}$. The result can be seen in Fig. 2 (right); it indicates that the influence of flawed data can be filtered out by our regularization approach. Furthermore, we are able to reduce the required number of pseudo-time-stepping iterations by a factor of around 15 due to the smoothing properties of the artificial diffusive term. Corresponding results for reconstruction from noisy surface elevation were obtained on refined grids. However, the regularization parameter ε has to be chosen much larger. In some cases, the steady-state convergence behavior can also be improved by decreasing the pseudo-time step. In our experience, promising results can be obtained even for $\varepsilon \to 0$.

5 Conclusion and Outlook

The main highlight of this work is a combined CG-DG finite element method for SWE-based reconstruction of bottom topography from surface elevation data. The use of a continuous finite element space $\mathbb{W}_h \subset \mathbb{V}_h$ for the bathymetry produces a realistic number of constraints and ensures compatibility to the DG scheme for the hyperbolic forward problem. A regularization term is added to the discretized continuity equation of the inverse problem to obtain stable steady state solutions. The presented numerical examples demonstrate the potential of the proposed methodology. Further work is required to study the sensitivity of results to the choice of the regularization parameter and definition of the artificial diffusion operator. These studies may involve theoretical investigations, as well as applications to rivers with well-explored bathymetry and/or comparison to laboratory experiments.

Acknowledgement This research was supported by the German Research Association (DFG) under grant AI 117/2-1 (KU 1530/12-1).

References

1. Aizinger, V., Dawson, C.: A discontinuous Galerkin method for two-dimensional flow and transport in shallow water. Adv. Water Resour. **25**(1), 67–84 (2002)
2. Becker, R., Hansbo, P.: A simple pressure stabilization method for the Stokes equation. Int. J. Numer. Methods Biomed. Eng. **24**(11), 1421–1430 (2008)
3. Brezzi, F., Pitkäranta, J.: On the stabilization of finite element approximations of the Stokes equations. In: Hackbusch, W. (ed.) Efficient Solutions of Elliptic Systems. Notes on Numerical Fluid Mechanics, vol. 10, pp. 11–19. Vieweg+Teubner Verlag, Wiesbaden (1984)
4. Chan, T.F., Golub, G.H., Mulet, P.: A nonlinear primal-dual method for total variation-based image restoration. SIAM J. Sci. Comput. **20**(6), 1964–1977 (1999)
5. Cushman-Roisin, B., Beckers, J.-M.: Introduction to Geophysical Fluid Dynamics: Physical and Numerical Aspects. Academic, Cambridge (2011)
6. Frank, F., Reuter, B., Aizinger, V., Knabner, P.: FESTUNG: A MATLAB / GNU Octave toolbox for the discontinuous Galerkin method, part I: diffusion operator. Comput. Math. Appl. **70**(1), 11–46 (2015)
7. Gessese, A.F., Sellier, M.: A direct solution approach to the inverse shallow-water problem. Math. Probl. Eng. **2012**, 18 (2012)
8. Gessese, A.F., Sellier, M., Van Houten, E., Smart, G.: Reconstruction of river bed topography from free surface data using a direct numerical approach in one-dimensional shallow water flow. Inverse Prob. **27**(2), 025001 (2011)
9. Gessese, A.F., Smart, G., Heining, C., Sellier, M.: One-dimensional bathymetry based on velocity measurements. Inverse Prob. Sci. Eng. **21**(4), 704–720 (2013)
10. Gessese, A.F., Wa, K.M., Sellier, M.: Bathymetry reconstruction based on the zero-inertia shallow water approximation. Theor. Comput. Fluid Dyn. **27**(5), 721–732 (2013)
11. Gottlieb, S., Shu, C.-W., Tadmor, E.: Strong stability-preserving high-order time discretization methods. SIAM Rev. **43**(1), 89–112 (2001)
12. Hajduk, H.: Numerical investigation of direct bathymetry reconstruction based on a modified shallow-water model. Master's thesis, University Erlangen-Nuremberg (2017)
13. Hajduk, H., Hodges, B.R., Aizinger, V., Reuter, B.: Locally filtered transport for computational efficiency in multi-component advection-reaction models. Environ. Model. Softw. **102**, 185–198 (2018)
14. Hilldale, R.C., Raff, D.: Assessing the ability of airborne lidar to map river bathymetry. Earth Surf. Process. Landf. **33**(5), 773–783 (2008)
15. Jaust, A., Reuter, B., Aizinger, V., Schütz, J., Knabner, P.: FESTUNG: A MATLAB/GNU Octave toolbox for the discontinuous Galerkin method, part III: hybridized discontinuous Galerkin (HDG) formulation. Comput. Math. Appl. **75**(12), 4505–4533 (2018). https://doi.org/10.1016/j.camwa.2018.03.045
16. Plant, W.J., Branch, R., Chatham, G., Chickadel, C.C., Hayes, K., Hayworth, B., Horner-Devine, A., Jessup, A., Fong, D.A., Fringer, O.B., Giddings, S.N., Monismith, S., Wang, B.: Remotely sensed river surface features compared with modeling and in situ measurements. J. Geophys. Res. Oceans **114**(C11002) (2009)
17. Reuter, B., Aizinger, V., Wieland, M., Frank, F., Knabner, P.: FESTUNG: A MATLAB / GNU Octave toolbox for the discontinuous Galerkin method. Part II: advection operator and slope limiting. Comput. Math. Appl. **72**(7), 1896–1925 (2016)
18. Roberts, A.C.B.: Shallow water bathymetry using integrated airborne multi-spectral remote sensing. Int. J. Remote Sens. **20**(3), 497–510 (1999)
19. Vreugdenhil, C.B.: Numerical Methods for Shallow-Water Flow. Springer Science & Business Media, Berlin (1994)
20. Westaway, R.M., Lane, S.N., Hicks, D.M.: The development of an automated correction procedure for digital photogrammetry for the study of wide, shallow, gravel-bed rivers. Earth Surf. Process. Landf. **25**(2), 209–226 (2000)

Enabling Scalable Multifluid Plasma Simulations Through Block Preconditioning

Edward G. Phillips, John N. Shadid, Eric C. Cyr, and Sean T. Miller

Abstract Recent work has demonstrated that block preconditioning can scalably accelerate the performance of iterative solvers applied to linear systems arising in implicit multiphysics PDE simulations. The idea of block preconditioning is to decompose the system matrix into physical sub-blocks and apply individual specialized scalable solvers to each sub-block. It can be advantageous to block into simpler segregated physics systems or to block by discretization type. This strategy is particularly amenable to multiphysics systems in which existing solvers, such as multilevel methods, can be leveraged for component physics and to problems with disparate discretizations in which scalable monolithic solvers are rare. This work extends our recent work on scalable block preconditioning methods for structure-preserving discretizatons of the Maxwell equations and our previous work in MHD system solvers to the context of multifluid electromagnetic plasma systems. We argue how a block preconditioner can address both the disparate discretization, as well as strongly-coupled off-diagonal physics that produces fast time-scales (e.g. plasma and cyclotron frequencies). We propose a block preconditioner for plasma systems that allows reuse of existing multigrid solvers for different degrees of freedom while capturing important couplings, and demonstrate the algorithmic scalability of this approach at time-scales of interest.

Sandia National Laboratories is a multimission laboratory managed and operated by National Technology & Engineering Solutions of Sandia, LLC, a wholly owned subsidiary of Honeywell International Inc., for the U.S. Department of Energy's National Nuclear Security Administration under contract DE-NA0003525. This paper describes objective technical results and analysis. Any subjective views or opinions that might be expressed in the paper do not necessarily represent the views of the U.S. Department of Energy or the United States Government.

E. G. Phillips (✉)
Plasma Theory and Simulation Department, Sandia National Laboratories, Albuquerque, NM, USA
e-mail: egphill@sandia.gov

J. N. Shadid · E. C. Cyr · S. T. Miller
Center for Computing Research, Sandia National Laboratories, Albuquerque, NM, USA

H. van Brummelen et al. (eds.), *Numerical Methods for Flows*,
Lecture Notes in Computational Science and Engineering 132,
https://doi.org/10.1007/978-3-030-30705-9_21

Keywords Multiphysics · Block preconditioning · Mixed discretizations

1 Introduction

This work considers the development of a robust preconditioning strategy for linear systems arising as a result of discretization and linearization (e.g. Newton' or fixed-point iteration) of implicit multifluid continuum plasma models. The model considered includes a set of Euler equations for each tracked species–which may include electrons and various ion and neutral species–coupled together through electromagnetic source terms. The electromagnetics are governed by the Maxwell equations with current defined by the fluid momenta [6, 16, 36]. The equations are nonlinear and may be strongly coupled, manifesting in dynamics over numerous length- and time-scales which may range over many orders of magnitude in practice [8]. Consequently, time-scales of interest in a simulation can be much longer than those dictated by explicit time-stepping methods, and in these cases implicit methods may be the only feasible way to obtain timely solutions. When implicit methods are applied, operators corresponding to fast time-scales with respect to the discrete time-step give rise to stiff modes in the linear system, and linear solvers must account for these time-scales to obtain scalable, robust performance. To further complicate the situation, fast time-scales in the plasma system often come from off-diagonal interactions between electromagnetic and fluid degrees of freedom. Additionally, edge and face discretizations may be employed for the electromagnetics in order to enforce Gauss' law ($\nabla \cdot \mathbf{B} = 0$) to machine precision [20, 27]. With fluids discretized by nodal finite elements, this mixed discretization poses an additional challenge to linear solvers, as monolithic preconditioners [1, 25, 34, 35] are difficult to adapt to disparate discretizations.

Block preconditioners, which segregate the system matrix by physical degrees of freedom, are well-suited to address the particular difficulties of multifluid plasma systems. They allow segregation based on discretization type such that scalable solvers for particular discretizations may be leveraged. By additionally segregating according species, block preconditioning allows for the additional flexibility of treating large disparities in mass ratio as would be the case for electrons, ions and heavy neutral species. Finally, important off-diagonal coupling can be captured in Schur complement operators on the block diagonal of the preconditioner. By devising effective and inexpensive (in terms of CPU time and memory) approximations to Schur complements, stiff off-diagonal physics can be represented in a partitioned way. Block preconditioners have been shown to be an effective tool in solving linear systems arising from mixed discretizations of the Navier-Stokes equations [4, 5, 13] and the Maxwell equations [2, 15, 32], as well as multiphysics systems such as magnetohydrodynamics [3, 9–11, 26, 30, 31].

We proceed by first defining a simplified PDE model for multifluid electromagnetic plasmas that we analyze here, followed by a brief discussion of the compatible discretization. We then propose a block preconditioner for the resulting linear systems, analyzing it in terms of time-scales and keeping in mind the need for performant sub-solves. This is followed by some results demonstrating the robustness and scalability of the preconditioning methodology and conclusions.

2 Governing Equations and Discretization

The complex structure of multifluid plasma systems is manifest even in simplified models. For simplicity of presentation, we restrict this work to consider collisionless two-fluid plasmas. The governing equations for such plasmas can be written as

$$\frac{\partial(\mathbf{M}_\alpha)}{\partial t} + \nabla\cdot(\mathbf{M}_\alpha\otimes\mathbf{u}_\alpha + p_\alpha I) - \frac{q_\alpha}{m_\alpha}\rho_\alpha(\mathbf{E}+\mathbf{u}_\alpha\times\mathbf{B}) = \mathbf{0}, \tag{1a}$$

$$\frac{\partial\rho_\alpha}{\partial t} + \nabla\cdot(\mathbf{M}_\alpha) = 0, \tag{1b}$$

$$\frac{\partial(\epsilon\mathbf{E})}{\partial t} - \nabla\times\left(\frac{1}{\mu}\mathbf{B}\right) + \sum_\alpha \frac{q_\alpha}{m_\alpha}\mathbf{M}_\alpha = \mathbf{0}, \tag{1c}$$

$$\frac{\partial\mathbf{B}}{\partial t} + \nabla\times\mathbf{E} = \mathbf{0}, \tag{1d}$$

where $\alpha = e, i$ represent electrons and ions [6, 16, 36]. The unknowns are density ρ_α, momentum $\mathbf{M}_\alpha = \rho_\alpha\mathbf{u}_\alpha$, electric field $\mathbf{E}$, and magnetic induction $\mathbf{B}$. Physical parameters include species mass m_α, species charge q_α, permittivity of free space ϵ, and permeability of free space μ. For this study, a simplifying assumption of an isentropic equation of state is made such that the pressure is defined as a function of density as

$$p_\alpha = \rho_\alpha^\gamma, \tag{2}$$

where the heat capacity ratio γ is a constant. More complex equations of state may be used which introduce an energy equation and additional time-scales, but with the isentropic simplification, key complexities of the multifluid plasma model are retained in a reduced set of equations. Note that if the involutions

$$\nabla\cdot\mathbf{B} = 0, \qquad \nabla\cdot\mathbf{E} = \rho_c = \sum_\alpha \frac{q_\alpha}{m_\alpha}\rho_\alpha, \tag{3}$$

are satisfied at time $t = 0$, then they are implicitly satisfied by Eqs. (1) for all time.

While these equations do not involve the collisional time-scales of a full 5-moment model, there are still several time-scales represented as summarized in Table 1. The light wave time-scale arises from the coupling between $\mathbf{E}$ and $\mathbf{B}$. The advection time-scales are produced by a diagonal coupling/interaction related to the operator $\nabla\cdot(\mathbf{M}_\alpha\otimes\mathbf{u}_\alpha)$. Sound waves arise due to the coupling of momentum

Table 1 Time-scales in the isentropic, collisionless model and their associated CFLs

Physical phenomenon	CFL notation	CFL definition
Light wave	CFL_{EM}	$c\frac{\Delta t}{\Delta x}$
Species advection	$CFL_{\mathbf{u}_\alpha}$	$\frac{\Delta t}{\Delta x}\|\|\bar{\mathbf{u}}_\alpha\|\|$
Species sound wave	CFL_{s_α}	$\frac{\Delta t}{\Delta x}\sqrt{\gamma\bar{\rho}_\alpha^{\gamma-1}}$
Species plasma oscillation	$CFL_{\omega_{p,\alpha}}$	$\Delta t\sqrt{\frac{\bar{\rho}_\alpha q_e{}^2}{\epsilon m_\alpha{}^2}}$
Species cyclotron frequency	$CFL_{\omega_{c,e}}$	$\Delta t\frac{\|\|\bar{\mathbf{B}}\|\|q_\alpha}{m_\alpha}$

and density through the density gradient. The plasma oscillation time-scales come from the coupling between momentum and electric field. Finally, the cyclotron frequencies arise from the operator $\frac{q_\alpha}{m_\alpha}\mathbf{M}_\alpha \times \mathbf{B}$. Each of these time-scales has an associated approximate explicit stability bound (CFL) which can be obtained from physical parameters, discretization parameters (mesh size Δx and time-step Δt), and linearized solution fields in a Newton iteration ($\bar{\rho}_\alpha$, $\bar{\mathbf{u}}_\alpha$, $\bar{\mathbf{B}}$). In general, electron time-scales tend to be faster than ion time-scales due to the larger mass of ions, and the separation of these time-scales is directly related to the mass ratio $\frac{m_i}{m_e}$. Interesting to note is that the light wave and advection CFLs are proportional to $\frac{1}{\Delta x}$ whereas the plasma oscillation and cyclotron CFLs are independent of the spatial discretization. This means that the relative speed of different physics is dependent on mesh refinement as well as physical constants. It is clear then that time-scales may range over many orders of magnitude and that they may be ordered differently depending on the problem and the discretization, and a robust preconditioner must be resilient to these changes.

Consider a finite element discretization of Eqs. (1) in which equal order Lagrangian nodal elements are used for momenta and densities, Nédélec edge elements [28, 29] for the electric field, and Raviart-Thomas face elements [33] for the magnetic induction. When discretized in time using a multistage or multistep implicit integrator and applying a Newton linearization, a linear system of the form

$$\begin{pmatrix} F_e & G_{\rho_e}^{\mathbf{M}_e} & 0 & 0 & Q_{\mathbf{E}}^{\mathbf{M}_e} & Q_{\mathbf{B}}^{\mathbf{M}_e} \\ B & Q_\rho & 0 & 0 & 0 & 0 \\ 0 & 0 & F_i & G_{\rho_i}^{\mathbf{M}_i} & Q_{\mathbf{E}}^{\mathbf{M}_i} & Q_{\mathbf{B}}^{\mathbf{M}_i} \\ 0 & 0 & B & Q_\rho & 0 & 0 \\ Q_{\mathbf{M}_e}^{\mathbf{E}} & 0 & Q_{\mathbf{M}_i}^{\mathbf{E}} & 0 & Q_{\mathbf{E}} & -\hat{K}^t \\ 0 & 0 & 0 & 0 & K & Q_{\mathbf{B}} \end{pmatrix} \begin{pmatrix} \Delta\mathbf{M}_e \\ \Delta\rho_e \\ \Delta\mathbf{M}_i \\ \Delta\rho_i \\ \Delta\mathbf{E} \\ \Delta\mathbf{B} \end{pmatrix} = \begin{pmatrix} R_{\mathbf{M}_e} \\ R_{\rho_e} \\ R_{\mathbf{M}_i} \\ R_{\rho_i} \\ R_{\mathbf{E}} \\ R_{\mathbf{B}} \end{pmatrix} \quad (4)$$

must be solved at each Newton step, where the component linear operators are defined in Table 2.

Table 2 Discrete operators appearing in the time discretized Newton linearization of the multi-fluid plasma equations and the corresponding continuous operators

Discrete operator	Semi-discrete operator
$F_\alpha \mathbf{M}_\alpha$	$\frac{1}{\Delta t}\mathbf{M}_\alpha + \nabla\cdot(\mathbf{M}_\alpha \otimes \bar{\mathbf{u}}_\alpha + \bar{\mathbf{u}}_\alpha \otimes \mathbf{M}_\alpha) - \frac{q_\alpha}{m_\alpha}\mathbf{M}_\alpha \times \bar{\mathbf{B}}$
$G_\alpha \rho_\alpha$	$\nabla\cdot(\rho_\alpha \bar{\mathbf{u}}_\alpha \otimes \bar{\mathbf{u}}_\alpha) + \gamma \bar{\rho}_\alpha^{\gamma-1}\nabla\rho_\alpha + \gamma\nabla\left(\bar{\rho}_\alpha^{\gamma-1}\right)\rho_\alpha - \frac{q_\alpha}{m_\alpha}\rho_\alpha \bar{\mathbf{E}}$
$Q_{\mathbf{E}}^{\mathbf{M}_\alpha}\mathbf{E}$	$-\frac{q_\alpha}{m_\alpha}\bar{\rho}_\alpha \mathbf{E}$
$Q_{\mathbf{B}}^{\mathbf{M}_\alpha}\mathbf{B}$	$-\frac{q_\alpha}{m_\alpha}\bar{\mathbf{M}}_\alpha \times \mathbf{B}$
$B\mathbf{M}_\alpha$	$\nabla\cdot\mathbf{M}_\alpha$
$Q_\rho \rho_\alpha$	$\frac{1}{\Delta t}\rho_\alpha$
$Q_{\mathbf{M}_\alpha}^{\mathbf{E}}\mathbf{M}_\alpha$	$\frac{q_\alpha}{m_\alpha}\mathbf{M}_\alpha$
$Q_{\mathbf{E}}\mathbf{E}$	$\frac{\epsilon}{\Delta t}\mathbf{E}$
$-\hat{K}^t\mathbf{B}$	$-\nabla\times\left(\frac{1}{\mu}\mathbf{B}\right)$
$K\mathbf{E}$	$\nabla\times\mathbf{E}$
$Q_{\mathbf{B}}\mathbf{B}$	$\frac{1}{\Delta t}\mathbf{B}$

Bars indicate the value of the solution at the previous Newton step

3 Block Preconditioner Definition

As noted above, one advantage of block preconditioning is that degrees of freedom may be segregated based on discretization type. As such, we define the vector of nodal degrees of freedom that consists of the hydrodynamic unknowns $\mathbf{f} = (\mathbf{M}_e, \rho_e, \mathbf{M}_i, \rho_i)$. Then, system (4) can be rewritten as

$$\begin{pmatrix} F_{\mathbf{f}} & Q_{\mathbf{E}}^{\mathbf{f}} & Q_{\mathbf{B}}^{\mathbf{f}} \\ Q_{\mathbf{f}}^{\mathbf{E}} & Q_{\mathbf{E}} & -\hat{K}^t \\ 0 & K & Q_{\mathbf{B}} \end{pmatrix} \begin{pmatrix} \Delta\mathbf{f} \\ \Delta\mathbf{E} \\ \Delta\mathbf{B} \end{pmatrix} = \begin{pmatrix} R_{\mathbf{f}} \\ R_{\mathbf{E}} \\ R_{\mathbf{B}} \end{pmatrix}, \tag{5}$$

or $\mathcal{A}\mathbf{x} = \mathbf{b}$. This 3×3 block structure has been selected to allow optimal multilevel solvers to be applied to the nodal, edge, and face degrees of freedom. The $F_{\mathbf{f}}$ block can be sub-blocked according to species and/or momentum versus density, or a monolithic nodal multigrid method can be applied to $F_{\mathbf{f}}$ as a whole.

Strategically taking advantage of the block structure $\mathbf{B}$ is eliminated from both the electric field ($\mathbf{E}$) and hydrodynamics ($\mathbf{f}$) equations. This is achieved discretely by adding $\hat{K}^t Q_{\mathbf{B}}^{-1}$ times the $\mathbf{B}$ equation to the $\mathbf{E}$ equation, and $-Q_{\mathbf{B}}^{\mathbf{f}} Q_{\mathbf{B}}^{-1}$ times the

B equation to the **F** equation. At the level of the preconditioner, this is analogous to observing that

$$\mathcal{A} = \begin{pmatrix} I & 0 & Q_{\mathbf{E}}^{\mathbf{f}} Q_{\mathbf{B}}^{-1} \\ 0 & I & -\hat{K}^t Q_{\mathbf{B}}^{-1} \\ 0 & 0 & I \end{pmatrix} \begin{pmatrix} F_{\mathbf{f}} & Q_{\mathbf{E}}^{\mathbf{f}} - Q_{\mathbf{E}}^{\mathbf{f}} Q_{\mathbf{B}}^{-1} K & 0 \\ Q_{\mathbf{f}}^{\mathbf{E}} & S_{\mathbf{E}} & 0 \\ 0 & K & Q_{\mathbf{B}} \end{pmatrix}, \tag{6}$$

where the electric field Schur complement is defined as

$$S_{\mathbf{E}} = Q_{\mathbf{E}} + \hat{K}^t Q_{\mathbf{B}}^{-1} K. \tag{7}$$

This is exactly the Schur complement associated with a compatible discretization of the Maxwell equations, and several specialized solvers have been developed for operators of this type [7, 18, 19, 22, 23, 32]. Note that it is impractical to use the exact inverse $Q_{\mathbf{B}}^{-1}$ in $Q_{\mathbf{E}}^{\mathbf{f}} Q_{\mathbf{B}}^{-1} K$ and $S_{\mathbf{E}}$, so, in practice, we use a diagonal approximation of $Q_{\mathbf{B}}^{-1}$. The use of a diagonal approximation to mass matrices is motivated by the fact that nodal mass matrices are spectrally equivalent to their diagonal [37] and that edge and face mass matrices have approximately the same magnitude as their diagonal [32]. Looking at the rightmost matrix in (6), one can associate each of the time-scales in Table 1 with an operator. The light wave is captured in the Schur complement $S_{\mathbf{E}}$, advection is contained in $F_{\mathbf{f}}$, and plasma frequencies come about through the 2×2 interaction of the **f** and **E** operators. It is also evident that cyclotron dynamics are not manifest in the off-diagonal operators, but only in the $F_{\mathbf{f}}$ operator. As noted above a further decoupling of the fluid species by a mass ratio ordering is also possible since the lower mass electron species will have much less inertia and react much faster to a strong electric and/or magnetic field.

The final step in devising the preconditioner is to account for the plasma oscillation time-scales. This can be achieved by eliminating **E** from the **F** equation, via

$$\mathcal{A} = \begin{pmatrix} I & S_{\mathbf{E}}^{-1}(Q_{\mathbf{E}}^{\mathbf{f}} - Q_{\mathbf{E}}^{\mathbf{f}} Q_{\mathbf{B}}^{-1} K) & Q_{\mathbf{E}}^{\mathbf{f}} Q_{\mathbf{B}}^{-1} \\ 0 & I & -\hat{K}^t Q_{\mathbf{B}}^{-1} \\ 0 & 0 & I \end{pmatrix} \begin{pmatrix} S_{\mathbf{f}} & 0 & 0 \\ Q_{\mathbf{f}}^{\mathbf{E}} & S_{\mathbf{E}} & 0 \\ 0 & K & Q_{\mathbf{B}} \end{pmatrix}, \tag{8}$$

where the fluid Schur complement is

$$S_{\mathbf{f}} = F_{\mathbf{f}} - Q_{\mathbf{f}}^{\mathbf{E}} S_{\mathbf{E}}^{-1} (Q_{\mathbf{E}}^{\mathbf{f}} - Q_{\mathbf{E}}^{\mathbf{f}} Q_{\mathbf{B}}^{-1} K). \tag{9}$$

The difficulty with $S_{\mathbf{f}}$ lies in the embedded inverse of $S_{\mathbf{E}}$ which is not well-approximated by a diagonal if CFL_{EM} is large. If $CFL_{EM} \ll 1$, then it can be shown that $S_{\mathbf{E}} \approx Q_{\mathbf{E}}$ and that $Q_{\mathbf{E}}^{\mathbf{f}} Q_{\mathbf{B}}^{-1} K$ is negligible compared to $Q_{\mathbf{E}}^{\mathbf{f}}$. In that case

$$S_{\mathbf{f}} \approx F_{\mathbf{f}} - Q_{\mathbf{f}}^{\mathbf{E}} Q_{\mathbf{E}}^{-1} Q_{\mathbf{E}}^{\mathbf{f}}, \tag{10}$$

which is the Schur complement obtained from the electrostatic coupling between only **f** and **E**, and thus represents the plasma oscillation time-scales. It can be shown that, for a particular species α, $-Q_{\mathbf{M}_\alpha}^{\mathbf{E}} Q_{\mathbf{E}}^{-1} Q_{\mathbf{E}}^{\mathbf{M}_\alpha} \mathbf{M}_\alpha$ is a discretization of

$$\frac{\Delta t}{\epsilon} \frac{q_\alpha{}^2}{m_\alpha{}^2} \bar{\rho}_\alpha \mathbf{M}_\alpha, \tag{11}$$

a semi-discrete operator directly capturing the plasma frequency. Thus, the additional terms in (9) can be viewed as perturbations of the linear plasma oscillation when light waves are not resolved. Furthermore, the magnitudes of operators can be analyzed to show that

$$||Q_{\mathbf{M}_\alpha}^{\mathbf{E}} S_{\mathbf{E}}^{-1} Q_{\mathbf{E}}^{\mathbf{M}_\alpha}|| \approx \frac{\Delta t}{\epsilon} \frac{q_\alpha{}^2}{m_\alpha{}^2} \bar{\rho}_\alpha \left[1 + CFL_{EM}^2\right]^{-1}, \tag{12}$$

meaning that as CFL_{EM} increases, the relative contribution of the perturbed plasma oscillation operator to the fluid Schur complement should decrease. In this sense, $S_{\mathbf{f}}$ is well approximated by (10) when CFL_{EM} is small, and the perturbation to $F_{\mathbf{f}}$ need not be approximated well when CFL_{EM} is large. Given this analysis, a diagonal approximation of $S_{\mathbf{E}}$ is justified inside of $S_{\mathbf{f}}$. That way, for small CFL_{EM}, where $S_{\mathbf{E}}$ is approximately the mass operator $Q_{\mathbf{E}}$, the approximation is good, and when CFL_{EM} is large, the approximation is lower fidelity, but the magnitude of $Q_{\mathbf{f}}^{\mathbf{E}} S_{\mathbf{E}}^{-1}(Q_{\mathbf{E}}^{\mathbf{f}} - Q_{\mathbf{E}}^{\mathbf{f}} Q_{\mathbf{B}}^{-1} K)$ is much smaller.

In summary, we use an approximation of the lower triangular factor in (8) as the preconditioner, i.e.

$$\mathcal{P} = \begin{pmatrix} \hat{S}_{\mathbf{f}} & 0 & 0 \\ Q_{\mathbf{f}}^{\mathbf{E}} & \hat{S}_{\mathbf{E}} & 0 \\ 0 & K & Q_{\mathbf{B}} \end{pmatrix}, \qquad \begin{aligned} \hat{S}_{\mathbf{E}} &= Q_{\mathbf{E}} + \hat{K}^t \bar{Q}_{\mathbf{B}}^{-1} K, \\ \hat{S}_{\mathbf{f}} &= F_{\mathbf{f}} - Q_{\mathbf{f}}^{\mathbf{E}} \bar{S}_{\mathbf{E}}^{-1} (Q_{\mathbf{E}}^{\mathbf{f}} - Q_{\mathbf{E}}^{\mathbf{f}} \bar{Q}_{\mathbf{B}}^{-1} K), \end{aligned} \tag{13}$$

where bars over linear operators indicate diagonal approximations. For sub-block solvers, a basic multigrid method with an inexpensive smoother suffices for the mass matrix $Q_{\mathbf{B}}$, and we use the augmentation-based Maxwell solver described in [32] for $\hat{S}_{\mathbf{E}}$. While this method was designed specifically for the case of large CFL_{EM}, any of the multigrid methods described in [7, 18, 19, 22, 23] may be used for $\hat{S}_{\mathbf{E}}$. Many options are available for solving the system $\hat{S}_{\mathbf{f}}$. In many cases, it may be advantageous to use a block method for this subsystem as this allows treatment of fast species differently from slow species. For collisional plasmas, additional terms coupling the different species appear within $\hat{S}_{\mathbf{f}}$, and a carefully defined Schur complement can handle these effects. For warm plasmas, energy is another required degree of freedom in the **f** vector, and off-diagonal acoustic time-scales appear. Monolithic multilevel methods [25, 34, 35] are applicable to $\hat{S}_{\mathbf{f}}$ because all degrees of freedom are co-located. These methods are also appealing because they do not require physics-specific analysis to define effective Schur complements. For the proof of concept in this work, a monolithic AMG is used in a fairly black box fashion to solve the $\hat{S}_{\mathbf{f}}$ system.

4 Computational Results

In this section, we demonstrate the effectiveness and scalability of the proposed preconditioner on two test problems. The dynamics of interest in the first test problem operates at time-scales close to the speed of light with stiff modes associated with plasma and cyclotron frequencies. The second test problem is driven by ion-acoustic dynamics, and the limiting time-scale is associated with light waves. For both cases, the nonlinear stopping criterion is a relative reduction of 10^{-3} of the nonlinear residual. Preconditioned GMRES is used with a relative residual stopping tolerance of 10^{-4}. The physics driver is the application code Drekar [35] which is built on the Trilinos framework[17]. In particular, Nédélec edge elements and nodal elements are provided by the discretization package Intrepid and degrees of freedom are managed by the Panzer package. The Aztec [21] implementation of GMRES is used as the linear solver. Block preconditioners are implemented using the Teko package [12], and the ML package [14] is employed for all multilevel subsolves. One V-cycle is used for each AMG solve. Four Gauss-Seidel smoother sweeps are used for mass matrix solves. As in [32], an overlapping domain decomposition smoother with an ILU(0) on each subdomain was needed for the augmented edge solve used in approximating $\hat{S}_{\mathbf{E}}^{-1}$. A monolithic AMG [25, 34, 35] was applied to approximate $\hat{S}_{\mathbf{f}}^{-1}$. In this case, the eight fluid degrees of freedom ($\mathbf{M}_{ex}, \mathbf{M}_{ey}, \mathbf{M}_{ez}, \rho_e, \mathbf{M}_{ix}, \mathbf{M}_{iy}, \mathbf{M}_{iz}, \rho_i$) are co-located allowing a single coarse grid hierarchy to be constructed for all of them. An overlapping domain decomposition ILU(0) smoother was also applied for this routine.

4.1 Current Pulse Test Problem

This test problem represents a cold, collisionless two-fluid plasma driven by an external current pulse. The domain is $[-3000, 3000] \times [87000, 93000]$ with two elements in the z-direction. The current pulse is Gaussian in space and time in the z-direction, defined as

$$\mathbf{J}_z = J_0 e^{-||\boldsymbol{x}-\boldsymbol{x}_0||^2/\ell} e^{-||t-t_0||/\tau}, \tag{14}$$

where $J_0 = 5.0 \times 10^4$ is the magnitude of the current, $\boldsymbol{x}_0 = (0, 90000, 0)$ is the center of the pulse, $\ell = 15{,}625$ governs the spatial width of the pulse, $t_0 = 1.5 \times 10^{-6}$ is the time of the pulse peak, and $\tau = 5.0 \times 10^{-7}$ governs the pulse width in time. These parameters define a strong, fast pulse over a large length-scale.

Additionally, a background magnetic field and a density gradient in the y-direction are supplied as initial conditions

$$\mathbf{B} = \begin{pmatrix} 2.9 \times 10^{-4} \\ 1.2 \times 10^{-4} \\ 0 \end{pmatrix}, \qquad \begin{aligned} &\rho_e = m_e N(y), \quad \rho_i = m_i N(y), \\ &N(y) = 8.8 \times 10^{14} + 8.7 \times 10^{14} \sin\left(\frac{\pi}{2}\frac{y-94500}{10500}\right), \end{aligned} \tag{15}$$

such that solutions are not grid aligned. Realistic values $q_e = -1.6 \times 10^{-19}$ and $m_e = 9.1 \times 10^{-31}$ are used with $q_i = -q_e$ and a mass ratio $m_i/m_e = 1000$. For a cold plasma the pressure is set to zero, such that there are no sound speed dynamics. A snapshot of the electron and ion momenta is plotted in Fig. 1. It can be seen that the plasma expands out from the current pulse in the center of the domain following the magnetic field.

A fixed time-step of $\Delta t = 5.0 \times 10^{-8}$ is used such that the current pulse is well resolved in time. A sequence of uniform meshes with $\Delta x = 240$ on the coarsest mesh and $\Delta x = 7.5$ on the finest mesh was used to obtain the following results. Given these parameters, the various CFL conditions for this problem are summarized in Table 3. When CFL values fall in ranges, they are largest for the finest grid. It can be seen that for this problem, the stiffest time-scale is always the electron plasma frequency. The ion plasma frequency and electron cyclotron frequency are also stiff, but not to the same degree. Because of the large length-scale of this problem, the speed of light and electron advection are resolved on almost all meshes. A weak scaling study of the block preconditioner proposed above is summarized in Table 4. For this study, the number of processors was increased keeping 1250 elements on each processor. The problem was run for 50 time-steps,

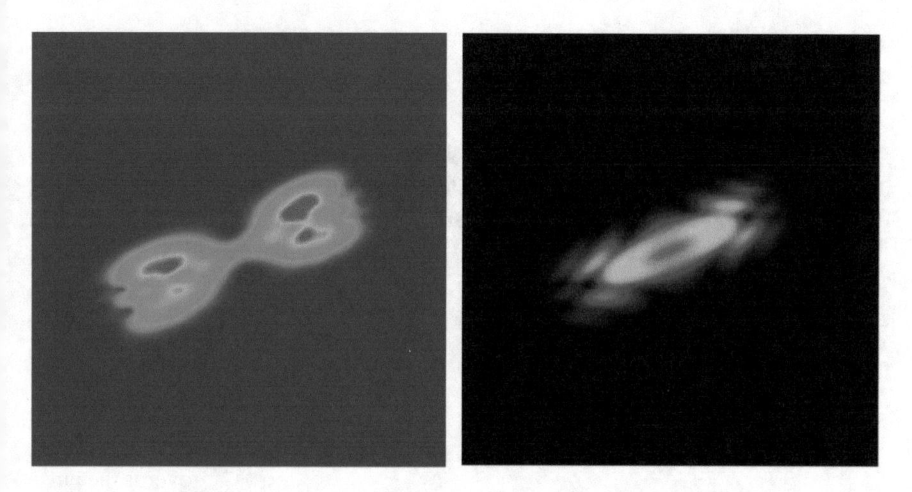

Fig. 1 Magnitude of electron and ion momenta for current pulse test problem at $t = 2.5 \times 10^{-6}$

Table 3 Time-scales in the test problems

Time-scale	Pulse problem	Soliton problem
CFL_{EM}	$[6.25 \times 10^{-2}, 2.0]$	62.5
$CFL_{\mathbf{u}_e}$	$[3.75 \times 10^{-2}, 1.2]$	0.075
$CFL_{\mathbf{u}_i}$	$[3.75 \times 10^{-5}, 1.2 \times 10^{-3}]$	0.05
CFL_{s_e}	0	0.05
CFL_{s_i}	0	1.0
$CFL_{\omega_{p,e}}$	1.2×10^2	[0.78,30.0]
$CFL_{\omega_{p,i}}$	3.8	[0.16,6.0]
$CFL_{\omega_{c,e}}$	2.7	[0.094,3.0]
$CFL_{\omega_{c,i}}$	2.7×10^{-3}	[0.0038,0.12]

Table 4 Preconditioner weak scaling for current pulse test problem

Processors	Total DOFs	Average GMRES iterations	Average setup time	Average solve time
1	18,618	30.36	3.3	3.4
4	72,218	35.93	5.3	5.0
16	284,418	29.75	7.2	9.0
64	1,128,818	28.81	7.5	9.4
256	4,497,618	18.93	8.0	7.0
1024	17,955,218	18.84	8.6	8.0

and averages are taken over all linear solves. It can be seen that preconditioner performance improves in terms of iteration count as the problem size increases. Computation time, in terms of both preconditioner setup and apply time increases slowly with problem size. Both of these increase by a factor of less than 3 when the problem size is increased by a factor of 1024.

4.2 Soliton-Like Test Problem

The second test problem operates at time-scales comparable to those found in MHD simulations in which the dynamics of interest are much slower than the speed of light. This problem is a 2D isentropic variation on the soliton problem defined in [24]. The domain is $[0, 12]^2$ and an initial perturbation in the densities

$$\rho_\alpha = m_\alpha \left(1 + e^{-10||\boldsymbol{x}-\boldsymbol{x}_0||^2}\right) \tag{16}$$

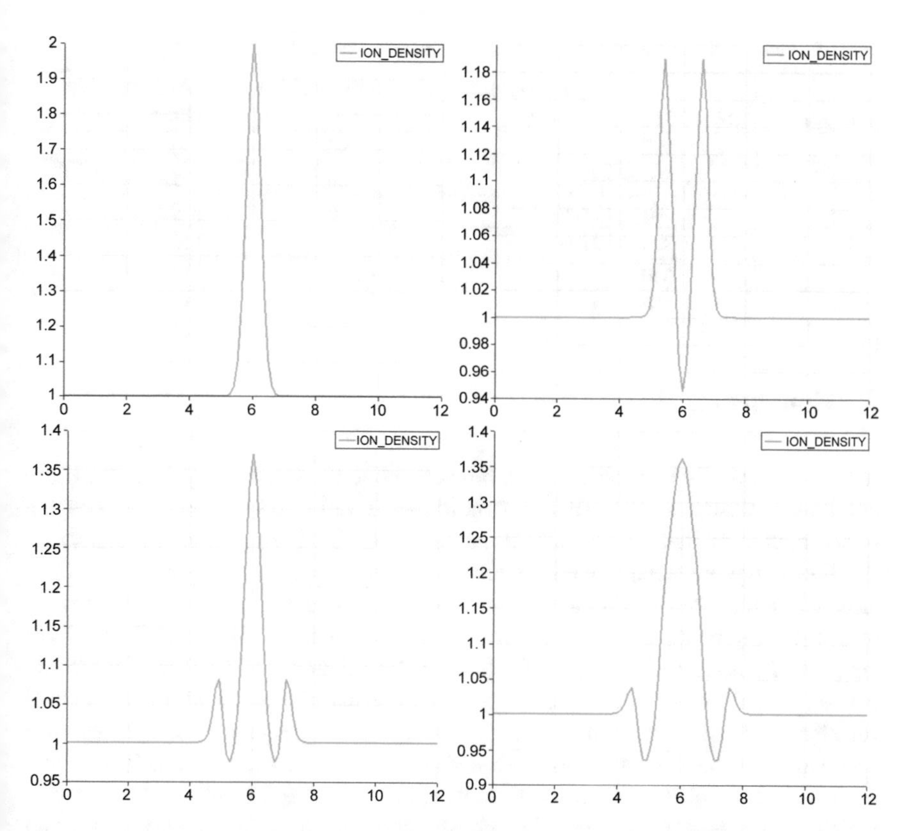

Fig. 2 Ion density plotted along $y = 0$ at $t = 0, 0.3, 0.6, 0.9$ for the soliton-like problem

drives the dynamics, where $\boldsymbol{x} = (6, 6)$. The masses are set to $m_e = 0.04$ and $m_i = 1$. Physical parameters are set such that the speed of light is $c = 250$ and the plasma frequencies are $\omega_{p,e} = 25$ and $\omega_{p,i} = 1$. The adiabatic index is set to $\gamma = 2$ such that the electron speed of sound is 0.2 and the ion speed of sound is 4.0. Snapshots of the ion density are plotted in Fig. 2. We consider a sequence of meshes with $\Delta x = 0.48$ on the coarsest mesh and $\Delta x = 0.015$ on the finest mesh.

The dynamics of this problem are dominated by the ion-acoustic wave, and as such we resolve these dynamics by fixing $CFL_{s_i} = 1$, resulting in the CFLs reported in Table 3. It can be seen that the light wave is always stiff, and that the electron plasma frequency CFL is almost always greater than 1. The results of a weak scaling study are presented in Table 5. Again, we can see very good scaling in terms of iteration count and a slight increase in computation time as the mesh is refined.

Table 5 Preconditioner weak scaling for soliton-like test problem

Processors	Total DOFs	Average GMRES iterations	Average setup time	Average solve time
1	17,500	40.81	3.6	4.7
4	70,000	32.77	6.6	5.4
16	280,000	28.46	7.6	9.3
64	1,120,000	31.64	8.0	10.9
256	4,480,000	31.31	8.9	12.5
1024	17,920,000	33.45	10.0	16.4

5 Conclusion

In this work, we have proposed a new block preconditioning strategy for a multifluid continuum plasma model. This preconditioner is informed by an analysis of the linear operators that couple electromagnetics to fluid dynamics. By analyzing the linear algebra as it relates to physical time-scales in the system, we have argued for the use of particular Schur complement approximations, allowing us to use the augmentation based Maxwell solver detailed in [32] as well as a fully coupled algebraic multilevel solver for the fluid degrees of freedom. While the plasma model considered neglects collisional terms, the preconditioner accounts for electromagnetic, plasma frequency, and cyclotron frequency time-scales. Our computational results have demonstrated the effectiveness and scalability of the proposed preconditioner in two different regimes: the large length-scale regime in which light waves are resolved but plasma and cyclotron frequency are fast and approaching the MHD limit in which the electromagnetics result in the stiffest modes. Future work will extend this method to collisional regimes and models including more complex equations of state and provide a more comprehensive set of numerical examples. This work may also be expanded by treating different species separately, thus allowing for cheaper fluid solves that are more informed by the time-scales of particular plasma species.

References

1. Adler, J., Benson, T.R., Cyr, E., MacLachlan, S.P., Tuminaro, R.S.: Monolithic multigrid methods for two-dimensional resistive magnetohydrodynamics. SIAM J. Sci. Comput. **38**(1), B1–B24 (2016)
2. Adler, J., Hu, X., Zikatanov, L.: Robust solvers for Maxwell's equations with dissipative boundary conditions. SIAM J. Sci. Comput. **39**(5), S3–S23 (2016)
3. Badia, S., Martín, A., Planas, R.: Block recursive LU preconditioners for the thermally coupled incompressible inductionless MHD problem. J. Comput. Phys. **274**, 562–591 (2014)
4. Benzi, M., Olshanskii, M.: An augmented Lagrangian-based approach to the Oseen problem. SIAM J. Sci. Comput. **28**, 2095–2113 (2006)

5. Benzi, M., Olshanskii, M., Wang, Z.: Modified augmented lagrangian preconditioners for the incompressible navier-stokes equations. Int. J. Numer. Meth. Fluids **66**, 486–508 (2011)
6. Bittencourt, J.: Fundamentals of Plasma Physics. Springer, Berlin (2004)
7. Bochev, P., Hu, J., Siefert, C., Tuminaro, R.: An algebraic multigrid approach based on a compatible gauge reformulation of Maxwell's equations. SIAM J. Sci. Comput. **31**, 557–583 (2008)
8. Bonoli, P., Curfman McInnes, L.: Report on workshop of the integrated simulations for magnetic fusion energy science. Technical report, DOE Office of Science ASCR and DOE Office of Fusion Energy Sciences (2015)
9. Chacón, L.: An optimal, parallel, fully implicit Newton-Krylov solver for three-dimensional viscoresistive magnetohydrodynamics. Phys. Plasmas **15**, 056103 (2008)
10. Chacón, L.: Scalable parallel implicit solvers for 3D magnetohydrodynamics. J. Phys. Conf. Ser. **125**, 012041 (2008)
11. Cyr, E.C., Shadid, J., Tuminaro, R., Pawlowski, R., Chacón, L.: A new approximate block factorization preconditioner for 2D incompressible (reduced) resistive MHD. SIAM J. Sci. Comput. **35**, B701–B730 (2013)
12. Cyr, E.C., Shadid, J.N., Tuminaro, R.S.: Teko: a block preconditioning capability with concrete example applications in Navier-Stokes and MHD. SIAM J. Sci. Comput. **38**(5), S307–S331 (2016)
13. Elman, H., Howle, V., Shadid, J., Shuttleworth, R., Tuminaro, R.: A taxonomy and comparison of parallel block multi-level preconditioners for the incompressible Navier-Stokes equations. J. Comput. Phys. **227**, 1790–1808 (2008)
14. Gee, M., Siefert, C., Hu, J., Tuminaro, R., Sala, M.: ML 5.0 smoother aggregration user's guide. Technical Report SAND2006-2649, Sandia National Laboratories (2006)
15. Greif, C., Schötzau, D.: Preconditioners for the discretized time-harmonic Maxwell equations in mixed form. Numer. Linear Algebra Appl. **14**, 281–297 (2007)
16. Hakim, A.: Extended MHD modelling with the ten-moment equations. J. Fusion Energ. **27**, 36–43 (2008)
17. Heroux, M., Bartlett, R., Howle, V., Hoekstra, R., Hu, J., Kolda, T., Lehoucq, R., Long, K., Pawlowski, R., Phipps, E., Salinger, A., Thornquist, H., Tuminaro, R., Willenbring, J., Williams, A., Stanley, K.: An overview of the Trilinos project. ACM Trans. Math. Softw. **31**, 397–423 (2005)
18. Hiptmair, R., Xu, J.: Nodal auxiliary space preconditioning in H(Curl) and H(Div) spaces. SIAM J. Numer. Anal. **45**, 2483–2509 (2007)
19. Hu, J., Tuminaro, R., Bochev, P., Garasi, C., Robinson, A.: Toward an h-independent algebraic multigrid method for Maxwell's equations. SIAM J. Sci. Comput. **27**, 1669–1688 (2006)
20. Hu, K., Ma, Y., Xu, J.: Stable finite element methods preserving $\nabla \cdot B = 0$ exactly for MHD models. Numer. Math. (2016). https://doi.org/10.1007/s00211-016-0803-4
21. Hutchinson, S., Prevost, L., Shadid, J.N., Tong, C., Tuminaro, R.S.: Aztec user's guide version 2.0. Technical Report Sand99-8801J, Sandia National Laboratories (1999)
22. Kolev, T., Vassilevski, P.: Parallel auxiliary space AMG for H(curl) problems. J. Comput. Math. **27**(5), 604–623 (2009)
23. Kolev, T., Pasciak, J., Vassilevski, P.: H(curl) auxiliary mesh preconditioning. Numer. Linear Algebra Appl. **15**, 455–471 (2008)
24. Kumar, H., Mishra, S.: Entropy stable numerical schemes for two-fluid plasma equations. J. Sci. Comput. **52**, 401–425 (2012)
25. Lin, P., Shadid, J., Tuminaro, R., Sala, M., Hennigan, G., Pawlowski, R.: A parallel fully-coupled algebraic multilevel preconditioner applied to multiphysics PDE applications: drift-diffusion, flow/transport/reaction, resistive MHD. Int. J. Numer. Meth. Fluids **64**, 1148–1179 (2010)
26. Ma, Y., Hu, K., Hu, X., Xu, J.: Robust preconditioners for incompressible MHD models. J. Comp. Phys. **316**, 721–746 (2016)
27. Monk, P.: Finite Element Methods for Maxwell's Equations. Oxford University Press, Oxford (2003)

28. Nédélec, J.-C.: Mixed finite elements in $\mathbb{R}^3$. Numer. Math. **35**, 315–341 (1980)
29. Nédélec, J.-C.: A new family of mixed finite elements in $\mathbb{R}^3$. Numer. Math. **50**, 57–81 (1986)
30. Phillips, E., Elman, H., Cyr, E.C., Shadid, J., Pawlowski, R.: A block preconditioner for an exact penalty formulation for stationary MHD. SIAM J. Sci. Comput. **36**, B930–B951 (2014)
31. Phillips, E., Shadid, J., Cyr, E., Elman, H., Pawlowski, R.: Block preconditioners for stable mixed nodal and edge finite element representations of incompressible resistive MHD. SIAM J. Sci. Comput. **38**, B1009–B1031 (2016)
32. Phillips, E., Shadid, J., Cyr, E.: Scalable preconditioners for structure preserving discretizations of Maxwell equations in first order form. SIAM J. Sci. Comput. **40**(3), B723–B742 (2018)
33. Raviart, P., Thomas, J.: A mixed finite element method for second order elliptic problems. Lect. Notes Math. **606**, 292–315 (1977)
34. Shadid, J., Pawlowski, R., Banks, J., Chacón, L., Lin, P., Tuminaro, R.: Towards a scalable fully-implicit fully-coupled resistive MHD formulation with stabilized FE methods. J. Comput. Phys. **229**, 7649–7671 (2010)
35. Shadid, J.N., Pawlowski, R.P., Cyr, E.C., Tuminaro, R.S., Chacon, L., Weber, P.D.: Scalable implicit incompressible resistive MHD with stabilized FE and fully-coupled Newton-Krylov-AMG. Comput. Methods Appl. Mech. Eng. **304**, 1–25 (2016)
36. Shumlak, U., Loverich, J.: Approximate Riemann solver for the two-fluid plasma model. J. Comp. Phys. **187**, 620–638 (2003)
37. Wathen, A.: Realistic eigenvalue bounds for the Galerkin mass matrix. IMA J. Numer. Anal. **7**, 449–457 (1987)

The Effort of Increasing Reynolds Number in Projection-Based Reduced Order Methods: From Laminar to Turbulent Flows

Saddam Hijazi, Shafqat Ali, Giovanni Stabile, Francesco Ballarin, and Gianluigi Rozza

Abstract We present in this double contribution two different reduced order strategies for incompressible parameterized Navier-Stokes equations characterized by varying Reynolds numbers. The first strategy deals with low Reynolds number (laminar flow) and is based on a stabilized finite element method during the offline stage followed by a Galerkin projection on reduced basis spaces generated by a greedy algorithm. The second methodology is based on a full order finite volume discretization. The latter methodology will be used for flows with moderate to high Reynolds number characterized by turbulent patterns. For the treatment of the mentioned turbulent flows at the reduced order level, a new POD-Galerkin approach is proposed. The new approach takes into consideration the contribution of the eddy viscosity also during the online stage and is based on the use of interpolation. The two methodologies are tested on classic benchmark test cases.

Keywords Stabilised RB methods · SUPG · Turbulence modelling · Projection based model reduction · Data driven model reduction · Viscous flows

1 Introduction

Nowadays we see an increasing need for numerical simulation of fluid dynamics problems with high Reynolds number. These problems come from different types of applications and fields. This pushes the scientific community to offer new techniques and approaches which can meet also the demand of industry to simulate higher Reynolds number fluid problems [11]. Today, in several situations, there is a need to perform simulations in a multi-query contest (e.g. optimization, uncertainty quantification) with an extremely reduced computational time as a requirement

S. Hijazi (✉) · S. Ali · G. Stabile · F. Ballarin · G. Rozza
mathLab, Mathematics Area, SISSA, Trieste, Italy
e-mail: shijazi@sissa.it; sali@sissa.it; gstabile@sissa.it; fballarin@sissa.it; grozza@sissa.it

H. van Brummelen et al. (eds.), *Numerical Methods for Flows*, Lecture Notes in Computational Science and Engineering 132,
https://doi.org/10.1007/978-3-030-30705-9_22

(real-time control). Therefore, in such situations, the resolution of the governing PDEs using standard discretization techniques may become unaffordable. Hence, reduced order modelling has become an important tool to reduce the computational complexity. This is a double contribution containing two related topics dealing with the efforts of increasing Reynolds number in viscous flows and being able to guarantee stable reduced order methods in parametric problems.

This chapter is organized as follows: in Sect. 2 we define the steady Navier-Stokes equations in strong formulation. In Sect. 3 we present residual based stabilized reduced basis method for parameterized Navier-Stokes problem characterized by low Reynolds number. Section 4 deals with POD-Galerkin reduction for parameterized Navier-Stokes problem in case of higher Reynolds number. In Sect. 5 we show some numerical results for both strategies. Finally in Sect. 6 we summarize the main outcomes of this chapter and we outline some perspectives.

2 Projection Based ROMs

In this section some basic notions of projection based ROMs [13] are recalled. Firstly, the mathematical problem deals with the steady Navier-Stokes equations, and reads as follows:

$$\begin{cases} (\mathbf{u}\cdot\nabla)\mathbf{u} - \nu\Delta\mathbf{u} + \nabla p = \mathbf{0} & \text{in } \Omega, \\ \nabla\cdot\mathbf{u} = 0 & \text{in } \Omega, \\ \mathbf{u} = \mathbf{U}_{in} & \text{on } \partial\Omega_{In}, \\ \mathbf{u} = \mathbf{0} & \text{on } \partial\Omega_0, \\ (\nu\nabla\mathbf{u} - p\mathbf{I})\mathbf{n} = \mathbf{0} & \text{on } \partial\Omega_{Out}, \end{cases} \tag{1}$$

where $\mathbf{u}(\mathbf{x})$ and $p(\mathbf{x})$ are the velocity and pressure fields respectively, $\Omega \subset \mathbb{R}^2$ is a bounded domain, while $\partial\Omega = \partial\Omega_{In} \cup \partial\Omega_0 \cup \partial\Omega_{Out}$ is the boundary of the domain formed by three parts $\partial\Omega_{In}$, $\partial\Omega_0$ and $\partial\Omega_{Out}$ which correspond to the inlet, the physical walls and the outlet respectively, $\mathbf{U}_{in}$ is the velocity at the inlet part of the boundary, and ν is the viscosity of the fluid. Then the problem reads find $\mathbf{u}(\mathbf{x})$ and $p(\mathbf{x})$ which satisfy (1) and lie respectively in the following spaces $\boldsymbol{V} = [\boldsymbol{H}^1(\Omega)]^d$, and $Q = L_0^2(\Omega)$. See [29] for more details.

In the context of this work, the main goal is studying how the flow fields change as a result of the variation of certain parameters. For this reason a parameterized version of (1) will be considered. The set of parameters is denoted by $\boldsymbol{\mu}$ where this vector of parameters lies in the parameter space $\mathbb{P}$, note that $\mathbb{P}$ is compact set in $\mathbb{R}^p$ with p being the length of the vector $\boldsymbol{\mu}$. The parameters can be geometrical or physical or a combination of them [7]. The objective is to be able to compute the velocity and pressure fields for every parameter value inside the parameter space. The cost of doing that operation resorting on full order methods can be prohibitive.

For this reason ROMs [7, 12, 18, 30] have been developed, as an approach to achieve the objective of computing efficiently and accurately the flow fields, when the input parameters are being varied.

One key assumption in ROMs is that the dynamics of the system under study is governed by a reduced number of dominant modes. In other words, the solution to the full order problem lies in a low dimensional manifold that is spanned by the previously mentioned modes [18]. Consequently the velocity and pressure fields can be approximated by decomposing them into linear combination of global basis functions $\boldsymbol{\phi}_i(\mathbf{x})$ and $\chi_i(\mathbf{x})$ (which do not depend on $\boldsymbol{\mu}$) multiplied by unknown coefficients $a_i(\boldsymbol{\mu})$ and $b_i(\boldsymbol{\mu})$, for velocity and pressure respectively, then this approximation reads as follows:

$$\mathbf{u}(\mathbf{x};\boldsymbol{\mu}) \approx \sum_{i=1}^{N_u} a_i(\boldsymbol{\mu})\boldsymbol{\phi}_i(\mathbf{x}), \qquad p(\mathbf{x};\boldsymbol{\mu}) \approx \sum_{i=1}^{N_p} b_i(\boldsymbol{\mu})\chi_i(\mathbf{x}). \tag{2}$$

The reduced basis spaces $\mathbb{V}_{rb} = \text{span}\{\boldsymbol{\phi}_i\}_{i=1}^{N_u}$ and $Q_{rb} = \text{span}\{\chi_i\}_{i=1}^{N_p}$ can be obtained either by Reduced Basis (RB) method with a greedy approach [18], using Proper Orthogonal Decomposition (POD) [35], by The Proper Generalized Decomposition [15], or by Dynamic Mode Decomposition [33]. In the next two sections we will consider RB and POD methods.

3 Stabilized Finite Element RB Reduced Order Method

In this section, we present a RB method for parameterized steady Navier-Stokes problem [30] which ensures stable solution [2]. Our focus in this section is to deal with flows at low Reynolds number with particular emphasis on inf-sup stability at reduced order level.

We know that the Galerkin projection on RB spaces does not guarantee the fulfillment of equivalent reduced inf-sup condition [31]. To fulfill this condition we have to enrich the RB velocity space with the solutions of a supremizer problem [5, 32]. In this work we propose a residual based stabilization technique which circumvents the inf-sup condition and guarantees stable RB solution. This approach consists in adding some stabilization terms into the Galerkin finite element formulation of (1) using equal order ($\mathbb{P}_k/\mathbb{P}_k$; $k = 1, 2$) velocity pressure interpolation, and than projecting onto RB spaces. As the results in Sect. 5 will show, residual based stabilization methods improves the stability of Galerkin finite element method without compromising the consistency.

We start with introducing two finite-dimensional subspaces $\mathbf{V}_h \subset \mathbf{V}$, $Q_h \subset Q$ of dimension $\mathcal{N}_u$ and $\mathcal{N}_p$, respectively, being h related to the computational mesh size. The Galerkin finite element approximation of the parameterized problem (1) with the addition of stabilization terms reads as follows: for a given parameter value $\mu \in \mathbb{P}$, we look for the full order solution $(\mathbf{u}_h(\mu), p_h(\mu)) \in \mathbf{V}_h \times Q_h$ such that

$$\begin{cases} a(\mathbf{u}_h, \mathbf{v}_h; \mu) + c(\mathbf{u}_h, \mathbf{u}_h, \mathbf{v}_h; \mu) + b(\mathbf{v}_h, p_h; \mu) = \xi_h(\mathbf{v}_h; \mu) & \forall \mathbf{v}_h \in \mathbf{V}_h, \\ b(\mathbf{u}_h, q_h; \mu) = \psi_h(q_h; \mu) & \forall q_h \in Q_h, \end{cases} \tag{3}$$

which we name as the stabilized Galerkin finite element formulation, where $a(.,.;\mu)$ and $b(.,.;\mu)$ are the bilinear forms related to diffusion and pressure-divergence operators, respectively and $c(.,.,.;\mu)$ is the trilinear form related to the convective term. The stabilization terms $\xi_h(\mathbf{v}_h;\mu)$ and $\psi_h(q_h;\mu)$ are defined as:

$$\begin{aligned} \xi_h(\mathbf{v}_h; \mu) &:= \delta \sum_K h_K^2 \int_K (-\nu \Delta \mathbf{u}_h + \mathbf{u}_h \cdot \nabla \mathbf{u}_h + \nabla p_h, -\gamma \nu \Delta \mathbf{v}_h + \mathbf{u}_h \cdot \nabla \mathbf{v}_h), \\ \psi_h(q_h; \mu) &:= \delta \sum_K h_K^2 \int_K (-\nu \Delta \mathbf{u}_h + \mathbf{u}_h \cdot \nabla \mathbf{u}_h + \nabla p_h, \nabla q_h), \end{aligned} \tag{4}$$

where K is an element of the domain, h_K is the diameter of K, δ is the stabilization coefficient such that, $0 < \delta \leq C$ (C is a suitable constant) needs to be chosen properly [9, 24]. For $\gamma = 0, 1, -1$, the stabilization (4) is respectively known as Streamline Upwind Petrov Galerkin (SUPG) [10], Galerkin least-squares (GLS) [20] and Douglas-Wang (DW) [14].

Next step is to construct the RB spaces $\mathbf{V}_{rb}$ and Q_{rb}, for velocity and pressure, respectively. These spaces are constructed using the greedy algorithm [18] and may or may not be enriched with supremizer [2]. In order to control the condition number of RB matrix, the basis functions $\boldsymbol{\phi}_i(\mathbf{x})$ and $\chi_i(\mathbf{x})$ for RB velocity and pressure, respectively are orthonormalized by using the Gram-Schmidt orthonormalization process [18].

Now we write the RB formulation, i.e, we perform a Galerkin projection of (3) onto the RB spaces. Therefore the reduced problem reads as follows: for any $\mu \in \mathbb{P}$, find $(\mathbf{u}_N(\mu), p_N(\mu)) \in \mathbf{V}_{rb} \times Q_{rb}$ such that

$$\begin{cases} a(\mathbf{u}_N, \mathbf{v}_N; \mu) + c(\mathbf{u}_N, \mathbf{u}_N, \mathbf{v}_N; \mu) + b(\mathbf{v}_N, p_N; \mu) = \xi_N(\mathbf{v}_N; \mu) & \forall \mathbf{v}_N \in \mathbf{V}_{rb}, \\ b(\mathbf{u}_N, q_N; \mu) = \psi_N(q_N; \mu) & \forall q_N \in Q_{rb}, \end{cases} \tag{5}$$

where $\xi_N(\mathbf{v}_N;\mu)$ and $\psi_N(q_N;\mu)$ are the reduced order counterparts of the stabilization terms defined in (4). We call (5) as the stabilized RB formulation.

The Galerkin projection of (3) onto RB spaces can also be performed without adding the stabilization terms in RB formulation. Therefore we have two options here [28]; the first option is the *offline-online stabilization*, where we apply the Galerkin projection on stabilized formulations in both the offline and the online stages, and the second option is *offline-only stabilization*, where we apply stabilization only in the offline stage and then we perform the online stage using the standard formulation. Finally, combining these two options with the supremizer enrichment [32], we come up with the following four options to discuss [2]:

- *offline-online stabilization* with supremizer;
- *offline-online stabilization* without supremizer;
- *offline-only stabilization* with supremizer;
- *offline-only stabilization* without supremizer.

In Sect. 5.2 the first three options are implemented. An extension of the work presented in this section to unsteady problems is currently in progress [3].

4 Finite Volume POD-Galerkin Reduced Order Model

In this section, the treatment of flow with high Reynolds number will be addressed. The starting point is with the POD-Galerkin projection method in the first subsection, and then the ROM for turbulent flows will be proposed in the second subsection.

4.1 POD-Galerkin Projection Method

POD is a very popular method for generating reduced order spaces. It is based on constructing a reduced order space which is optimal in the sense that it minimizes the projection error (the L^2 norm of the difference between the snapshots and their projection onto the reduced order basis). After generating the POD space one can project (1) into that space. This approach is called POD-Galerkin projection which has been widely used for building ROMs for variety of problems in Computational Fluid Dynamics (CFD) [1, 4, 8, 16, 21, 27].

The POD space is obtained by solving the following minimization problem:

$$\mathbb{V}_{POD} = \arg\min \frac{1}{N_s} \sum_{n=1}^{N_s} ||\mathbf{u}_n - \sum_{i=1}^{N_u} (\mathbf{u}_n, \boldsymbol{\phi}_i)_{L^2(\Omega)} \boldsymbol{\phi}_i ||^2_{L^2(\Omega)}, \quad (6)$$

where $\mathbf{u}_n$ is a general snapshot of the velocity field which is obtained for the sample μ_n and N_s is the total number of snapshots. The minimization problem can be solved by performing Singular Value Decomposition (SVD) on the matrix formed by the snapshots, or by computing a correlation matrix whose entries are the scalar product between the snapshots and then performing eigenvalue decomposition on that correlation matrix, for more details we refer the reader to [23, 36].

The next step in building the reduced order model is to project the momentum equation of (1) onto the POD space spanned by the velocity POD modes, namely:

$$(\boldsymbol{\phi}_i, (\mathbf{u} \cdot \nabla)\mathbf{u} - \nu \Delta \mathbf{u} + \nabla p)_{L^2(\Omega)} = 0. \tag{7}$$

Inserting the approximations (2) into (7) yields the following system:

$$\nu \mathsf{B}\mathbf{a} - \mathbf{a}^T \mathsf{C}\mathbf{a} - \mathsf{H}\mathbf{b} = \mathbf{0}, \tag{8}$$

where $\mathbf{a}$ and $\mathbf{b}$ are the vectors of coefficients $a_i(\boldsymbol{\mu})$ and $b_i(\boldsymbol{\mu})$, respectively, while the other terms are computed as follows:

$$B_{ij} = \left(\boldsymbol{\phi}_i, \Delta \boldsymbol{\phi}_j\right)_{L^2(\Omega)}, \tag{9}$$

$$C_{ijk} = \left(\boldsymbol{\phi}_i, \nabla \cdot (\boldsymbol{\phi}_j \otimes \boldsymbol{\phi}_k))\right)_{L^2(\Omega)}, \tag{10}$$

$$H_{ij} = \left(\boldsymbol{\phi}_i, \nabla \chi_j\right)_{L^2(\Omega)}. \tag{11}$$

For better understanding of the treatment of the nonlinearity introduced by the convective term the reader may refer to [36]. To close the system (8) an additional number of N_p equations is needed since there are just N_u equations but with N_u+N_p unknowns. The continuity equation cannot be directly used to close the system since the snapshots which are obtained using the full order solver are already divergence free, and the velocity POD modes which are obtained using those snapshots have the same property. This problem can be overcome by two possible approaches, the first one is to use Poisson equation for pressure to get the needed additional equations such that one can close the system. Poisson equation for pressure can be derived by just taking the divergence of the momentum equation and then exploiting the continuity equation. The second possible approach, is the supremizer stabilization method [5, 32] which has been already mentioned in Sect. 3. The latter approach has been developed for finite volume discretization method as well and one can refer to [35] for more details on that. The supremizer approach will ensure that the velocity modes are not all divergence free. One can project the continuity equation onto the space spanned by the pressure modes, which results in the following system:

$$\begin{cases} \nu \mathsf{B}\mathbf{a} - \mathbf{a}^T \mathsf{C}\mathbf{a} - \mathsf{H}\mathbf{b} = \mathbf{0}, \\ \mathsf{P}\mathbf{a} = \mathbf{0}, \end{cases} \tag{12}$$

where the new matrix is P, is computed as follows:

$$P_{ij} = \left(\chi_i, \nabla \cdot \boldsymbol{\phi}_j\right)_{L^2(\Omega)}. \tag{13}$$

Concerning the treatment of boundary conditions, a lifting function method is employed. A new set of snapshots with homogeneous boundary condition is created. For the selection of an appropriate lifting function, several options are available such as snapshots average or the solution to a linear problem. We decided here to rely on the latter approach. For more details one can refer to [36].

4.2 POD-Galerkin Reduced Order Model for Turbulent Flows

In this subsection, the main goal is to focus on flows which have higher Reynolds number than those considered in the Sect. 3. In these flows the turbulence phenomenon is present. The full order discretization technique used in this case for solving (1) is the Finite Volume Method (FVM) [26, 37] which is widely used in industrial applications. One advantage of the FVM is that the equations are written in conservative form, and therefore the conservation law is ensured at a local level.

The turbulence modelling is employed using $k-\omega$ turbulence model [25] which is a two equations model, is used to ensure the stability of the simulation. In this model the eddy viscosity ν_t depends algebraically on two variables k and ω respectively stand for the turbulent kinetic energy and the specific turbulent dissipation rate. The values of these two variables are computed solving two additional PDEs. The new set of equations to be solved is the Reynolds Averaged Navier-Stokes (RANS) equations which read as follow:

$$\begin{cases} (\boldsymbol{u}\cdot\nabla)\,\boldsymbol{u} = \nabla\cdot\left[-p\mathbf{I} + (\nu+\nu_t)\left(\nabla\boldsymbol{u} + (\nabla\boldsymbol{u})^T\right) - \frac{2}{3}k\mathbf{I}\right], \\ \nabla\cdot\boldsymbol{u} = 0, \\ \nu_t = f(k,\omega), & \text{in } \Omega \\ \text{Transport-Diffusion equation for } k, \\ \text{Transport-Diffusion equation for } \omega. \end{cases}$$

In order to build a reduced order model for the new set of equations one can extend the previous assumption (2) to the eddy viscosity field, namely:

$$\nu_t(\mathbf{x};\boldsymbol{\mu}) \approx \sum_{i=1}^{N_{\nu_t}} g_i(\boldsymbol{\mu})\eta_i(\mathbf{x}),$$

The eddy viscosity modes η_i are computed similarly to those of velocity and pressure. Following the procedure explained in Sect. 4.1 one can project the momentum equation onto the spatial bases of velocity. The continuity equation is projected onto the spatial bases of pressure with the use of a supremizer stabilization approach. In contrast, $k-\omega$ transport-diffusion equations are not used in the projection procedure, this makes the reduced order model general and independent of the turbulence model used in the full order simulations.

The resulting system is the following:

$$\begin{cases} \nu(\mathsf{B}+\mathsf{B_T})\mathbf{a} - \mathbf{a}^{\mathbf{T}}\mathsf{C}\mathbf{a} + \mathbf{g}^{\mathbf{T}}(\mathsf{C_{T1}} + \mathsf{C_{T2}})\mathbf{a} - \mathsf{H}\mathbf{b} = \mathbf{0}, \\ \mathsf{P}\mathbf{a} = 0, \end{cases} \tag{14}$$

Where the new terms with respect to the dynamical system in (12) are computed as follows:

$$B_{T_{ij}} = \left(\boldsymbol{\phi}_i, \nabla \cdot (\nabla \boldsymbol{\phi}_j^T)\right)_{L^2(\Omega)}, \tag{15}$$

$$C_{T1_{ijk}} = \left(\boldsymbol{\phi}_i, \eta_j \Delta \boldsymbol{\phi}_k)\right)_{L^2(\Omega)}, \tag{16}$$

$$C_{T2_{ijk}} = \left(\boldsymbol{\phi}_i, \nabla \cdot \eta_j (\nabla \boldsymbol{\phi}_k^T)\right)_{L^2(\Omega)}. \tag{17}$$

One can see that a new set of coefficients g_i has been introduced. These coefficients are used in the approximation of the eddy viscosity fields, and in order to compute them an interpolation procedure using Radial Basis Functions (RBF) [22] has been used in the online stage. After that one can solve the system (14) for the vectors of coefficients **a** and **b**.

In the remaining part of this subsection, the interpolation method used to compute the coefficients of the reduced viscosity will be explained in further details. The starting point consists of the set of samples used in the offline stage $X_\mu = \{\boldsymbol{\mu_1}, \boldsymbol{\mu_2}, \ldots, \boldsymbol{\mu_{N_s}}\}$. The associated outputs y_i are the coefficients resulted from the projection of the viscosity snapshots that correspond to each μ_i onto the viscosity spatial modes $[\eta_j]_{j=1}^{N_{\nu_t}}$. The goal is to interpolate the known coefficients by making the use of RBF ζ_i for $i = 1, \ldots, N_s$. One may assume that Y has the following form:

$$Y(x) = \sum_{j=1}^{N_s} w_j \zeta_j(\|x - x_j\|_2), \tag{18}$$

where w_j are some appropriate weights. In order to interpolate the known data, the following property is required:

$$Y(x_i) = y_i, \quad \text{for} \quad i = 1, 2, \ldots, N_s. \tag{19}$$

In other words,

$$\sum_{j=1}^{N_s} w_j \zeta_j(\|x_i - x_j\|_2) = y_i, \quad \text{for} \quad i = 1, 2, \ldots, N_s. \tag{20}$$

The latter system can be solved to find the weights. The procedure dealing with what concerns the use of RBF interpolation is summarized in the following box for both offline and online stages.

In the context of this work, RBF is used according to the following algorithm. The methodology has two parts, the first is within the offline stage in which the interpolant RBF is constructed. The second part, which takes place during the online stage, consists into the evaluation of the coefficients $[g_i]_{i=1}^{N_{\nu_t}}$ using the latter mentioned RBF methodology.

Offline Stage

Input: The set of samples for which the offline stage has been run $X_\mu = \{\boldsymbol{\mu}_1, \boldsymbol{\mu}_2, \ldots, \boldsymbol{\mu}_{N_s}\}$, with the corresponding eddy viscosity snapshots $\nu_{t1}, \nu_{t2}, \ldots, \nu_{tN_s}$, the number of eddy viscosity modes to be used in the reduction during online phase N_{ν_t} and finally $i = 1$ which is an index to be used during the stage.

Goal: for $i = 1, 2, \ldots, N_{\nu_t}$ construct $g_i(\boldsymbol{\mu}) = \sum_{j=1}^{N_s} w_{i,j}\zeta_{i,j}(\|\boldsymbol{\mu} - \boldsymbol{\mu}_j\|_2)$

Step 1

Compute the eddy viscosity modes $[\eta_k]_{k=1}^{N_{\nu_t}}$ using POD as mentioned before.

Step 2

Compute the following coefficients

$$g_{i,j} = (\nu_{tj}, \eta_i)_{L^2(\Omega)}, \quad \text{for} \quad j = 1, 2, \ldots N_s \quad . \tag{21}$$

Step 3

Solve the following linear system for the vector of weights $\mathbf{w}_i = [w_{i,j}]_{j=1}^{N_s}$

$$\sum_{j=1}^{N_s} w_{i,j}\zeta_{i,j}(\|\boldsymbol{\mu}_k - \boldsymbol{\mu}_j\|_2) = g_{i,k}, \quad \text{for} \quad k = 1, 2, \ldots, N_s. \tag{22}$$

Step 4

Store the weights $[w_{i,j}]_{j=1}^{N_s}$ and construct the scalar coefficients $g_i(\boldsymbol{\mu})$.

Step 5

If $i = N_{\nu_t}$ **terminate**, otherwise set $i = i + 1$ and go to **Step 2**.

Online Stage

As **Input** we have the new parameter value $\boldsymbol{\mu}^*$ and the goal is to compute $\mathbf{g}(\boldsymbol{\mu}^*) = [g_i(\boldsymbol{\mu}^*)]_{i=1}^{N_{\nu_t}}$

Which is done simply by computing $g_i(\boldsymbol{\mu}^*) = \sum_{j=1}^{N_s} w_{i,j}\zeta_{i,j}(\|\boldsymbol{\mu}^* - \boldsymbol{\mu}_j\|_2)$ for $i = 1, 2, \ldots, N_{\nu_t}$

After computing the coefficients of the viscosity reduced order solution $[g_i]_{i=1}^{N_{\nu_t}}$ then it will be possible to solve the reduced order system (14). Afterwards, one can compute the reduced order solution for both velocity and pressure using (2). From now on this approach will be referred to as *POD-Galerkin-RBF ROM*. The POD-Galerkin-RBF model will be tested on a simple benchmark test case of the backstep in steady setting, with the offline phase being done with a RANS approach. For the application of this model on more complex cases involving unsteady RANS simulations the reader may refer to [19].

5 Numerical Results

In this section we present numerical results for both reduced order modelling strategies presented in the previous sections. In Sect. 5.1 we present the numerical results for low Reynolds number using stabilized RB method developed in Sect. 3 for steady Navier-Stokes equations. Section 5.2 is based on the results for POD-Galerkin-RBF on a backward facing step problem. In both cases we consider only physical parameters.

5.1 Stabilized Finite Element Based ROM Results

In this test case, we apply the stabilized RB model developed in Sect. 3 for the Navier-Stokes problem to the *lid driven-cavity* problem with only one physical parameter μ which denotes the Reynolds number. We consider only the first three options and we have done several test cases to compare the three options. Fourth option is the worst option and is not reported here. The stabilization option that we consider here is the SUPG stabilization, corresponding to $\gamma = 0$ in (4). The computational domain is shown in Fig. 1 and the boundary conditions are

$$u_1 = 1, u_2 = 0 \text{ on } \partial\Omega_{In} \text{ and } \mathbf{u} = \mathbf{0} \text{ on } \partial\Omega_0 \tag{23}$$

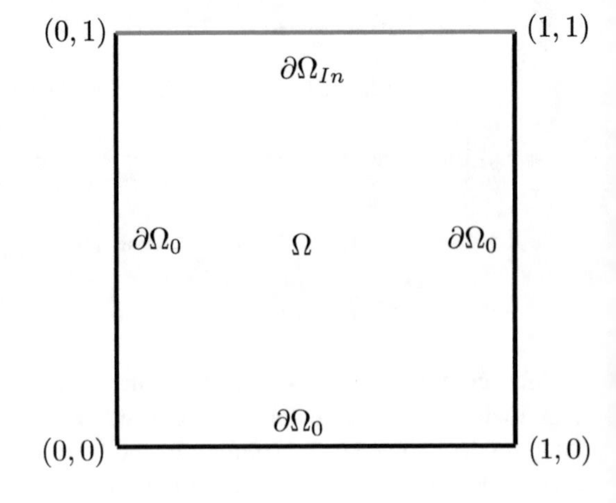

Fig. 1 Unit cavity domain Ω for RB problem with boundaries identified

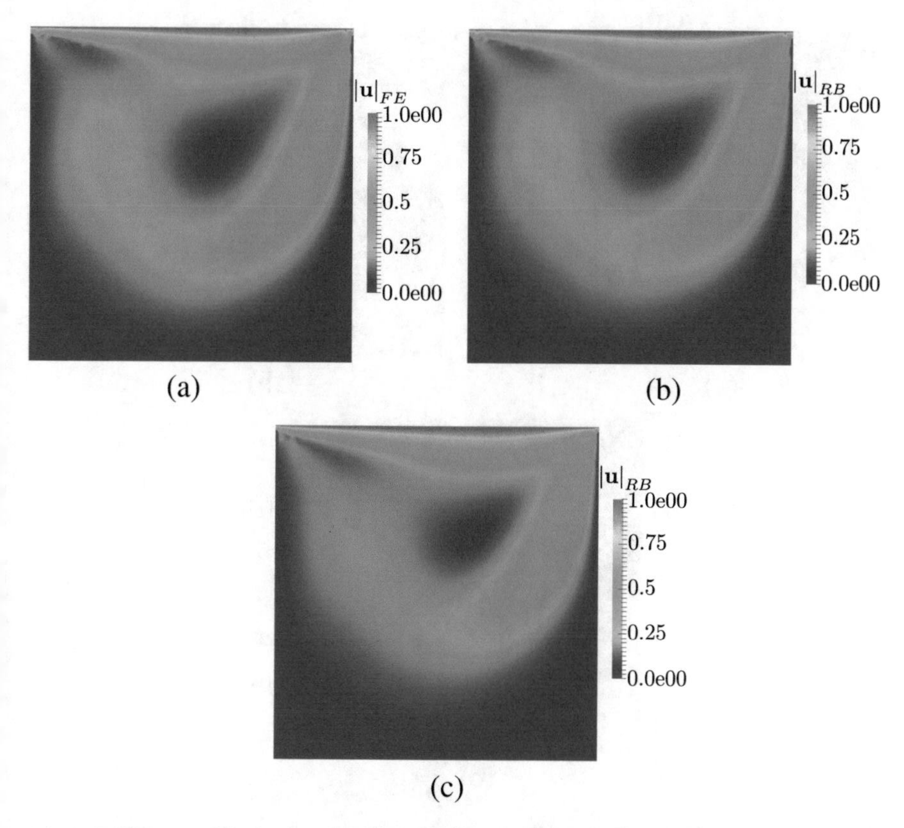

Fig. 2 SUPG stabilization: FE and RB solutions for velocity at $Re = 200$. (**a**) FE velocity. (**b**) RB velocity (*offline-online*). (**c**) RB velocity (*offline-only*)

The mesh of this problem is non-uniform with 3794 triangles and 1978 nodes, whereas the minimum and maximum size elements are $h_{min} = 0.0193145$ and $h_{max} = 0.0420876$, respectively. All the numerical simulations for this case are performed using FreeFem++ [17] and RBniCS [6].

Figure 2 shows the FE velocity (left), RB velocity obtained using *offline-online stabilization* (center), and the RB velocity obtained for *offline-only stabilization* (right). From these solutions we see that the FE and RB solutions are similar.

Figure 3 plots the FE pressure (left), RB pressure obtained using *offline-online stabilization* (center), and the RB pressure obtained for *offline-only stabilization* (right). These results show that the RB pressure with *offline-online stabilization* is stable but RB solution obtained by *offline-only stabilization* is highly oscillatory even with the supremizer enrichment. All these solutions are obtained for equal order linear velocity pressure interpolation $\mathbb{P}_1/\mathbb{P}_1$. Similar results can be shown for $\mathbb{P}_2/\mathbb{P}_2$ [2].

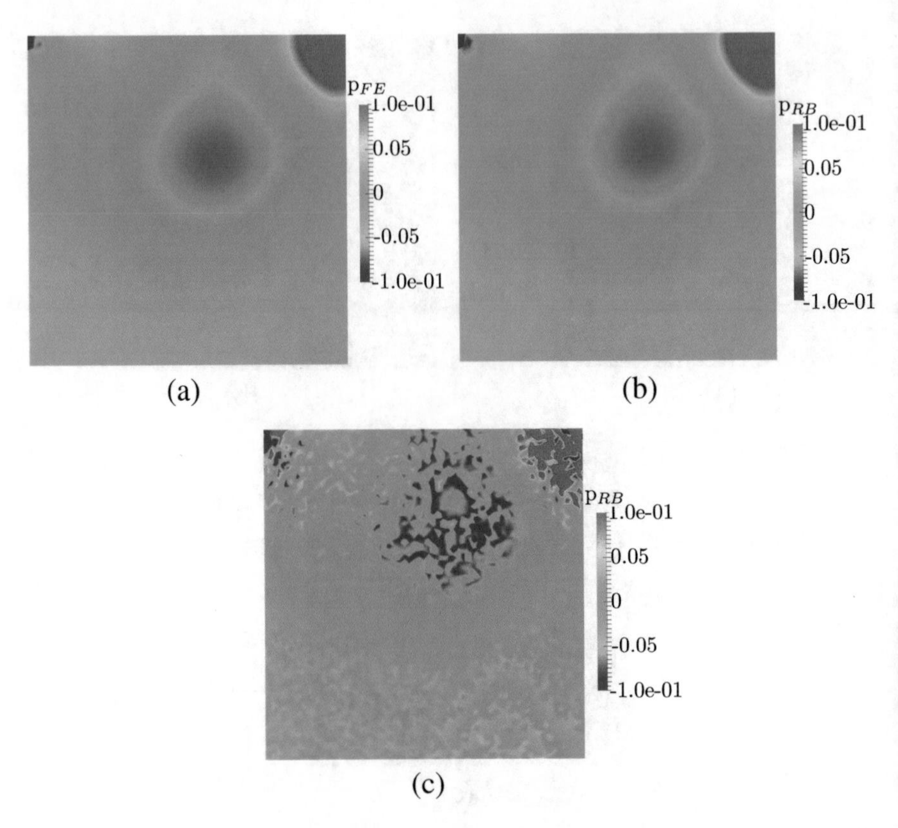

Fig. 3 SUPG stabilization: FE and RB solutions for pressure at $Re = 200$. (**a**) FE pressure. (**b**) RB pressure (*offline-online*). (**c**) RB pressure (*offline-only*)

Figure 4 illustrates the error between FE and RB solutions for velocity (left) and pressure (right). We show the comparison between *offline-online stabilization* with/without supremizer and *offline-only stabilization* with supremizer. These comparison shows that the *offline-online stabilization* is the most appropriate way to stabilize and the enrichment of RB velocity space with supremizer may not be necessary. We are getting even a better approximation of the velocity without the supremizer, which is polluted a little bit by the supremizer. However in case of pressure, supremizer is improving the accuracy in the case of *offline-online stabilization*. All the results here are presented for equal order linear velocity pressure interpolation $\mathbb{P}_1/\mathbb{P}_1$.

In Table 1 we summarize the computational cost of offline and online stage for different choices of FE spaces, parameter detail, FE and RB dimensions. From this table we can see that the *offline-online stabilization* without supremizer is less expensive as compared to *offline-online stabilization* with supremizer.

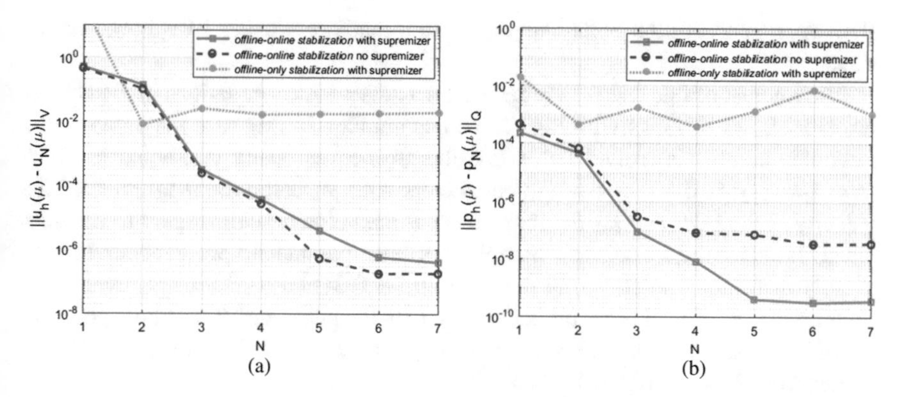

Fig. 4 Error between FE and RB solutions: velocity (left) and pressure (right), obtained by different options using SUPG stabilization. (**a**) Velocity errors. (**b**) Pressure errors

Table 1 Computational details of steady Navier-Stokes problem with physical parameter only

Physical parameter	μ (Reynolds number)
Range of μ	[100,500]
Online μ	200
FE degrees of freedom	13,218 ($\mathbb{P}_1/\mathbb{P}_1$)
	52,143 ($\mathbb{P}_2/\mathbb{P}_2$)
RB dimension	$N_u = N_s = N_p = 7$
Offline time ($\mathbb{P}_1/\mathbb{P}_1$)	1182 s (*offline-online stabilization* with supremizer)
	842 s (*offline-online stabilization* without supremizer)
Offline time ($\mathbb{P}_2/\mathbb{P}_2$)	2387 s (*offline-online stabilization* with supremizer)
	2121 s (*offline-online stabilization* without supremizer)
Online time ($\mathbb{P}_1/\mathbb{P}_1$)	74 s (with supremizer)
	65 s (without supremizer)
Online time ($\mathbb{P}_2/\mathbb{P}_2$)	131 s (with supremizer)
	108 s (without supremizer)

5.2 *Finite Volume POD-Galerkin-RBF ROM Results*

In this subsection the numerical results for the reduced order model obtained using the POD-Galerkin-RBF approach are shown. The finite volume C++ library OpenFOAM® (OF) [38] is used as the numerical solver at the full order level. At the reduced order level the reduction is done using the library ITHACA-FV [34] which is based on C++.

We have tested the proposed model on the benchmark case of the backstep see Fig. 5. The test is performed in steady state setting, the two considered parameters are both physical and consist into the magnitude of the velocity at the inlet and the inclination of the velocity with respect to the inlet. In addition a comparison is presented between the results obtained using the newly developed POD-Galerkin-

RBF approach with the POD-Galerkin option that is not using RBF in the online stage.

The interest of this test is in reducing the Navier-Stokes equations in the case of turbulent flows or flows with high Reynolds number. In this case the value of the Reynolds number is around 10^4, while the physical viscosity ν is equal to 10^{-3}. $\boldsymbol{\mu} = [\mu_1, \mu_2]$ is the vector of the parameters with μ_1 being the magnitude of the velocity at the inlet and μ_2 the inclination of the velocity with respect to the inlet which is measured in degrees. Samples for both parameters are generated as 20 equally distributed points in the ranges of $[0.18, 0.3]$ and $[0, 30]$ respectively. The training of the reduced order model is done with the generated 400 sample points in the offline stage. The Reynolds number as mentioned before is of order of 10^4 and ranges from 9.144×10^3 to 1.524×10^4.

In Fig. 5 one can see the computational domain that has been used in this work. The characteristic length d and is equal to 50.8 m. In the full order problem the boundary conditions for velocity and pressure are set as reported in Table 2.

In the reduced order model the supremizer approach has been used to stabilize pressure. In Table 3 one can see the cumulative eigenvalues for velocity, pressure, supremizer (which is denoted by **S**) and viscosity.

During the online phase another set of samples has been used to check the reduced order model which is a cross validation test of the model. The value of the parameter vector given in the online phase is denoted by $\boldsymbol{\mu}_i^*$ where $i = 1, \ldots, N_{online-samples}$. The samples which were used in the cross validation have been chosen inside the ranges of the samples used in the offline stage. For the sake of better evaluation of the model the online samples have been chosen such that they are as far as possible in the parameter space from those used to educate the model. After taking that criterion into consideration the samples used in the online phase happened to be equally distributed in the ranges of $[0.20826, 0.28405]$ and $[5.5368, 29.221]$ for μ_1 and μ_2, respectively. Seven samples were used for μ_1 and six for μ_2.

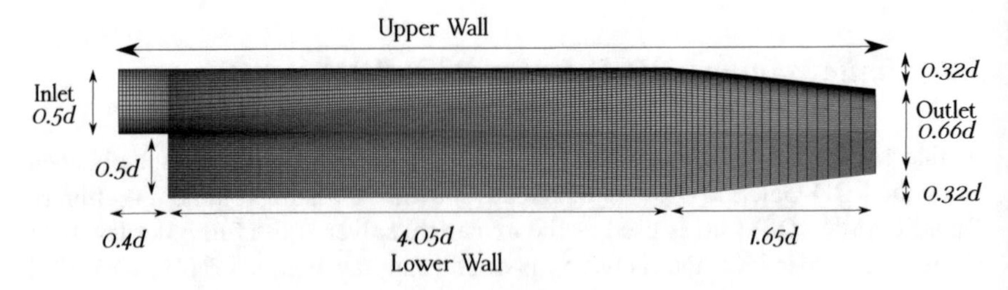

Fig. 5 The computational domain used in the numerical simulations, d is equal to 50.8 m

Table 2 Boundary conditions

	Inlet	Outlet	Lower and upper walls
u	$\mathbf{u_{in}} = [\mu_1 cos(\mu_2), \mu_1 sin(\mu_2)]$	$\nabla \mathbf{u} \cdot \mathbf{n} = \mathbf{0}$	$\mathbf{u} = \mathbf{0}$
p	$\nabla p \cdot \mathbf{n} = 0$	$p = 0$	$\nabla p \cdot \mathbf{n} = 0$

Table 3 Cumulative eigenvalues

N modes	**u**	p	**S**	ν_t
1	0.971992	0.868263	0.899488	0.985703
2	0.993017	0.998541	0.996392	0.998884
3	0.997589	0.999915	0.999767	0.999673
4	0.999196	0.999963	0.999929	0.999880
5	0.999545	0.999985	0.999965	0.999926
6	0.999828	0.999997	0.999988	0.999971
7	0.999914	0.999999	0.999996	0.999986
8	0.999952	0.999999	0.999998	0.999992
9	0.999978	1.000000	0.999999	0.999995
10	0.999986	1.000000	0.999999	0.999997

Recall that in this case the parameters were basically the two components of the velocity at the inlet. Therefore two lifting fields were computed which correspond to the full order solution for the velocity field with unitary boundary condition. The first and second lifting fields are the steady state solutions with the velocity at the inlet being $U = (1, 0)$ and $U = (0, 1)$, respectively. These two fields are added to the velocity modes.

The RBF functions for the turbulent viscosity are chosen to be Gaussian functions. The system (14) has been solved for each online sample $\boldsymbol{\mu}_i^*$ in the online phase and the fields have been constructed. The ROM fields obtained by solving (14) have been compared to those resulted from solving the POD-Galerkin system (12), which does not take into consideration the contribution of the eddy viscosity.

In Fig. 6 one can see the velocity fields obtained by the full order solver, ROM velocity field obtained by the POD-Galerkin approach and ROM velocity obtained by the new POD-Galerkin-RBF model. In Fig. 7 there is the same comparison but for the pressure fields. In both figures the online sample which has been introduced to both reduced order models is the one with $\boldsymbol{\mu}^* = (0.22089, 24.484)$, which corresponds to the velocity vector at the inlet to be $\mathbf{U} = (0.20103, 0.091548)$. The reduction has been made with seven modes for velocity, pressure, supremizer and eddy viscosity (just considered in the POD-Galerkin-RBF model). One can see that the POD-Galerkin-RBF model is able to capture the dynamics efficiently. It has successfully reconstructed the full order solution from both qualitative and quantitative aspects. On the other hand, it is quite clear that the classical POD-Galerkin model, which does not consider the contribution of the eddy viscosity in its formulation fail to give an accurate reproduction of the full order solution, especially close to the top and to the outlet for the velocity field and at the inlet for the pressure field.

Looking on the results from a quantitative point of view, in the POD-Galerkin-RBF model we have values of 0.00612 and 0.02957 for the relative error in L^2 norm for velocity and pressure, respectively, while the POD-Galerkin model has errors of 0.37967 and 2.2296. Table 4 shows a comparison between the two models in terms of the error over all the samples used in the online phase (average and maximum

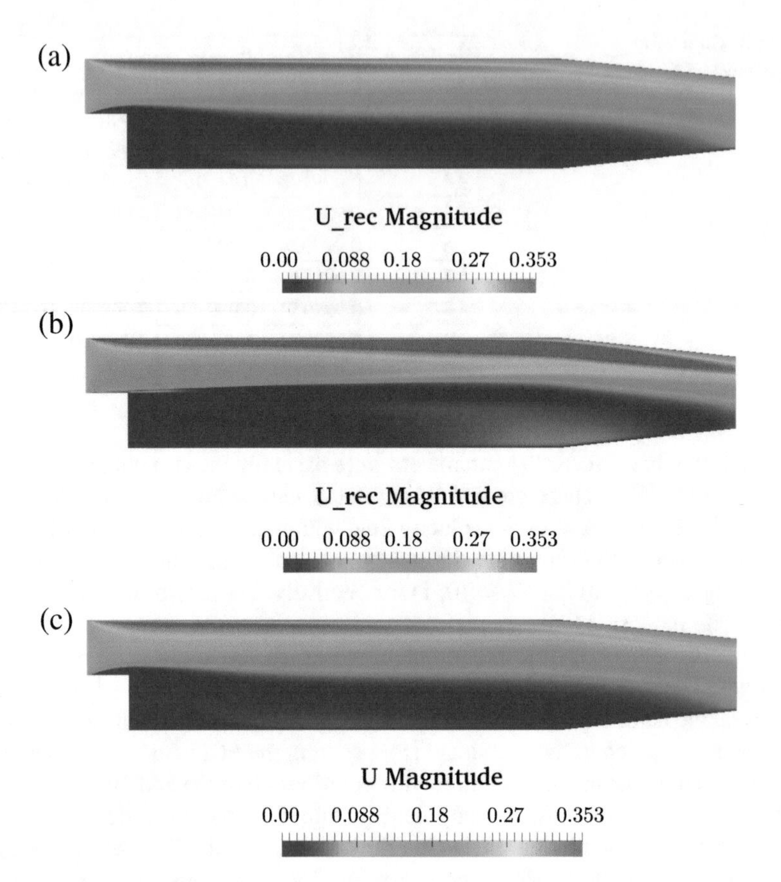

Fig. 6 Velocity fields: (**a**) shows the ROM Velocity obtained by POD-Galerkin-RBF ROM model, while in (**b**) one can see the ROM Velocity (without viscosity incorporated in ROM), and finally in (**c**) we have the FOM Velocity

value). Figures 8 and 9 show the error as function of the two parameters when one of them is fixed and the other is varied.

6 Conclusion and Perspectives

In this chapter we have proposed two different ROM strategies for the incompressible parameterized Navier-Stokes equations to deal from low to higher Reynolds number, respectively. In case of low Reynolds number, we have used a stabilized FE discretization technique at the full order level and then we performed Galerkin projection onto RB spaces, obtained by a greedy algorithm. We have compared the *offline-online stabilization* approach with supremizer enrichment in context of RB inf-sup stability. Based on numerical results, we conclude that a residual based

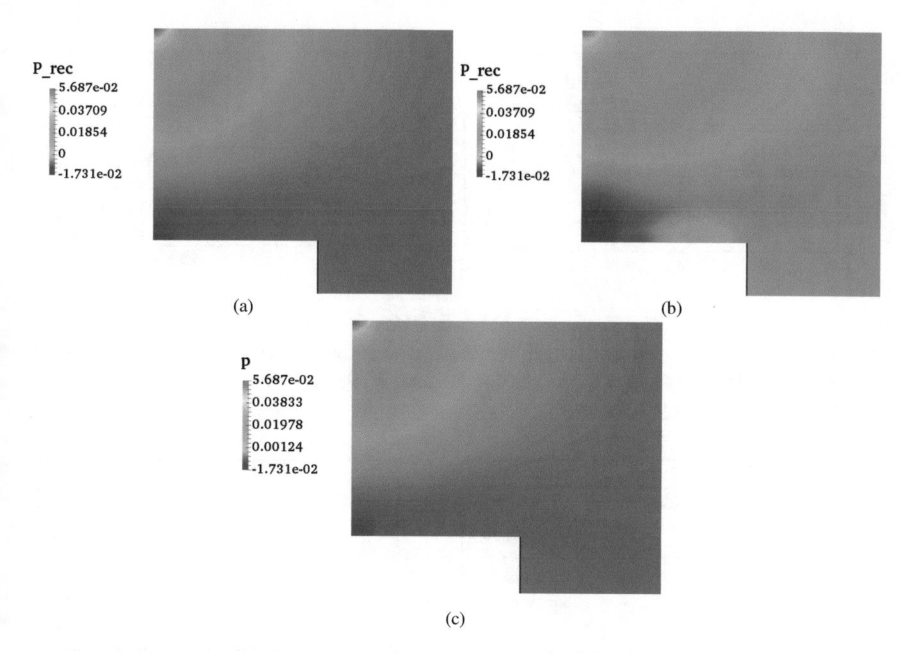

Fig. 7 Pressure fields: (**a**) shows the ROM Pressure obtained by POD-Galerkin-RBF ROM model, while in (**b**) one can see the ROM Pressure (without viscosity incorporated in ROM), and finally in (**c**) we have the FOM Pressure

Table 4 Relative L^2 error for velocity and pressure fields: *Average* is taken over all samples used in the online phase, while *maximum* represents the worse case among the samples

	$\mathbf{u}$ with RBF	p with RBF	$\mathbf{u}$ without RBF	p without RBF
Average relative error	0.0073	0.0276	0.2592	1.5412
Maximum relative error	0.0104	0.0475	0.3810	2.3616

POD-Galerkin-RBF model results are compared to those of the normal POD-Galerkin one

stabilization technique, if applied in both offline and online stage (*offline-online stabilization*), is sufficient to ensure a stable RB solution and therefore we can avoid the supremizer enrichment which consequently reduces the online computation cost. Supremizer may help in improving the accuracy of pressure approximation. We also conclude that a stable RB solution is not guaranteed if we stabilize the offline stage and not the online stage (*offline-only stabilization*) even with supremizer enrichment.

For higher Reynolds number, the test case was the backstep benchmark test case, we have used the FV discretization technique at the full order level. At the reduced order level, we have used a POD-Galerkin projection approach taking into consideration the contribution of the eddy viscosity. The newly proposed approach involves the usage of radial basis functions interpolation in the online stage. The model has been tested on the benchmark case of the backstep, the results showed

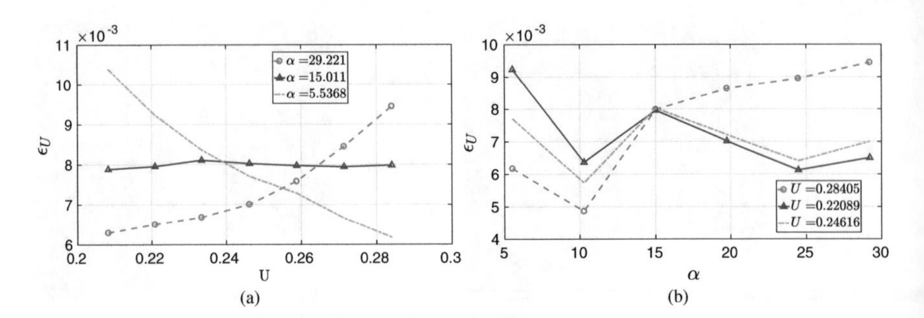

Fig. 8 The L^2 relative error for velocity fields as function of the parameters. In (**a**) the error is plotted versus the inclination of the velocity at the inlet. While in (**b**) the error is plotted versus the magnitude of the velocity at the inlet

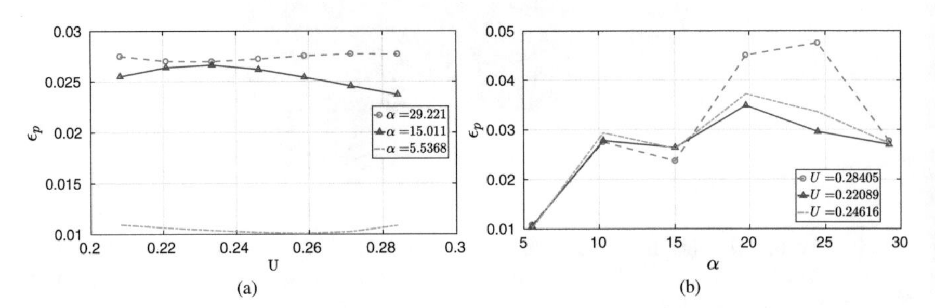

Fig. 9 The L^2 relative error for pressure fields as function of the parameters. In (**a**) the error is plotted versus the inclination of the velocity at the inlet. While in (**b**) the error is plotted versus the magnitude of the velocity at the inlet

that the proposed model has successfully reduced RANS equations. On the other hand, the classical POD-Galerkin approach has not been able to reduce the equations accurately in the same study case.

For the future work, we aim to extend the POD-Galerkin-RBF approach to work also on unsteady Navier-Stokes equations. In addition, one important goal is to reduce problems where the offline phase is simulated with LES.

Acknowledgements We acknowledge the support provided by the European Research Council Consolidator Grant project Advanced Reduced Order Methods with Applications in Computational Fluid Dynamics—GA 681447, H2020-ERC COG 2015 AROMA-CFD, and INdAM-GNCS.

References

1. Akhtar, I., Nayfeh, A.H., Ribbens, C.J.: On the stability and extension of reduced-order Galerkin models in incompressible flows. Theor. Comput. Fluid Dyn. **23**(3), 213–237 (2009). https://doi.org/10.1007/s00162-009-0112-y

2. Ali, S., Ballarin, F., Rozza, G.: Stabilized reduced basis methods for parametrized steady Stokes and Navier-Stokes equations (2018, submitted). https://arxiv.org/abs/2001.00820
3. Ali, S., Ballarin, F., Rozza, G.: Unsteady stabilized reduced basis methods for parametrized Stokes and Navier-Stokes equations (2019, submitted)
4. Baiges, J., Codina, R., Idelsohn, S.R.: Reduced-order modelling strategies for the finite element approximation of the incompressible Navier-Stokes equations. In: Computational Methods in Applied Sciences, pp. 189–216. Springer International Publishing, New York (2014). https://doi.org/10.1007/978-3-319-06136-8_9
5. Ballarin, F., Manzoni, A., Quarteroni, A., Rozza, G.: Supremizer stabilization of POD-Galerkin approximation of parametrized steady incompressible Navier-Stokes equations. Int. J. Numer. Methods Eng. **102**(5), 1136–1161 (2014). https://doi.org/10.1002/nme.4772
6. Ballarin, F., Sartori, A., Rozza, G.: RBniCS – reduced order modelling inFEniCS. http://mathlab.sissa.it/rbnics (2016)
7. Benner, P., Ohlberger, M., Patera, A., Rozza, G., Urban, K. (eds.): Model Reduction of Parametrized Systems, vol. 17. Springer International Publishing, New York (2017). https://doi.org/10.1007/978-3-319-58786-8
8. Bergmann, M., Bruneau, C.H., Iollo, A.: Enablers for robust POD models. J. Comput. Phys. **228**(2), 516–538 (2009). https://doi.org/10.1016/j.jcp.2008.09.024
9. Braack, M., Burman, E., John, V., Lube, G.: Stabilized finite element methods for the generalized oseen problem. Comput. Methods Appl. Mech. Eng. **196**(4), 853–866 (2007). https://doi.org/10.1016/j.cma.2006.07.011
10. Brooks, A.N., Hughes, T.J.: Streamline Upwind/Petrov-Galerkin formulations for convection dominated flows with particular emphasis on the incompressible Navier-Stokes equations. Comput. Methods Appl. Mech. Eng. **32**(1–3), 199–259 (1982). https://doi.org/10.1016/0045-7825(82)90071-8
11. Chacón Rebollo, T., Delgado Ávila, E., Mármol Gómez, M., Ballarin, F., Rozza, G.: On a certified Smagorinsky reduced basis turbulence model. SIAM J. Numer. Anal. **55**(6), 3047–3067 (2017). https://doi.org/10.1137/17M1118233
12. Chinesta, F., Huerta, A., Rozza, G., Willcox, K.: Model reduction methods. Encyclopedia of Computational Mechanics, 2nd edn., pp. 1–36 (2017)
13. David, A., Charbel, F.: Stabilization of projection-based reduced-order models. Int. J. Numer. Methods Eng. **91**(4), 358–377 (2012). https://doi.org/10.1007/s11071-012-0561-5
14. Douglas, J.J., Wang, J.: An absolutely stabilized finite element formulation for the Stokes problem. Math. Comput. **52**(186), 495–508 (1989). https://doi.org/10.1090/S0025-5718-1989-0958871-X
15. Dumon, A., Allery, C., Ammar, A.: Proper general decomposition (PGD) for the resolution of Navier–Stokes equations. J. Comput. Phys. **230**(4), 1387–1407 (2011). https://doi.org/10.1016/j.jcp.2010.11.010
16. Giere, S., Iliescu, T., John, V., Wells, D.: SUPG reduced order models for convection-dominated convection-diffusion-reaction equations. Comput. Methods Appl. Mech. Eng. **289**, 454–474 (2015). https://doi.org/10.1016/j.cma.2015.01.020
17. Hecht, F.: New development in freefem++. J. Numer. Math. **20**(3–4), 251–266 (2013). https://doi.org/10.1515/jnum-2012-0013
18. Hesthaven, J.S., Rozza, G., Stamm, B.: Certified Reduced Basis Methods for Parametrized Partial Differential Equations. Springer International Publishing, New York (2016). https://doi.org/10.1007/978-3-319-22470-1
19. Hijazi, S., Stabile, G., Rozza, G.: Data-driven POD-Galerkin reduced order model for turbulent flows (2018, submitted). https://arxiv.org/abs/1907.09909
20. Hughes, T.J., Franca, L.P., Hulbert, G.M.: A new finite element formulation for computational fluid dynamics: VIII. The Galerkin/least-squares method for advective-diffusive equations. Comput. Methods Appl. Mech. Eng. **73**(2), 173–189 (1989). https://doi.org/10.1016/0045-7825(89)90111-4

21. Kunisch, K., Volkwein, S.: Galerkin proper orthogonal decomposition methods for a general equation in fluid dynamics. SIAM J. Numer. Anal. **40**(2), 492–515 (2002). https://doi.org/10.1137/s0036142900382612
22. Lazzaro, D., Montefusco, L.B.: Radial basis functions for the multivariate interpolation of large scattered data sets. J. Comput. Appl. Math. **140**(1–2), 521–536 (2002). https://doi.org/10.1016/s0377-0427(01)00485-x
23. Lorenzi, S., Cammi, A., Luzzi, L., Rozza, G.: POD-Galerkin method for finite volume approximation of Navier-Stokes and RANS equations. Comput. Methods Appl. Mech. Eng. **311**, 151–179 (2016). http://dx.doi.org/10.1016/j.cma.2016.08.006
24. Lube, G., Rapin, G.: Residual-based stabilized higher-order FEM for a generalized oseen problem. Math. Models Methods Appl. Sci. **16**(07), 949–966 (2006)
25. Menter, F.R.: Two-equation eddy-viscosity turbulence models for engineering applications. AIAA J. **32**(8), 1598–1605 (1994). https://doi.org/10.2514/3.12149
26. Moukalled, F., Mangani, L., Darwish, M.: The Finite Volume Method in Computational Fluid Dynamics. Springer International Publishing, New York (2016). https://doi.org/10.1007/978-3-319-16874-6
27. Noack, B.R., Eckelmann, H.: A low-dimensional Galerkin method for the three-dimensional flow around a circular cylinder. Phys. Fluids **6**(1), 124–143 (1994). https://doi.org/10.1063/1.868433
28. Pacciarini, P., Rozza, G.: Stabilized reduced basis method for parametrized advection-diffusion PDEs. Comput. Methods Appl. Mech. Eng. **274**, 1–18 (2014). https://doi.org/10.1016/j.cma.2014.02.005
29. Quarteroni, A.: Numerical Models for Differential Problems, vol. 2. Springer, Berlin (2009)
30. Quarteroni, A., Rozza, G.: Numerical solution of parametrized Navier–Stokes equations by reduced basis methods. Numer. Methods Partial Differential Equations **23**(4), 923–948 (2007). https://doi.org/10.1002/num.20249
31. Rovas, D.: Reduced-basis output bound methods for parametrized partial differential equations. Ph.D. Thesis, Massachusetts Institute of Technology (2003)
32. Rozza, G., Veroy, K.: On the stability of the reduced basis method for Stokes equations in parametrized domains. Comput. Methods Appl. Mech. Eng. **196**(7), 1244–1260 (2007). https://doi.org/10.1016/j.cma.2006.09.005
33. Schmid, P.J.: Dynamic mode decomposition of numerical and experimental data. J. Fluid Mech. **656**, 5–28 (2010). https://doi.org/10.1017/s0022112010001217
34. Stabile, G., Rozza, G.: ITHACA-FV – In real Time Highly Advanced Computational Applications for Finite Volumes. http://www.mathlab.sissa.it/ithaca-fv. Accessed 30 Jan 2018
35. Stabile, G., Rozza, G.: Finite volume POD-galerkin stabilised reduced order methods for the parametrised incompressible Navier–Stokes equations. Comput. Fluids **173**, 273–284 (2018). https://doi.org/10.1016/j.compfluid.2018.01.035
36. Stabile, G., Hijazi, S., Mola, A., Lorenzi, S., Rozza, G.: POD-Galerkin reduced order methods for CFD using Finite Volume Discretisation: vortex shedding around a circular cylinder. Commun. Appl. Ind. Math. **8**(1) (2017). https://doi.org/10.1515/caim-2017-0011
37. Versteeg, H.K., Malalasekera, W.: An Introduction to Computational Fluid Dynamics: The Finite Volume Method. Pearson Education, London (2007)
38. Weller, H.G., Tabor, G., Jasak, H., Fureby, C.: A tensorial approach to computational continuum mechanics using object-oriented techniques. Comput. Phys. **12**(6), 620 (1998). https://doi.org/10.1063/1.168744

Optimization Based Particle-Mesh Algorithm for High-Order and Conservative Scalar Transport

Jakob M. Maljaars, Robert Jan Labeur, Nathaniel A. Trask, and Deborah L. Sulsky

Abstract A particle-mesh strategy is presented for scalar transport problems which provides diffusion-free advection, conserves mass locally (i.e. cellwise) and exhibits optimal convergence on arbitrary polyhedral meshes. This is achieved by expressing the convective field naturally located on the Lagrangian particles as a mesh quantity by formulating a dedicated particle-mesh projection based via a PDE-constrained optimization problem. Optimal convergence and local conservation are demonstrated for a benchmark test, and the application of the scheme to mass conservative density tracking is illustrated for the Rayleigh–Taylor instability.

Keywords Lagrangian-Eulerian · Particle-mesh · Advection equation · PDE-constraints · Conservation · Hybridized discontinuous Galerkin

---

Sandia National Laboratories is a multimission laboratory managed and operated by National Technology & Engineering Solutions of Sandia, LLC, a wholly owned subsidiary of Honeywell International Inc., for the U.S. Department of Energy's National Nuclear Security Administration under contract DE-NA0003525. This paper describes objective technical results and analysis. Any subjective views or opinions that might be expressed in the paper do not necessarily represent the views of the U.S. Department of Energy or the United States Government.

J. M. Maljaars (✉) · R. J. Labeur
Delft University of Technology, Delft, The Netherlands
e-mail: j.m.maljaars@tudelft.nl; r.j.labeur@tudelft.nl

N. A. Trask
Sandia National Laboratories, Albuquerque, NM, USA
e-mail: natrask@sandia.gov

D. L. Sulsky
The University of New Mexico, Albuquerque, NM, USA
e-mail: sulsky@math.unm.edu

H. van Brummelen et al. (eds.), *Numerical Methods for Flows*, Lecture Notes in Computational Science and Engineering 132,
https://doi.org/10.1007/978-3-030-30705-9_23

1 Introduction

Tracing back to the particle-in-cell (PIC) method developed by Harlow et al. [1], hybrid particle-mesh methods attempt to combine a particle-based approach with a mesh-based approach, exploiting the distinct advantages of each framework. Hence, Lagrangian particles are conveniently used in the convective part of the problem, whereas a mesh is particularly efficient to account for the dynamic interactions between particles.

Despite many successful applications to model, e.g., dense particulate flows [2], history-dependent materials [3], and free-surface flows [4–6], some fundamental issues remain pertaining to such a hybrid particle-mesh strategy. In particular, formulating an accurate and conservative coupling between the scattered particle data and the mesh is a non-trivial issue. Existing approaches generally either compromise conservation in favor of accuracy [3] or vice versa [7–9].

This contribution outlines a particle-mesh algorithm which fundamentally overcomes the aforementioned issue as it conserves the transported quantity both globally and locally (i.e. cellwise), while preserving extensions to arbitrary order accuracy. Key to the approach is the formulation of the particle-mesh projection in terms of a PDE-constrained minimization problem in such a way that, from a mesh-perspective, the transported Lagrangian particle field weakly satisfies an advection equation. The formulation for this optimization problem relies on the use of a hybridized discontinuous Galerkin (HDG) method.

For brevity, we present our method for a scalar hyperbolic conservation law on closed (i.e. no inflow or outflow through the boundaries) or periodic domains. Making combined use of particles and a mesh for this problem has the distinct advantage in that it allows handling the advection term free of any artificial diffusion. Forthcoming work will present the method in a particle-mesh operator splitting context for the advection-diffusion equation and the incompressible Navier–Stokes equations, also including in- and outflow boundaries [10].

The remainder is organized as follows. Section 2 introduces the governing equations, some definitions, and states the problem. Presenting the PDE-constrained particle mesh interaction and proving conservation constitute the main part of Sect. 3. In Sect. 4, we demonstrate the performance of the scheme in terms of accuracy and conservation, and apply the scheme for mass conservative density tracking in multiphase flows. Finally, Sect. 5 summarizes our findings.

2 Governing Equations and Problem Statement

2.1 Governing Equations

We now define the hyperbolic conservation law on the space-time domain $\Omega \times I$ for a scalar quantity ψ_h. Under the simplifying assumption that the solenoidal advective field $\mathbf{a}: \Omega \times I \rightarrow \mathbb{R}^d$ has a vanishing normal component at the boundary (i.e. $\mathbf{a} \cdot \mathbf{n} = 0$ on $\partial\Omega$), this problem reads: given the initial condition $\psi^0 : \Omega \rightarrow \mathbb{R}$, find the scalar quantity $\psi : \Omega \times I \rightarrow \mathbb{R}$ such that

$$\frac{\partial \psi}{\partial t} + \nabla \cdot \mathbf{a}\psi = 0 \qquad \text{in} \quad \Omega \times I, \tag{1a}$$

$$(\mathbf{a} \cdot \mathbf{n})\psi = 0 \qquad \text{on} \quad \Gamma_N^0 \times I, \tag{1b}$$

$$\psi(\mathbf{x}, t^0) = \psi^0 \qquad \text{in} \quad \Omega, \tag{1c}$$

where the notation Γ_N^0 reflects that in the scope of this paper we only consider boundaries with vanishing normal velocity (i.e. $\mathbf{a} \cdot \mathbf{n} = 0$) for the sake of brevity. Hence, note that Γ_N^0 coincides with the domain boundary $\partial\Omega$. For the more generic case, including inflow and outflow boundaries, reference is made to upcoming work [10].

Problem Eq. (1) is solved using a set of scattered, Lagrangian particles, and our aim is to express fields as flux degrees of freedom on an Eulerian background mesh from this set of moving particles in an accurate and physically correct manner. To state this problem mathematically in Sect. 2.3, we first introduce some notation related to the Lagrangian particles and the Eulerian mesh.

2.2 Definitions

Let $\mathcal{X}_t$ define the configuration of Lagrangian particles in the domain Ω at a time instant t

$$\mathcal{X}_t := \{\mathbf{x}_p(t) \in \Omega\}_{p=1}^{N_p}, \tag{2}$$

in which $\mathbf{x}_p$ denotes the spatial coordinates of particle p, and N_p is the number of particles.
Furthermore, a Lagrangian scalar field on the particles is defined as

$$\Psi_t := \left\{\psi_p(t) \in \mathbb{R}\right\}_{p=1}^{N_p}, \tag{3}$$

where ψ_p denotes the scalar quantity associated with particle p.

Next, we define an Eulerian mesh as the triangulation $\mathcal{T} := \{K\}$ of Ω into open, non-overlapping cells K. A measure of the cell size is denoted by h_K, and the outward pointing unit normal vector on the boundary ∂K of a cell is denoted by $\mathbf{n}$. Adjacent cells K_i and K_j $(i \neq j)$ share a common facet $F = \partial K_i \cap \partial K_j$. The set of all facets (including the exterior boundary facets $F = \partial K \cap \partial\Omega$) is denoted by $\mathcal{F}$.

The following scalar finite element spaces are defined on $\mathcal{T}$ and $\mathcal{F}$:

$$W_h := \left\{ w_h \in L^2(\mathcal{T}),\ w_h|_K \in P_k(K)\ \forall\ K \in \mathcal{T} \right\}, \tag{4}$$

$$T_h := \left\{ \tau_h \in L^2(\mathcal{T}),\ \tau_h|_K \in P_l(K)\ \forall\ K \in \mathcal{T} \right\}, \tag{5}$$

$$\bar{W}_h := \left\{ \bar{w}_h \in L^2(\mathcal{F}),\ \bar{w}_h|_F \in P_k(F)\ \forall\ F \in \mathcal{F} \right\}, \tag{6}$$

in which $P(K)$ and $P(F)$ denote the spaces spanned by Lagrange polynomials on K and F, respectively, and $k \geq 1$ and $l = 0$ indicating the polynomial order. The latter is chosen to keep the discussion concise, and reference is made to [10] for the more generic case $l \geq 0$. Also, note that $\bar{W}_h$ is continuous inside cell facets and discontinuous at their borders.

Importantly, we henceforth distinguish between Lagrangian particle data and Eulerian mesh fields by using the subscripts p and h, respectively.

2.3 Problem Statement

We now formulate the two core components comprising our algorithm: solving Eq. (1) in a Lagrangian, particle-based framework, and projecting the Lagrangian quantities to a locally conservative Eulerian mesh field via a particle mesh-projection.

In a Lagrangian, particle-based frame of reference, the advection problem Eq. (1) is solved straightforwardly by decomposing the problem into two ordinary differential equations for the particle scalar quantity and the particle position, given by

$$\dot{\psi}_p(t) = 0, \tag{7a}$$

$$\dot{\mathbf{x}}_p(t) = \mathbf{a}(\mathbf{x}_p(t), t), \tag{7b}$$

where $\dot{\psi}_p(t)$ and $\dot{\mathbf{x}}_p(t)$ are the total derivatives at time t of the scalar quantity and the position of particle p, respectively. From Eq. (7a) it readily follows that the scalar valued particle property remains constant over time, i.e. $\psi_p = \psi_p(0) = \psi^0(\mathbf{x}_p)$. Furthermore, any appropriate time integration method can be used to integrate Eq. (7b) in time, which will not be subject of further discussion. Finally, as a result of our simplifying assumption that $\mathbf{a} \cdot \mathbf{n} = 0$ at the exterior boundary Γ_N^0, we do not consider the inflow and outflow of particles through exterior boundaries, and we refer to [10] for a further discussion of this topic.

Instead, we focus on the reconstruction of a locally conservative mesh field $\psi_h \in W_h$ from the scattered particle data $\psi_p \in \Psi_t$ in a subsequent particle-mesh projection step. Abstractly, this projection $\mathcal{P}_E : \Psi_t \rightarrow W_h$ can be denoted as

$$\psi_h(\mathbf{x}, t) = \mathcal{P}_E\left(\psi_p(t)\right). \tag{8}$$

Our specific aim is to formulate the projection operator $\mathcal{P}_E$ in such a way that local conservation is guaranteed, in the sense that the integral of ψ_h over each element is invariant.

3 PDE-Constrained Particle-Mesh Interaction

3.1 Formulation

In order to define the projection operator $\mathcal{P}_E$, we take as our starting point a local least squares minimization problem [9]

$$\min_{\psi_h \in W_h} J = \sum_p \frac{1}{2}\left(\psi_h(\mathbf{x}_p(t), t) - \psi_p\right)^2. \tag{9}$$

With W_h a discontinuous function space, this approach allows for an efficient cellwise implementation, and gives accurate results provided that the particle configuration satisfies unisolvency (Definition 2.6 in [11]) with respect to W_h. The latter requirement practically implies that the particle locations in a cell are not collinear, and the number of particles in a cell is bounded from below by the number of local basis functions. In the remainder of this work we assume that this criterion is met, so as to focus on a more important issue concerning Eq. (9) in that conservation of the quantity ψ_h cannot be guaranteed a priori.

In order to achieve conservation, Eq. (9) is extended by imposing the additional constraint that the projection has to satisfy a hyperbolic conservation law in a weak sense. To cast this into an optimization problem, the functional in Eq. (9) is augmented with terms multiplying Eq. (1a) with a Lagrange multiplier $\lambda_h \in T_h$. After integration by parts and exploiting that gradients of the Lagrange multiplier vanish on K for $l = 0$, the minimization problem may be stated: given a particle field $\psi_p \in \Psi_t$, and a solenoidal velocity field $\mathbf{a}$, find the stationary points of the Lagrangian functional

$$\mathcal{L}(\psi_h, \bar{\psi}_h, \lambda_h) = \sum_p \frac{1}{2}\left(\psi_h(\mathbf{x}_p(t), t) - \psi_p(t)\right)^2 + \sum_K \oint_{\partial K} \frac{1}{2}\beta\left(\bar{\psi}_h - \psi_h\right)^2 \mathrm{d}\Gamma$$
$$+ \int_\Omega \frac{\partial \psi_h}{\partial t}\lambda_h \mathrm{d}\Omega + \sum_K \oint_{\partial K \setminus \Gamma_N^0} \mathbf{a}\cdot\mathbf{n}\bar{\psi}_h\lambda_h \mathrm{d}\Gamma + \oint_{\Gamma_N^0} \mathbf{a}\cdot\mathbf{n}\psi_h\lambda_h \mathrm{d}\Gamma, \tag{10}$$

The first two terms at the right-hand side in this equation are recognized as a regularized least squares projection, and the last three terms constitute a weak form of the advection problem Eq. (1), with the Lagrange multiplier $\lambda_h \in T_h$ as the weight function. Furthermore, the unknown facet-based field $\bar{\psi}_h \in \bar{W}_h$, resulting from integration by parts, determines the interface flux, and is crucial in providing the required optimality control. The additional term containing $\beta > 0$ penalizes the jumps between ψ_h and $\bar{\psi}_h$ on cell interfaces, thereby avoiding the problem of becoming singular in cases with vanishing normal velocity $\mathbf{a} \cdot \mathbf{n}$.

Equating the variations of Eq. (10) with respect to the three unknowns $(\psi_h, \lambda_h, \bar{\psi}_h) \in (W_h, T_h, \bar{W}_h)$ to zero yields the semi-discrete optimality system. An in-depth derivation can be found in [10]. Here we suffice to present the resulting fully-discrete system, thus assuming that the particle field $\psi_p \in \Psi_t$, the particle positions $\mathbf{x}_p^{n+1} \in \mathcal{X}_t$ after the Lagrangian advection, and the mesh field at the previous time level $\psi_h^n \in W_h$ are known.

Variation with respect to the scalar field $\psi_h^{n+1} \in W_h$ yields the co-state equation

$$\sum_p \left(\psi_h^{n+1}(\mathbf{x}_p^{n+1}) - \psi_p\right) w_h(\mathbf{x}_p^{n+1}) - \sum_K \oint_{\partial K} \beta \left(\bar{\psi}_h^{n+1} - \psi_h^{n+1}\right) w_h \mathrm{d}\Gamma$$

$$+ \int_\Omega \frac{w_h}{\Delta t} \lambda_h^{n+1} \mathrm{d}\Omega + \oint_{\Gamma_N^0} \mathbf{a} \cdot \mathbf{n} \lambda_h^{n+1} w_h \mathrm{d}\Gamma = 0 \quad \forall w_h \in W_h. \tag{11a}$$

Variation with respect to the Lagrange multiplier $\lambda_h^{n+1} \in T_h$ yields the discrete state equation

$$\int_\Omega \frac{\psi_h^{n+1} - \psi_h^n}{\Delta t} \tau_h \mathrm{d}\Omega + \sum_K \oint_{\partial K \setminus \Gamma_N^0} \mathbf{a} \cdot \mathbf{n} \bar{\psi}_h^{n+1} \tau_h \mathrm{d}\Gamma + \oint_{\Gamma_N^0} \mathbf{a} \cdot \mathbf{n} \psi_h^{n+1} \tau_h \mathrm{d}\Gamma = 0 \quad \forall \tau_h \in T_h. \tag{11b}$$

And variation with respect to the facet variable $\bar{\psi}_h^{n+1} \in W_h$ results in the optimality condition

$$\sum_K \oint_{\partial K} \mathbf{a} \cdot \mathbf{n} \lambda_h^{n+1} \bar{w}_h \mathrm{d}\Gamma + \sum_K \oint_{\partial K} \beta \left(\bar{\psi}_h^{n+1} - \psi_h^{n+1}\right) \bar{w}_h \mathrm{d}\Gamma = 0 \quad \forall \bar{w}_h \in \bar{W}_h. \tag{11c}$$

Solving Eq. (11) for $(\psi_h^{n+1}, \lambda_h^{n+1}, \bar{\psi}_h^{n+1}) \in (W_h, T_h, \bar{W}_h)$ yields the reconstructed field $\psi_h^{n+1} \in W_h$.

3.2 Conservation

Next, we will show that from the perspective of the Eulerian field the particle-mesh projection via Eq. (11) indeed conserves mass, both in a global and a local sense. To this end, consider the discrete state equations (Eq. (11b)) and set $\tau_h = 1$. Exploiting the single-valuedness of the facet flux variable $\bar{\psi}$ at the facets $F \in \mathcal{F}$, we readily obtain

$$\int_{\Omega} \frac{\psi_h^{n+1} - \psi_h^n}{\Delta t} \mathrm{d}\Omega = -\sum_K \oint_{\partial K \setminus \Gamma_N^0} \mathbf{a} \cdot \mathbf{n} \bar{\psi}_h^{n+1} \mathrm{d}\Gamma - \oint_{\Gamma_N^0} \mathbf{a} \cdot \mathbf{n}\, \psi_h^{n+1} \mathrm{d}\Gamma = 0, \tag{12}$$

where we made use of the fact that the flux term vanishes at opposing sides of interior facets, and the flux at facets on the exterior boundary vanishes due to our earlier simplification that $\mathbf{a} \cdot \mathbf{n} = 0$ on Γ_N^0.

Local mass conservation follows when setting $\tau_h = 1$ on an interior cell K and $\tau_h = 0$ on $\Omega \setminus K$, resulting in

$$\int_K \frac{\psi_h^{n+1} - \psi_h^n}{\Delta t} \mathrm{d}\Omega = -\oint_{\partial K} \mathbf{a} \cdot \mathbf{n} \bar{\psi}_h^{n+1} \mathrm{d}\Gamma. \tag{13}$$

Thus, the storage over an element balances the net ingoing advective flux through the cell boundary ∂K which proves local conservation in terms of the numerical flux on $\mathcal{F}$.

3.3 Numerical Implementation

The optimality system Eq. (11) leads to a seemingly large global system. However, this system is amenable to an efficient implementation via static condensation by eliminating the unknowns local to a cell, i.e. $(\psi_h, \lambda_h) \in (W_h, T_h)$, in favor of the global control variable $\bar{\psi}_h^{n+1} \in \bar{W}_h$, leading to a much smaller global system which is to be solved for $\bar{\psi}_h^{n+1}$ only. The local unknowns ψ_h^{n+1} and λ_h^{n+1} can be found in a subsequent backsubstitution step [9, 12].

We emphasize that our PDE-constrained particle-mesh projection hinges on the single-valued facet flux variable $\bar{\psi}_h$, acting as the control variable to our optimization procedure. This imperative ingredient is naturally provided by employing a HDG-framework (see, e.g., [12–14]).

4 Numerical Examples

In Sect. 4.1, the convergence and conservation of the scheme is studied for a benchmark test for which an analytical solution is available. Section 4.2 illustrates how the scheme can be applied as a tool for mass conservative density tracking in multiphase flows.

4.1 Convergence Study: Translation of Periodic Pulse

Following LeVeque [15], the translation of a sinusoidal profile $\psi(\mathbf{x}, 0) = \sin 2\pi x \sin 2\pi y$ on the bi-periodic unit square is considered. The velocity field $\mathbf{a} = [1, 1]^\top$ is used, so that at $t = 1$ the initial data should be recovered. The β-parameter is set to 1e-6, and a simple Euler scheme suffices for exact particle advection. Using different polynomial orders $k = 1, 2, 3$, the accuracy of the method is assessed at $t = 1$ by refining the mesh and the time step. We assign approximately a safe number of 28 particles per cell initially in order to comply with the unisolvency criterion. Furthermore, the time step corresponds to a CFL-number of approximately 1. The errors as well as the convergence rates are tabulated in Table 1. Optimal convergence rates of order $k + 1$ are observed, thus revealing the accuracy of our approach.

Table 1 also shows the local mass conservation error at $t = 1$, with this error for a time level $n + 1$ being defined as

$$\epsilon_{\Delta\phi_K} = \left(\sum_K \left(\int_K \frac{\psi_h^{n+1} - \psi_h^n}{\Delta t} \mathrm{d}\Omega + \oint_{\partial K} \mathbf{a} \cdot \mathbf{n} \bar{\psi}_h^{n+1} \mathrm{d}\Gamma \right)^2 \right)^{1/2}. \tag{14}$$

As expected, mass is conserved locally in terms of the facet flux. Global mass conservation is readily verified by noting that the facet flux term cancels at opposing sides of the facets.

Table 1 Translating pulse: overview of model runs with the associated L^2-error $\|\psi - \psi_h\|$, the convergence rate and the local mass conservation error $\epsilon_{\Delta\phi_K}$ at time $t = 1$

		$(k, l) = (1, 0)$			$(k, l) = (2, 0)$			$(k, l) = (3, 0)$		
Cells	Δt	Error	Rate	$\epsilon_{\Delta\phi_K}$	Error	Rate	$\epsilon_{\Delta\phi_K}$	Error	Rate	$\epsilon_{\Delta\phi_K}$
128	0.1	6.0e−2	–	1.3e−15	4.3e−3	–	4.5e−15	3.3e−4	–	1.3e−15
512	0.05	1.6e−2	1.9	7.7e−16	5.5e−4	3.0	3.1e−16	2.1e−5	4.0	7.1e−16
2048	0.025	3.9e−3	2.0	4.3e−16	6.9e−5	3.0	2.5e−16	1.3e−6	4.0	4.4e−16
8192	0.0125	9.8e−4	2.0	2.7e−16	8.6e−6	3.0	2.3e−16	8.2e−8	4.0	3.5e−16

4.2 Application: Mass Conservative Rayleigh–Taylor Instability

We next illustrate how the above presented scheme can be used for a mass conservative multiphase scheme in which particles are used for the tracking of sharp interfaces. As an example, we take the Rayleigh–Taylor instability test from [16] with an Atwood number of 0.5 and a Reynolds number of 256. In addition to a PDE-constrained particle-mesh strategy for tracking the density fields, we also track specific momentum at the particle level and enforce incompressibility and viscous forces via a Stokes step at the mesh. Details of such a particle-mesh operator splitting approach for the Navier–Stokes equations can be found in [10]. A regular and symmetric mesh with $60 \times 240 \times 4$ cells is used. Initially, approximately 20 particles per cell are assigned, and we note that advecting the particles through a pointwise divergence free velocity field obviates the need for a particle reseeding strategy [9, 10]. Furthermore, we use polynomial orders $(k, l) = (1, 0)$ and use a timestep $\Delta t = 1\text{e}{-3}$. Particles are advected using an explicit, three-stage Runge–Kutta scheme [17]. The evolution of the initial perturbation is visually assessed in Fig. 1. The sharp interface between the two phases is maintained and the interface shape is qualitatively in good agreement with [16]. Most importantly, computations confirm that the total mass remains constant to machine precision.

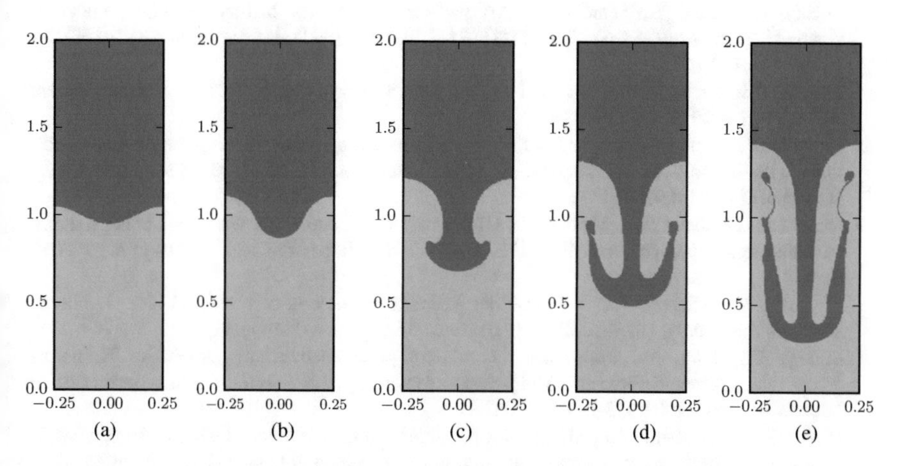

Fig. 1 Rayleigh–Taylor: time evolution of density field at particle level for Re = 256. (**a**) t = 0. (**b**) t = 0.25. (**c**) t = 0.5. (**d**) t = 0.75. (**e**) t = 1.0

5 Conclusions

We outlined a particle-mesh projection which enables the reconstruction of high-order accurate and diffusion-free mesh fields from a set of scattered Lagrangian particles. By casting the problem as a PDE-constrained optimization problem discrete conservation principles can be derived. Importantly, in the presented optimization strategy the advective flux was expressed in terms of a flux variable at the facet which provides the required optimality control. Such a facet function is typical to an HDG approach, and comes with the additional benefit that the resulting scheme can be implemented efficiently via static condensation. The scheme was assessed in terms of accuracy and conservation, and we highlighted a potential application to multiphase flows.

References

1. Evans, M., Harlow, F., Bromberg, E.: The Particle-in-Cell Method for Hydrodynamic Calculations. Technical Report. Los Alamos Scientific Laboratory, Mexico (1957)
2. Snider, D.: An incompressible three-dimensional multiphase particle-in-cell model for dense particle flows. J. Comput. Phys. **170**(2), 523–549 (2001). https://doi.org/10.1006/jcph.2001.6747
3. Sulsky, D., Chen, Z., Schreyer, H.: A particle method for history-dependent materials. Comput. Meth. Appl. Mech. Eng. **118**(1–2), 179–196 (1994). https://doi.org/10.1016/0045-7825(94)90112-0
4. Zhu, Y., Bridson, R.: Animating sand as a fluid. ACM Trans. Graph. **24**(3), 965 (2005). https://doi.org/10.1145/1073204.1073298
5. Kelly, D.M., Chen, Q., Zang, J.: PICIN: a particle-in-cell solver for incompressible free surface flows with two-way fluid-solid coupling. SIAM J. Sci. Comput. **37**(3), 403–424 (2015). https://doi.org/10.1137/140976911
6. Maljaars, J., Labeur, R.J., Möller, M., Uijttewaal, W.: A numerical wave tank using a hybrid particle-mesh approach. Proc. Eng. **175**, 21–28 (2017). https://doi.org/10.1016/j.proeng.2017.01.007
7. Edwards, E., Bridson, R.: A high-order accurate particle-in-cell method. Int. J. Numer. Methods Eng. **90**(9), 1073–1088 (2012). https://doi.org/10.1002/nme.3356
8. Sulsky, D., Gong, M.: Improving the material-point method. In: Innovative Numerical Approaches for Multi-Field and Multi-Scale Problems, pp. 217–240. Springer, Berlin (2016). https://doi.org/10.1007/978-3-319-39022-2_10
9. Maljaars, J.M., Labeur, R.J., Möller, M.: A hybridized discontinuous Galerkin framework for high–order particle–mesh operator splitting of the incompressible Navier–Stokes equations. J. Comput. Phys. **358**, 150–172 (2018). https://doi.org/10.1016/j.jcp.2017.12.036
10. Maljaars, J.M., Labeur, R.J., Trask, N., Sulsky, D.: Conservative, high-order particle-mesh scheme with applications to advection-dominated flows. Comput. Methods Appl. Mech. Eng. **348**, 443–465 (2019). ISSN 0045-7825. https://doi.org/10.1016/J.CMA.2019.01.028
11. Wendland, H.: Scattered Data Approximation, vol. 17. Cambridge University Press, Cambridge (2004)
12. Rhebergen, S., Wells, G.N.: Analysis of a hybridized/interface stabilized finite element method for the stokes equations. SIAM J. Numer. Anal. **55**(4), 1982–2003 (2017). https://doi.org/10.1137/16M1083839

13. Wells, G.N.: Analysis of an interface stabilized finite element method: the advection-diffusion-reaction equation. SIAM J. Numer. Anal. **49**(1), 87–109 (2011). https://doi.org/10.1137/090775464
14. Labeur, R.J., Wells, G.N.: Energy stable and momentum conserving hybrid finite element method for the incompressible Navier–Stokes equations. SIAM J. Sci. Comput. **34**(2), 889–913 (2012). https://doi.org/10.1137/100818583
15. LeVeque, R.J.: High-resolution conservative algorithms for advection in incompressible flow. SIAM J. Numer. Anal. **33**(2), 627–665 (1996). https://doi.org/10.1137/0733033
16. He, X., Chen, S., Zhang, R.: A lattice boltzmann scheme for incompressible multiphase flow and its application in simulation of Rayleigh–Taylor instability. J. Comput. Phys. **152**(2), 642–663 (1999). https://doi.org/10.1006/jcph.1999.6257
17. Ralston, A.: Runge–Kutta methods with minimum error bounds. Math. Comput. **16**(80), 431–437 (1962). https://doi.org/10.2307/2003133

Krylov Smoothing for Fully-Coupled AMG Preconditioners for VMS Resistive MHD

Paul T. Lin, John N. Shadid, and Paul H. Tsuji

Abstract This study explores the use of a Krylov iterative method (GMRES) as a smoother for an algebraic multigrid (AMG) preconditioned Newton–Krylov iterative solution approach for a fully-implicit variational multiscale (VMS) finite element (FE) resistive magnetohydrodynamics (MHD) formulation. The efficiency of this approach is critically dependent on the scalability and performance of the AMG preconditioner for the linear solutions and the performance of the smoothers play an essential role. Krylov smoothers are considered an attempt to reduce the time and memory requirements of existing robust smoothers based on additive Schwarz domain decomposition (DD) with incomplete LU factorization solves on each subdomain. This brief study presents three time dependent resistive MHD test cases to evaluate the method. The results demonstrate that the GMRES smoother can be faster due to a decrease in the preconditioner setup time and a reduction in outer GMRESR solver iterations, and requires less memory (typically 35% less memory for global GMRES smoother) than the DD ILU smoother.

Keywords Multigrid · Krylov smoother · Preconditioner · Newton–Krylov · Finite element method · Magnetohydrodynamics

---

P. T. Lin (✉)
Sandia National Laboratories, Albuquerque, NM, USA
e-mail: paullin@lbl.gov

J. N. Shadid
Center for Computing Research, Sandia National Laboratories, Albuquerque, NM, USA
e-mail: jnshadi@sandia.gov

P. H. Tsuji
Lawrence Livermore National Laboratory, Livermore, CA, USA

H. van Brummelen et al. (eds.), *Numerical Methods for Flows*,
Lecture Notes in Computational Science and Engineering 132,
https://doi.org/10.1007/978-3-030-30705-9_24

1 Introduction

The resistive magnetohydrodynamics (MHD) model describes the dynamics of charged fluids in the presence of electromagnetic fields and is used as a base-level continuum plasma model. Resistive MHD models find application in fundamental plasma physics phenomena, fusion science technology applications and astrophysical phenomena [12]. The MHD system is strongly coupled, highly nonlinear and characterized by coupled physical phenomena that induce a very large range of time-scales in the response of the system. These characteristics make the scalable, robust, accurate, and efficient computational solution of these systems extremely challenging. In this context fully-implicit formulations, coupled with effective robust iterative solution methods, become attractive, as they have the potential to provide stable, higher-order time-integration of these complex multiphysics systems when long dynamical time-scales are of interest (see e.g. [7, 8, 20, 21]).

Krylov iterative linear solver algorithms are among the fastest and most robust iterative solvers for a wide variety of applications [13, 18]. The key factor influencing the robustness and efficiency of these solution methods is the choice of preconditioner. Among current preconditioning techniques multilevel type methods (e.g. two-level domain decomposition, multigrid and algebraic multigrid (AMG)) have been demonstrated to provide scalable solutions to a wide range of challenging linear systems [23]. Multigrid scalability and performance is critically dependent on both the projection and the smoothers. This study focuses on the latter, specifically the performance of smoothers based on a Krylov type iterative method (e.g. GMRES) applied to the fully-coupled Newton–Krylov algebraic multigrid preconditioned solution approach described in [21]. In this context the solution of the discrete system developed by a fully-implicit backward differentiation (BDF) type formulation of a variational multiscale (VMS) finite element spatial discretization of the resistive MHD system is considered [21].

In the context of fully-coupled direct-to-steady-state solution methods for VMS CFD and MHD large linear non-symmetric systems, experience has indicated that robust, and therefore more expensive, smoothing methods are required [19–21] and local incomplete LU factorizations (ILU) have been shown to be an effective smoother [14, 19, 21]. This robustness however comes at a price since these methods are expensive and require large setup time and larger memory requirements to compute the ILU factors. We consider the case of transient resistive MHD problems and carry out an initial assessment of AMG preconditioned Krylov methods based on a few well known standard stationary iterative methods (e.g. Jacobi, Gauss-Seidel, etc.) and the recursive application of a Krylov method used as a smoother.

Previous work employing Krylov smoothers with multigrid for SPD problems has considered scalar elliptic problems and elasticity problems [1, 3, 4, 16]. There has been considerably less previous work employing Krylov smoothers for Helmholtz problems [10] and for nonsymmetric systems; for example [2] has considered Krylov smoothers for the convection-diffusion equation. Additionally and independently we have considered Krylov smoothers for the MHD equations [15]; the present study is an extension of this work.

2 Resistive MHD Model Equations and Discretization

The governing equations considered in this study are the 3D resistive iso-thermal MHD equations including dissipative terms for the momentum and magnetic induction equations [12]. This model provides a continuum description of charged fluids in the presence of electromagnetic fields. The system of equations in residual form:

$$\mathbf{R}_m = \frac{\partial(\rho\mathbf{u})}{\partial t} + \nabla\cdot[\rho\mathbf{u}\otimes\mathbf{u} - \frac{1}{\mu_0}\mathbf{B}\otimes\mathbf{B} + (P + \frac{1}{2\mu_0}\|\mathbf{B}\|^2)\mathbf{I} - \mu[\nabla\mathbf{u} + \nabla\mathbf{u}^T]] = \mathbf{0}$$

$$R_P = \frac{\partial\rho}{\partial t} + \nabla\cdot[\rho\mathbf{u}] = 0$$

$$\mathbf{R}_I = \frac{\partial\mathbf{B}}{\partial t} + \nabla\cdot\left[\mathbf{u}\otimes\mathbf{B} - \mathbf{B}\otimes\mathbf{u} - \frac{\eta}{\mu_0}\left(\nabla\mathbf{B} - (\nabla\mathbf{B})^T\right) + \psi\mathbf{I}\right] = \mathbf{0}.$$

Here $\mathbf{u}$ is the plasma velocity; ρ is the ion mass density; P is the plasma pressure; $\mathbf{B}$ is the magnetic induction (also termed the magnetic field) that is subject to the divergence-free involution $\nabla\cdot\mathbf{B} = 0$. In this formulation the Lagrange multiplier, ψ is introduced to allow numerical enforcement of the divergence involution as a constraint, $R_\psi = \nabla\cdot\mathbf{B} = 0$ [9, 21]. This study focuses on the variable density low-Mach-number compressible case. A finite element (FE) discretization of the stabilized equations gives rise to a system of coupled, nonlinear, non-symmetric algebraic equations, the numerical solution of which can be very challenging. These equations are linearized using an inexact form of Newton's method. The result of stabilization is that two weak form Laplacians, a discrete "pressure Laplacian" and "Lagrange multiplier Laplacian" are introduced on the block diagonals for the continuity equation and the solenoidal constraint on the magnetic field (Lagrange multiplier equation). These non-zero blocks allow approximate inversion of the discrete algebraic system with ILU and Gauss-Seidel type methods for transient problems (see [21] for a more detailed description).

3 Fully-Coupled Newton–Krylov Multigrid Preconditioned Solution Approach

Numerical discretization of the governing equations produces a large sparse, strongly-coupled nonlinear system. Although fully-coupled Newton–Krylov techniques [5], where a Krylov solver is used to solve the linear system generated by a Newton's method, are robust, efficient solution of the large sparse linear system that must be solved for each nonlinear iteration is challenging [13]. The performance, efficiency and scalability of the preconditioner is critical [13]. It is well known in the literature that Schwarz domain decomposition preconditioners do not scale due to lack of global coupling [6, 17]. Multigrid methods are one of the most efficient techniques for solving large linear systems [23]. As we have described our Newton–Krylov preconditioned by algebraic multigrid solution method in detail in our previous work [14, 21], we provide only a very brief description here. We employ a nonsmoothed aggregation multigrid approach with uncoupled aggregation. For systems of partial differential equations (PDEs), aggregation is performed on the graph where all the PDEs per mesh node is represented by a single vertex. The discrete equations are projected to the coarser level employing a Galerkin fashion with a triple matrix product, $A_{\ell+1} = R_\ell A_\ell P_\ell$, where R_ℓ restricts the residual from level ℓ to level ℓ+1, A_ℓ is the discretization matrix on level ℓ and P_ℓ prolongates the correction from level ℓ+1 to ℓ. We typically employ both pre- and post-smoothing on each level of the multigrid V-cycle.

The Trilinos framework provides the preconditioned Newton–Krylov method and preconditioners used for this work. Krylov methods are provided by the Aztec [24] library and the multigrid cycles and grid transfers are provided by ML [11].

As mentioned in the introduction, we were interested in evaluating Krylov smoothers compared with our standard ILU smoother for our MHD test cases. For our Newton–Krylov solution approach, a GMRES solver is employed that is preconditioned by multigrid with ILU smoother. Because our test cases matrices are nonsymmetric, the choice of GMRES [18] for the Krylov smoother for our initial evaluations is appropriate. When the Krylov/GMRES smoother is employed, there are two "levels" of Krylov methods and possibly two "levels" of preconditioners. Because the preconditioner is changing due to the GMRES smoother, it is necessary to employ a GMRES approach such as GMRESR [25] for the "outer" GMRES Krylov method that is then preconditioned by multigrid. Each level of the multigrid V-cycle employs a GMRES smoother, which is denoted as the "inner" GMRES. This "inner" GMRES can also be preconditioned, e.g. by a standard relaxation approach such as point or block Jacobi or Gauss-Seidel.

4 Results and Discussion

4.1 3D Taylor-Green MHD Turbulent Vortex Decay

The first transient MHD test case is a Taylor-Green vortex generalized to MHD as described in [22], with Reynolds number $Re = 1800$ and magnetic Reynolds number $Re_m = 1800$. We employ the full VMS formulation as described in [22]. Table 1 presents a weak scaling study for three different mesh sizes: 128^3, 256^3 and 512^3 elements cubed (16.8M, 134M and 1.1 billion DOFs) run on 256, 2048 and 16,384 cores respectively (one MPI process per core) of a linux cluster that consists of dual-socket Intel Xeon 2.6 GHz 8-core Sandy Bridge processors. 20

Table 1 Taylor-Green turbulent MHD vortex decay test case weak scaling for 16.8M, 134M and 1.1B DOFs run on 256, 2048 and 16,384 cores of an Intel Xeon linux cluster

Mesh	Smoother		GMRESR	Time(s)			Memory
		Prec	iterations/Δt	Prec setup	Solve	Linear solve	(MB)
128^3 elem	SGS		87.7	18	730	748	1050
(16.8M DOFs):	ILU(0)overlap=1		14.2	263	97	360	1440
256 cores	GMRES	noprec	15.4	22.7	267	290	917
		bkJac	13.1	35	238	273	927
		ptGS	13.6	24	413	437	917
		bkGS	12.0	34	539	573	930
256^3 elem	SGS		Failed				
(134M DOFs):	ILU(0)overlap=1		20.2	311	134	445	1440
2048 cores	GMRES	noprec	18.7	23.5	328	352	920
		bkJac	15.7	41.3	306	348	933
		ptGS	16.9	26.8	516	542	920
		bkGS	17.1	40.9	954	995	936
512^3 elem	ILU(0)overlap=1		37.3	414	276	690	1520
(1.1B DOFs):	GMRES	noprec	31.0	34	590	624	1000
16,384 cores		bkJac	21.9	51	482	533	1020
		ptGS	20.0	47	1176	1223	1023

Columns 4–8: GMRESR iterations per time step (sum of “outer” GMRESR iterations over the Newton steps in a time step), total preconditioner setup time (“prec time”), total linear system solve time (“solve”), sum of total preconditioner setup and solve time (“linear solve”) which is the metric for comparing the different smoothers, and maximum high water memory over MPI processes

time steps were used for all the simulations, and weak scaling was performed with fixed CFL (CFL $\approx$ 0.5). A BDF3 time integration approach was employed. We refer to the linear system solve time or preconditioned iteration time (i.e. not including preconditioner setup) as the “solve” time. The “linear solve” time is the sum of this “solve” time and the preconditioner setup time. Results are presented for various smoothers: sub-domain Symmetric Gauss-Seidel (SGS) with no overlap, ILU(0) with overlap of 1 (“ILU(0)overlap=1”) and GMRES smoother with no preconditioner (“noprec”) as well as block Jacobi (“bkJac”), point Gauss-Seidel (“ptGS”), and block Gauss-Seidel (“bkGS”). The fourth column is the outer GMRESR iterations per time step, and is the sum of the GMRESR iterations over the Newton steps within the time step. The eighth column is the maximum high watermark memory usage over the MPI processes. In general, standard relaxation smoothers are not sufficiently robust for our MHD test cases, and we seldom employ them. For the 16.8M DOFs case, the outer GMRESR iterations per time for SGS smoother is considerably higher than the other smoothers, which makes it uncompetitive compared with the standard ILU smoother (more than double the linear solve time). When the mesh is uniformly refined to 134M DOFs, with SGS smoother the outer GMRESR Krylov solver no longer converges (convergence stalls). Therefore we do not consider the SGS smoother for the 1.1B DOFs case or for any further test cases in this work. For the standard ILU smoother, the cost for smoother setup (time to compute the ILU factors) is very expensive, and considerably larger than the solve time. For the GMRES smoothers, while the setup time is inexpensive compared to ILU, the solve time is considerably more expensive compared to the ILU smoother. GMRES smoother with either no preconditioner or block Jacobi preconditioner is 20 and 25% faster (for linear solve time) than the standard ILU smoother respectively, while requiring only 66% of the memory. For all the cases involving the GMRES smoother, five iterations of GMRES are employed. Other than for the SGS smoother for which the outer GMRESR Krylov solver no longer converges (convergence stalls), the trends for the 134M DOFs case are similar to the 16.8M DOFs case.

For the 1.1B DOFs on 16,384 cores case, the GMRES smoother with either no preconditioner or block Jacobi preconditioner is 10 and 23% faster (for linear solve time) than the standard ILU smoother respectively, while requiring only 66% of the memory. For ILU smoother, the factorization is expensive, but once the factors are obtained, applying the factors is considerably less expensive. For the GMRES smoother, the preconditioner setup is inexpensive, but the solve time is expensive. It is a trade-off between the expensive ILU factorization for setup versus the expensive solve for GMRES smoother. The GMRES smoother can considerably lower the number of outer GMRESR iterations, but cost per iteration is not inexpensive.

4.2 3D Island Coalescence

The island coalescence problem follows the unstable evolution of two 3D current tubes (in the cross plane—islands) embedded in a sheared magnetic field Harris sheet and is described in detail in [21].

Table 2 presents a weak scaling study for the test case described in [21] with Lundquist number 10^3 for 64^3 element cube mesh (2.1M DOFs), 128^3 element cube mesh (16.9M DOFs) and 256^3 element cube mesh (135M DOFs) run on 64, 512 and 4096 cores of an Intel Xeon 2.6 GHz Sandy Bridge linux cluster. For this study, the time step is fixed at 0.1 and run to simulation time 4.0 (40 time steps). Each time the mesh is refined, the CFL is doubled. For the 256^3 element cube mesh the time scale for Alfven wave is associated with $CFL_A = (\frac{B}{\sqrt{\rho\mu_0}}\Delta t)/\Delta x \approx 15$. For the 64^3 element cube mesh, all the Krylov smoothers are faster than the standard ILU smoother, with the GMRES smoother with no preconditioner or block Jacobi preconditioner being 40–50% faster. Here the GMRES smoother is faster than the standard ILU smoother. As with the two coarser meshes, for the 135M DOFs case, the global GMRES smoother is faster than the standard ILU smoother (30–35% reduction in time) while only requiring 65% of the memory.

Table 2 Island coalescence test case weak scaling study for 2.1M, 16.9M and 135M DOFs run on 64, 512 and 4096 cores

Mesh	Smoother	Prec	GMRESR iterations/Δt	Time(s) Prec setup	Solve	Linear solve	Memory (MB)
64^3 elements	ILU(0)overlap=1		11.9	410	62	472	917
(2.1M DOFs):	GMRES	noprec	17.5	25	244	270	631
64 cores		bkJac	12.9	42	204	246	631
128^3 elements	ILU(0)overlap=1		15.1	455	96	552	907
(16.9M DOFs):	GMRES	noprec	21.6	30	378	407	645
512 cores		bkJac	14.6	50	362	412	632
256^3 elements	ILU(0)overlap=1		15.5	528	110	638	922
(135M DOFs):	GMRES	noprec	21.8	67	380	448	666
4096 cores		bkJac	15.5	88	317	406	653

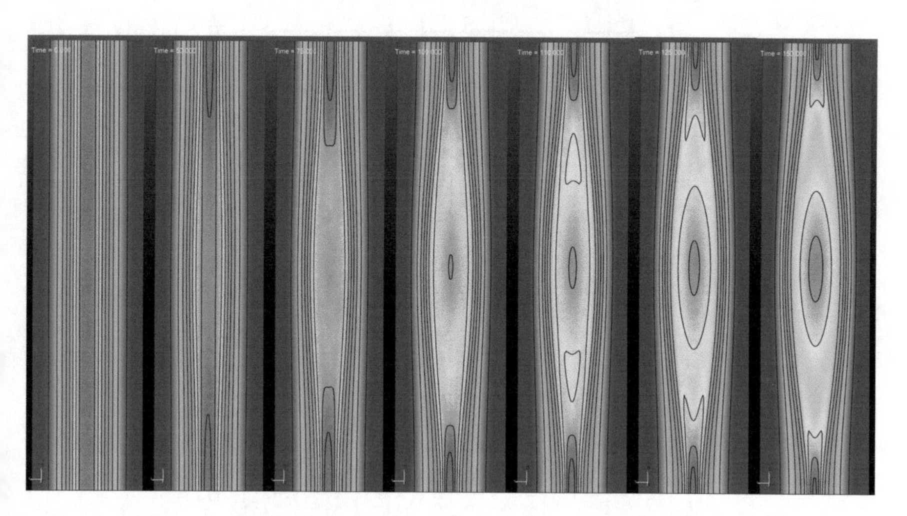

Fig. 1 2D compressible tearing mode test case. Colored contours and isolines of the current J_z are shown at times $t = 0, 50, 75, 100, 110, 125, 150$

4.3 2D Compressible Tearing Mode

The final test problem is a 2D low flow-Mach number compressible tearing mode simulation that follows the unstable evolution of a thin current sheet formed by a sheared magnetic field (Harris sheet) within a initially stationary velocity field [8]. The domain is a rectangle [0,4] × [0,1] and the computation is for a Lundquist number of 10^3. The thin current sheet becomes unstable and forms a magnetic island structure as presented in Fig. 1 with the x-axis oriented in the vertical direction.

Table 3 presents a weak scaling study for the 2D tearing mode for a sequence of four meshes (232k, 1.3M, 5.1M and 20.5M DOFs run on 36, 144, 576 and 2304 cores of an Intel Xeon 2.1 GHz 18-core Broadwell linux cluster). All were run to the

Table 3 2D tearing mode compressible test case weak scaling study run on an Intel Xeon Broadwell linux cluster

Mesh	Smoother		GMRESR	Time(s)				Memory
		Prec	iterations/Δt	Prec setup	Solve	Linear solve	(Lin sol)/Δt	(MB)
232k DOFs	ILU(2)overlap=1		112.5	1349	4017	5366	8.88	503
36 cores	GMRES	noprec	13.61	50.0	825.6	875.6	1.45	426
		bkJac	14.63	90.9	1069	1160	1.92	431
1.3M DOFs	ILU(2)overlap=1		274	2349	21,310	23,659	19.7	575
144 cores	GMRES	noprec	13.0	64.6	1410	1475	1.23	414
		bkJac	19.0	139	2568	2707	2.25	415
5.1M DOFs	ILU(2)overlap=1		Failed					
576 cores	GMRES	noprec	12.87	143.7	2698	2842	1.19	425
		bkJac	20.90	289.5	5559	5849	2.44	425
20.5M DOFs	ILU(2)overlap=1		Failed					
2304 cores	GMRES	noprec	12.94	543.3	5977	6520	1.36	444
		bkJac	22.60	859.3	13,350	14,209	2.97	447

Column 8, the average linear solve time per time step, is the metric for comparison between different meshes same simulation time of $t = 150$ with the fast magnetosonic CFL fixed at 75. The ILU(2) with overlap=1 smoother struggled for the 1.3M DOFs test case, and would not converge for subsequent refinements of the mesh. The GMRES smoother was considerably more robust, with good convergence for the sequence of meshes, with the no preconditioner case performing considerably better than the block Jacobi preconditioner (the former taking half the time per time step as the latter).

5 Conclusions

In this study, we evaluated the use of Krylov smoothers for multigrid as an alternative smoother to our robust but expensive in terms of time and memory standard ILU smoother for our fully-coupled Newton–Krylov algebraic preconditioned multigrid solution approach for large-scale VMS resistive MHD simulations. Our study considered three transient MHD simulations, but more test cases need to be considered. The GMRES smoother can be faster due to reduction in outer GMRESR solver iterations and requires less memory (typically 35% less memory for GMRES smoother) than our standard ILU smoother. The GMRES smoother was considerably more robust than the ILU smoother for the third test case. Our next steps are to evaluate the Krylov smoother at very large scales, explore other Krylov methods, and try to better analyze mathematically the behavior of the Krylov smoothers.

Acknowledgements The authors would like to thank R. Pawlowski and E. Cyr for their collaborative effort in developing the Drekar MHD code and D. Sondak and T. Smith for the collaborative development of the VMS LES turbulent MHD modeling capability. The authors gratefully acknowledge the generous support of the DOE NNSA ASC program and the DOE Office of Science AMR program at Sandia National Laboratories. Sandia National Laboratories is a multimission laboratory managed and operated by National Technology and Engineering Solutions of Sandia, LLC., a wholly owned subsidiary of Honeywell International, Inc., for the U.S. Department of Energy, National Nuclear Security Administration under contract DE-NA-0003525.

References

1. Bank, R.E., Douglas, C.C.: Sharp estimates for multigrid rates of convergence with general smoothing and acceleration. SIAM J. Numer. Anal. **22**, 617–633 (1983)
2. Birken, P., Bull, J., Jameson, A.: A study of multigrid smoothers used in compressible CFD based on the convection diffusion equations. In: Papadrakakis M., et al. (eds.) Proceedings of the VII ECCOMAS Congress 2016, Crete Island (2016)
3. Bornemann, F.A., Deuflhard, P.: The cascadic multigrid method for elliptic problems. Numerische Mathematik **75**, 135–152 (1996)
4. Braess, D.: On the combination of the multigrid method and conjugate gradients. In: Multigrid Methods II. Lecture Notes in Mathematics, vol. 1228, pp. 52–64. Springer, Berlin (2006)

5. Brown, P.N., Saad, Y.: Hybrid Krylov methods for nonlinear systems of equations. SIAM J. Sci. Stat. Comput. **11**, 450–481 (1990)
6. Cai, X.-C., Sarkis, M.: A restricted additive Schwarz preconditioner for general sparse linear systems. SIAM J. Sci. Comput. **21**, 792–797 (1999)
7. Chacón, L.: A non-staggered, conservative, $\nabla \cdot \mathbf{B} = 0$, finite-volume scheme for 3D implicit extended magnetohydrodynamics in curvilinear geometries. Comput. Phys. Comm. **163**, 143–171 (2004)
8. Chacón, L.: An optimal, parallel, fully implicit Newton-Krylov solver for three-dimensional visco-resistive magnetohydrodynamics. Phys. Plasmas **15**, 056103 (2008)
9. Codina, R., Hernández-Silva, N.: Stabilized finite element approximation of the stationary magneto-hydrodynamics equations. Comput. Mech. **38**, 344–355 (2006)
10. Elman, H.C., Ernst, O.G., O'Leary, D.P.: A multigrid method enhanced by Krylov subspace iteration for discrete Helmholtz equations. SIAM J. Sci. Commun. **23**, 1291–1315 (2001)
11. Gee, M.W., Siefert, C.M., Hu, J.J., Tuminaro, R.S., Sala, M.G.: ML 5.0 smoothed aggregation user's guide. Technical Report SAND2006-2649. Sandia National Laboratories, New Mexico (2006)
12. Goedbloed, H., Poedts, S.: Principles of Magnetohydrodynamics with Applications to Laboratory and Astrophysical Plasmas. Cambridge University Press, Cambridge (2004)
13. Knoll, D.A., Keyes, D.E.: Jacobian-free Newton–Krylov methods: a survey of approaches and applications. J. Comput. Phys. **193**, 357–397 (2004)
14. Lin, P.T., Sala, M., Shadid, J.N., Tuminaro, R.S.: Performance of fully coupled algebraic multilevel domain decomposition preconditioners for incompressible flow and transport. Int. J. Num. Meth. Eng. **67**(2), 208–225 (2006)
15. Lin, P., Shadid, J., Tsuji, P., Hu, J.J.: Performance of smoothers for algebraic multigrid preconditioners for finite element variational multiscale incompressible magnetohydrodynamics. In: Proceedings of SIAM PP16. SIAM, Philadelphia (2016)
16. Notay, Y., Vassilevski, P.: Recursive Krylov-based multigrid cycles. Numer. Linear Algebra Appl. **15**, 473–487 (2008)
17. Quarteroni, A., Valli, A.: Domain Decomposition Methods for Partial Differential Equations. Oxford University Press, Oxford (1999)
18. Saad, Y.: Iterative Methods for Sparse Linear Systems. SIAM, Philadelphia (2003)
19. Shadid, J.N., Tuminaro, R.S., Devine, K.D., Hennigan, G.L., Lin, P.T.: Performance of fully-coupled domain decomposition preconditioners for finite element transport/reaction simulations. J. Comput. Phys. **205**(1), 24–47 (2005)
20. Shadid, J.N., Pawlowski, R.P., Banks, J.W., Chacón, L., Lin, P.T., Tuminaro, R.S.: Towards a scalable fully-implicit fully-coupled resistive MHD formulation with stabilized FE methods. J. Comput. Phys. **229**(20), 7649–7671 (2010)
21. Shadid, J.N., Pawlowski, R.P., Cyr, E.C., Tuminaro, R.S., Chacon, L., Weber, P.D.: Scalable implicit incompressible resistive MHD with stabilized FE and fully-coupled Newton–Krylov–AMG. Comput. Methods Appl. Mech. Eng. **304**, 1–25 (2016)
22. Sondak, D., Shadid, J.N., Oberai, A.A., Pawlowski, R.P., Cyr, E.C., Smith, T.M.: A new class of finite element variational multiscale turbulence models for incompressible magnetohydrodynamics. J. Comput. Phys. **295**, 596–616 (2015)
23. Trottenberg, U., Oosterlee, C., Schüller, A.: Multigrid. Academic, London (2001)
24. Tuminaro, R.S., Heroux, M., Hutchinson, S.A., Shadid, J.N.: Aztec User's Guide–Version 2.1. Technical Report SAND99-8801J. Sandia National Laboratories, New Mexico (1999)
25. Van der Vorst, H.A., Vuik, C.: GMRESR: a family of nested GMRES methods. Numer. Linear Algebra Appl. **1**, 369–386 (1994)

Double Layer Potential Density Reconstruction Procedure for 3D Vortex Methods

Ilia K. Marchevsky and Georgy A. Shcheglov

Abstract A new approach is developed for the no-slip boundary condition in vortex methods. The procedure of double layer potential density reconstruction is considered, which consist of two steps. Firstly the integral equation with respect to vortex sheet intensity is solved, which expresses the equality between the tangential components of flow velocity limit value and the body surface velocity. It is solved by using a Galerkin approach. Secondly, the least-squares procedure is implemented, which permits to find nodal values of the potential. It is shown that the developed algorithm makes it possible to improve significantly the quality of solution for the bodies with very complicated geometry and low-quality surface meshes. It can be useful for CFD applications and visual effects reproducing in computer graphics.

Keywords Vortex method · Boundary integral equation · Double layer potential · Vortex sheet · Singularity exclusion · Least squares method

1 Introduction

Vortex methods are a well-known tool for unsteady incompressible flows simulation and coupled FSI-problems solution in number of engineering applications [1, 3]. These methods are also useful in computer graphics in visual effects simulation [9]. One of the key problems in vortex method development is connected to boundary condition satisfaction with high accuracy. The aim of this paper is to develop a new numerical approach to improve the existing numerical schemes of vortex methods.

I. K. Marchevsky (✉) · G. A. Shcheglov
Bauman Moscow State Technical University, Moscow, Russia
Ivannikov Institute for System Programming of the Russian Academy of Sciences, Moscow, Russia
e-mail: shcheglov_ga@bmstu.ru

H. van Brummelen et al. (eds.), *Numerical Methods for Flows*,
Lecture Notes in Computational Science and Engineering 132,
https://doi.org/10.1007/978-3-030-30705-9_25

2 Integral Equations Arising in Vortex Methods

The problem of 3D incompressible flow simulation around an immovable body is considered. The governing equations are the Navier–Stokes equations with no-slip boundary conditions on the body surface K and perturbation decay conditions.

It is well-known from physical point of view, that in order to take into account the presence of the body in the flow, it is possible to replace it with the vortex sheet of unknown intensity $\boldsymbol{\gamma}(\mathbf{r})$, placed on the body surface, $\mathbf{r} \in K$, which generates the velocity field $\mathbf{V}_\gamma(\mathbf{r})$. Then the summary velocity field is the superposition of the incident flow velocity $\mathbf{V}_\infty$, velocity field, generated by vorticity inside the flow domain $\mathbf{V}_\Omega(\mathbf{r})$, and the introduced field $\mathbf{V}_\gamma(\mathbf{r})$:

$$\mathbf{V}(\mathbf{r}) = \mathbf{V}_\infty + \mathbf{V}_\Omega(\mathbf{r}) + \mathbf{V}_\gamma(\mathbf{r}).$$

From mathematical point of view, the velocity $\mathbf{V}_\gamma$ potential can be expressed through unknown double layer potential density $g(\mathbf{r})$ [4]:

$$\Phi(\mathbf{r}) = \oint_K g(\boldsymbol{\xi}) \frac{\partial}{\partial \mathbf{n}(\boldsymbol{\xi})} \frac{1}{4\pi |\mathbf{r} - \boldsymbol{\xi}|} dS_\xi.$$

Note, that the velocity field, which corresponds to this potential

$$\mathbf{V}_\gamma(\mathbf{r}) = \nabla \Phi(\mathbf{r}) = \oint_K g(\boldsymbol{\xi}) \frac{\partial}{\partial \mathbf{n}(\mathbf{r})} \frac{\partial}{\partial \mathbf{n}(\boldsymbol{\xi})} \frac{1}{4\pi |\mathbf{r} - \boldsymbol{\xi}|} dS_\xi, \tag{1}$$

also can be written down in the following form [4]:

$$\mathbf{V}_\gamma(\mathbf{r}) = \nabla \Phi(\mathbf{r}) = \oint_K \frac{\boldsymbol{\gamma}(\boldsymbol{\xi}) \times (\mathbf{r} - \boldsymbol{\xi})}{4\pi |\mathbf{r} - \boldsymbol{\xi}|^3} dS_\xi, \tag{2}$$

where vector $\gamma(\mathbf{r}) = -\operatorname{Grad} g(\mathbf{r}) \times \mathbf{n}(\mathbf{r})$; Grad is surface gradient operator. One can notice, that the expression (2) coincides with the Biot–Savart law for incompressible flows. So the potential $g(\mathbf{r})$ is closely connected with vortex sheet intensity $\gamma(\mathbf{r})$. The velocity $\mathbf{V}(\mathbf{r})$ is discontinuous at the body surface; its limit value is

$$\mathbf{V}_-(\mathbf{r}) = \mathbf{V}(\mathbf{r}) - \frac{\operatorname{Grad} g(\mathbf{r})}{2} = \mathbf{V}(\mathbf{r}) - \frac{\boldsymbol{\gamma}(\mathbf{r}) \times \mathbf{n}(\mathbf{r})}{2}, \quad \mathbf{r} \in K.$$

Taking into account the no-slip boundary condition in the form $\mathbf{V}_- = \mathbf{0}$ at the body surface, we obtain form (1) and (2) two forms of the integral equation:

$$\oint_K g(\boldsymbol{\xi}) \frac{\partial}{\partial \mathbf{n}(\mathbf{r})} \frac{\partial}{\partial \mathbf{n}(\boldsymbol{\xi})} \frac{1}{4\pi |\mathbf{r} - \boldsymbol{\xi}|} dS_\xi - \frac{\operatorname{Grad} g(\mathbf{r})}{2} = -\big(\mathbf{V}_\infty + \mathbf{V}_\Omega(\mathbf{r})\big), \quad \mathbf{r} \in K, \tag{3}$$

or

$$\oint_K \frac{\boldsymbol{\gamma}(\boldsymbol{\xi}) \times (\mathbf{r} - \boldsymbol{\xi})}{4\pi |\mathbf{r} - \boldsymbol{\xi}|^3} dS_\xi - \frac{\boldsymbol{\gamma}(\mathbf{r}) \times \mathbf{n}(\mathbf{r})}{2} = -\big(\mathbf{V}_\infty + \mathbf{V}_\Omega(\mathbf{r})\big), \quad \mathbf{r} \in K. \tag{4}$$

It is proven in [2], that in order to satisfy these equations, it is enough to satisfy the corresponding equations, being projected either on surface normal unit vector or on tangential plane.

3 Double Layer Potential Density Direct Reconstruction

The most common approach to solve the problem is Eq. (3) projection on normal unit vector, that leads to the hypersingular integral equation with respect to the double layer potential. Its solution is normally found as piecewise-constant double layer density on surface mesh, which consists of polygonal panels. The efficient numerical formulae for the Hadamard principal values calculation of hypersingular integrals are suggested by Lifanov [4].

Note, that the i-th polygonal panel with double layer potential density $g_i = \text{const}$ put exactly the same contribution $\mathbf{V}_\gamma^{(i)}$ to the velocity field $\mathbf{V}_\gamma(\mathbf{r})$ as closed vortex filament, placed on the panel circumfery, with circulation $\Gamma_i = g_i$. So the vorticity on the body surface automatically becomes represented as closed vortex lines, that corresponds to the Helmholtz fundamental theorems [7].

Numerical experiments show that such approach works satisfactory for flow simulations around smooth airfoils of rather simple shape, when the surface mesh is close to uniform. However, even in this case the direction of vortex lines on the body surface is determined by the mesh, and can differ significantly from the true vorticity surface distribution. This can lead to significant error in velocity field reconstruction in the neighborhood of the body surface, especially in the case of unsteady flow simulation around the body being followed with vorticity generation on the surface (the so-called vorticity flux model) [1].

The mentioned problems can be overcome by closed vortex filament (vortex loop) reconstruction. Positions and circulations of such vortex loops can be found according to the following algorithm [8, 9]:

1. the double layer potential density values are calculated at the surface triangular mesh vertices; if the surface mesh consists of polygonal cells, they should be split into triangular sub-panels, maybe by introducing additional nodes;
2. the double layer potential surface distribution is reconstructed by FEM-type interpolation using 1-st order shape functions;
3. vortex loops are generated along the level lines of this potential; vortex loops circulations are determined by the difference between potential density values at the neighboring level lines.

Such an approach works perfect, for example, in computer graphics applications [9], where it is enough to provide only qualitative results and high accuracy is not required. Its usage for flow simulation and hydrodynamic forces calculation is restricted, again, to rather simple body geometries and uniform meshes [8].

4 Vortex Sheet Intensity Reconstruction

Another way to satisfy the boundary condition is developed in [2]. It supposes projection of (4) on the tangential plane, leading to the integral equation of the 2-nd kind

$$\mathbf{n}(\mathbf{r}) \times \left(\int\limits_K \frac{\boldsymbol{\gamma}(\boldsymbol{\xi}, t) \times (\mathbf{r} - \boldsymbol{\xi})}{4\pi |\mathbf{r} - \boldsymbol{\xi}|^3} \times \mathbf{n}(\mathbf{r}) dS_\xi \right) - \frac{\boldsymbol{\gamma}(\mathbf{r}, t) \times \mathbf{n}(\mathbf{r})}{2} = \mathbf{f}(\mathbf{r}, t), \quad \mathbf{r} \in K, \tag{5}$$

where the right-hand side $\mathbf{f}(\mathbf{r}, t)$ is a known vector function, which depends on the vortex wake influence and the incident flow velocity:

$$\mathbf{f}(\mathbf{r}, t) = -\mathbf{n}(\mathbf{r}) \times \Big(\big(\mathbf{V}_\infty + \mathbf{V}_\Omega(\mathbf{r}) \big) \times \mathbf{n}(\mathbf{r}) \Big).$$

Note, that the kernel of Eq (5) is unbounded when $|\mathbf{r} - \boldsymbol{\xi}| \to 0$, so in order to solve it numerically with rather high accuracy the following assumptions are introduced:

1. The body surface is discretized into N triangular panels K_i with areas A_i and unit normal vectors $\mathbf{n}_i$, $i = 1, \dots, N$.
2. The unknown vortex sheet intensity on the i-th panel is assumed to be a constant vector $\boldsymbol{\gamma}_i$, $i = 1, \dots, N$, which lies in the plane of the i-th panel, *i. e.*, $\boldsymbol{\gamma}_i \cdot \mathbf{n}_i = 0$.
3. The integral equation (5) is satisfied on average over the panel, or, the same, in a Galerkin sense: its residual is orthogonal to the basis function which is equal to the 1 on the j-th panel and equal to 0 on all other panels.

According to these assumptions the discrete analogue of (5) can be derived:

$$\frac{1}{A_i} \sum_{j=1}^{N} \int\limits_{K_i} \left(\int\limits_{K_j} \mathbf{n}_i \times \left(\frac{\boldsymbol{\gamma}_j \times (\mathbf{r} - \boldsymbol{\xi})}{4\pi |\mathbf{r} - \boldsymbol{\xi}|^3} \times \mathbf{n}_i \right) dS_\xi \right) dS_r - \frac{\boldsymbol{\gamma}_i \times \mathbf{n}_i}{2} = \tag{6}$$

$$= \frac{1}{A_i} \int\limits_{K_i} \mathbf{f}(\mathbf{r}, t) dS_r, \quad i = 1, \dots, N.$$

To write down (6) in the form of a linear algebraic system we choose a local orthonormal basis on every cell $(\boldsymbol{\tau}_i^{(1)}, \boldsymbol{\tau}_i^{(2)}, \mathbf{n}_i)$, where tangent vectors $\boldsymbol{\tau}_i^{(1)}, \boldsymbol{\tau}_i^{(2)}$ can be chosen arbitrarily (in the plane of the cell, orthogonal one to the other) and $\boldsymbol{\tau}_i^{(1)} \times \boldsymbol{\tau}_i^{(2)} = \mathbf{n}_i$, so

$$\boldsymbol{\gamma}_i = \gamma_i^{(1)}\boldsymbol{\tau}_i^{(1)} + \gamma_i^{(2)}\boldsymbol{\tau}_i^{(2)},$$

and we can project (6) for every i-th panel on directions $\boldsymbol{\tau}_i^{(1)}$ and $\boldsymbol{\tau}_i^{(2)}$ [5, 6].

The obtained algebraic system has an infinite set of solutions; in order to select the unique solution the additional condition for the total vorticity (the integral from the vorticity over the body surface) should be satisfied:

$$\int\limits_K \boldsymbol{\gamma}(\mathbf{r},\, t)dS_r = \mathbf{0},$$

which also should be written down in the discretized form.

The resulting algebraic system is overdetermined. It should be regularized similarly to [4] by introducing the regularization vector $\mathbf{R} = (R_1,\ R_2,\ R_3)^T$:

$$\begin{aligned}
&\frac{1}{A_i}\boldsymbol{\tau}_i^{(1)} \cdot \left(\sum_{j=1}^{N} \gamma_j^{(1)}\boldsymbol{\nu}_{ij}^{(1)} + \sum_{j=1}^{N} \gamma_j^{(2)}\boldsymbol{\nu}_{ij}^{(2)}\right) - \frac{\gamma_i^{(2)}}{2} - \mathbf{R}\cdot\boldsymbol{\tau}_i^{(2)} = \frac{b_i^{(1)}}{A_i},\\
&\frac{1}{A_i}\boldsymbol{\tau}_i^{(2)} \cdot \left(\sum_{j=1}^{N} \gamma_j^{(1)}\boldsymbol{\nu}_{ij}^{(1)} + \sum_{j=1}^{N} \gamma_j^{(2)}\boldsymbol{\nu}_{ij}^{(2)}\right) + \frac{\gamma_i^{(1)}}{2} + \mathbf{R}\cdot\boldsymbol{\tau}_i^{(1)} = \frac{b_i^{(2)}}{A_i},\\
&\sum_{j=1}^{N} A_j\left(\gamma_j^{(1)}\boldsymbol{\tau}_j^{(1)} + \gamma_j^{(2)}\boldsymbol{\tau}_j^{(2)}\right) = \mathbf{0}, \qquad i = 1,\, \dots,\, N.
\end{aligned} \tag{7}$$

The semi-analytical numerical algorithm is developed [5, 6] for the integrals calculation in the coefficients

$$\boldsymbol{\nu}_{ij}^{(k)} = \int\limits_{K_i}\left(\int\limits_{K_j} \frac{\boldsymbol{\tau}_j^{(k)} \times (\mathbf{r} - \boldsymbol{\xi})}{4\pi|\mathbf{r} - \boldsymbol{\xi}|^3} dS_\xi\right) dS_r, \quad b_i^{(k)} = \int\limits_{K_i} \boldsymbol{\tau}_i^{(k)} \cdot \mathbf{f}(\mathbf{r},\, t)dS_r,$$

Numerical experiments show that the developed algorithm permits to reconstruct surface vorticity distribution with rather high accuracy, even on coarse meshes and, practically even more important, on non-uniform meshes with refinements. The velocity field, generated by such vorticity, is rather smooth near to the body surface.

In order to use this approach in the above described in Sect. 3 algorithm of the vortex loops generation, it is necessary to reconstruct the double layer potential at the vertices of the surface mesh.

The solution of linear system (7) gives us a piecewise-constant vortex sheet intensity distribution over the panels. From the other side, the vortex sheet intensity is the surface gradient of the double layer potential density. It means, that the most convenient way to double layer potential density recovery is its approximation by a function, which is piecewise-linear at the panels. To do it, we consider the nodal values of the potential g_j, $j = 1, \dots, M$ to be unknown; M is number if vertices of the surface mesh. Then the potential density can be recovered by FEM-type interpolation using 1-st order shape functions. Let us denote the positions of all the vertices of the surface mesh as $\boldsymbol{\rho}_j$, $j = 1, \dots, M$. The vertices of the i-th triangular panel have indices p_i^k, $k = 1, 2, 3$. The shape functions, defined over the i-th panel, coincide with barycentric coordinates on the triangle:

$$\phi_i^{(k)}(\boldsymbol{\rho}) = \frac{|(\boldsymbol{\rho}_{p_i^l} - \boldsymbol{\rho}) \times (\boldsymbol{\rho}_{p_i^m} - \boldsymbol{\rho})|}{|(\boldsymbol{\rho}_{p_i^l} - \boldsymbol{\rho}_{p_i^k}) \times (\boldsymbol{\rho}_{p_i^m} - \boldsymbol{\rho}_{p_i^k})|}, \quad \boldsymbol{\rho} \in K_i,$$

where $(k, l, m) = (1, 2, 3)$, or $(2, 3, 1)$ or $(3, 1, 2)$, then the double layer density over the i-th panel is linear function with respect to $\boldsymbol{\rho}$ has the form

$$g(\boldsymbol{\rho}) = \sum_{k=1}^{3} g_{p_i^k} \phi_i^{(k)}(\boldsymbol{\rho}), \quad \boldsymbol{\rho} \in K_i.$$

The gradient of the approximate double layer on every i-th panel, multiplied by the normal unit vector $\mathbf{n}_i$ gives a constant vector, which in a physical sense provides an approximate value of the vortex sheet intensity at the corresponding panel:

$$\boldsymbol{\gamma}_i^* = -\sum_{k=1}^{3} g_{p_i^k} \left(\operatorname{Grad} \phi_i^{(k)} \times \mathbf{n}_i\right), \quad \boldsymbol{\rho} \in K_i,$$

where the surface gradients of the shape functions $\operatorname{Grad} \phi_i^{(k)}$ are constant vectors.

The unknown values g_j can be found from the least-squares procedure:

$$\Psi = \sum_{i=1}^{N} \left|\boldsymbol{\gamma}_i - \boldsymbol{\gamma}_i^*\right|^2 \to \min$$

Taking partial derivatives of Ψ with respect to g_j, $j = 1, \dots, M$, and making them equal to zero, we obtain a linear system of $M \times M$ size with a symmetric matrix.

This system is singular (in practice, due to the truncation errors, it is ill-conditioned), which follows from the fact that the value of the potential density can be chosen arbitrary at some arbitrary specified point. To be more specific, we assume $g_M = 0$. That means that the last row and last column in the least-squares matrix should be nullified, the diagonal coefficient can be chosen arbitrary (non-zero); the last coefficient in the right-hand side also should be nullified. The resulting regularized matrix is symmetric and positive definite.

5 Numerical Results

Let us consider the results of the double layer potential density reconstruction according to the "direct" method (see Sect. 3) and to the "indirect" one (through the vortex sheet intensity recovery intermediate step, see Sect. 4).

The results are shown in Fig. 1 for the sphere discretized into 446 triangular panels of nearly the same size. The incident flow is directed vertically (upward). Here the direct and indirect approaches lead to nearly the same results: the level-set lines are close to horizontal, that corresponds qualitatively to the exact solution.

The "indirect" algorithm permits to obtain rather good results also on coarse and non-uniform meshes. In Fig. 2 the same sphere is split into 342 panels when the ratio of the largest panel area to the smallest one is about 32. The quality of the level lines shape remains high for the "indirect" method, as opposed to the "direct" one.

Fig. 1 Level lines of the double layer potential density, obtained "directly" (left picture) and "indirectly" (right picture) on the uniform mesh

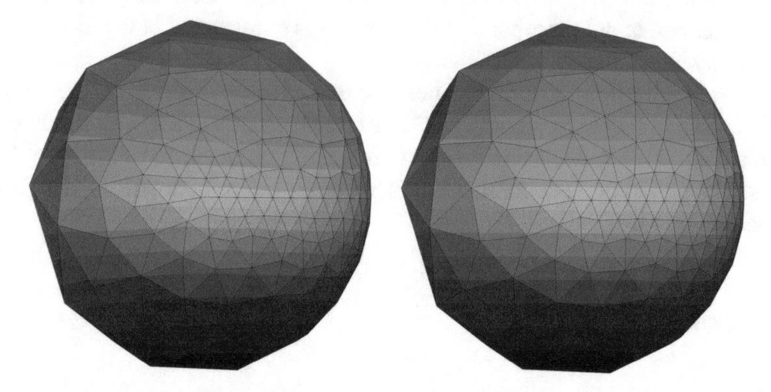

Fig. 2 Level lines of the double layer potential density, obtained "directly" (left picture) and "indirectly" (right picture) on the coarse non-uniform mesh

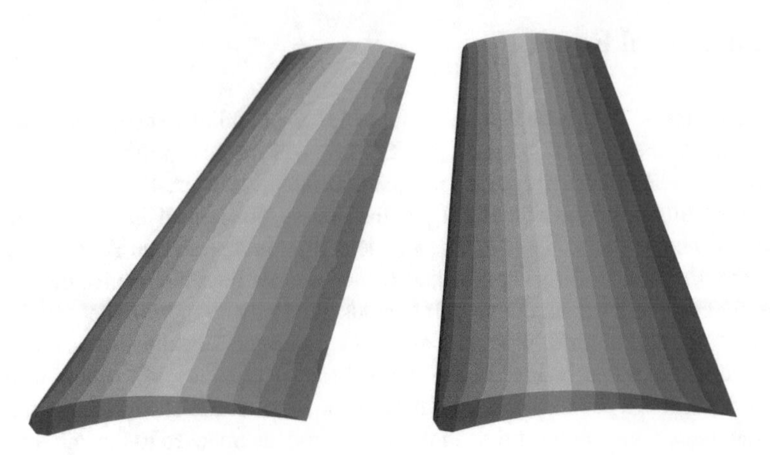

Fig. 3 Level lines of the double layer potential density, obtained "directly" (left picture) and "indirectly" (right picture) for the wing of finite span model

Fig. 4 Level lines of the double layer potential density for the fish stl-model, obtained by using the "direct" (left picture) and the "indirect" (right picture) method

The difference between the two approaches is more apparent for bodies of more complicated geometry. The example of flow around a wing of a finite span is shown in Fig. 3; the incident flow is directed from left to right. The triangular mesh with local refinement in neighborhood of the edges was constructed in the Salome open-source software and consists of 7040 panels.

For essentially non-uniform meshes, for example, obtained from the `stl`-file, which consist of large number of "bad" cells (elongated triangles with small angles) generated in some CAD software, the "indirect" approach makes it possible to reconstruct the solution with much better quality in comparison to the "direct" approach (Fig. 4).

6 Conclusion

The developed algorithm permits to improve significantly the quality of the double layer potential density reconstruction for the bodies with very complicated geometry and low-quality surface meshes. Its numerical complexity is higher than for the "direct" one due to solution of twice-larger linear system to recover vortex sheet intensity and the least-square problem. The developed approach can be useful for CFD applications and visual effects reproducing in computer graphics.

Acknowledgement The research is supported by Russian Science Foundation (proj. 17-79-20445).

References

1. Cottet, G.-H., Koumoutsakos, P.: Vortex Methods: Theory and Practice. Cambridge University Press, Cambridge (2000)
2. Kempka, S.N., et al.: Accuracy considerations for implementing velocity boundary conditions in vorticity formulations. SANDIA Report SAND96-0583 UC-700 (1996). https://doi.org/10.2172/242701
3. Lewis, R.I.: Vortex Element Methods for Fluid Dynamic Analysis of Engineering Systems. Cambridge University Press, Cambridge (2005)
4. Lifanov, I.K., Poltavskii, L.N., Vainikko, G.M.: Hypersingular Integral Equations and Their Applications. CRC Press, Boca Raton (2003)
5. Marchevsky, I.K., Shcheglov, G.A.: Efficient Semi-Analytical Integration of Vortex Sheet Influence in 3D Vortex Method. In: Wriggers P., et al. (eds.) 5th International Conference on Particle-Based Methods—Fundamentals and Applications, PARTICLES 2017, Hannower (2017)
6. Marchevsky, I.K., Shcheglov, G.A.: Semi-analytical influence computation for vortex sheet with piecewise constant intensity distribution in 3D vortex methods. In: Proceedings of the 6th European Conference on Computational Mechanics (ECCM 6), 7th European Conference on Computational Fluid Dynamics (ECFD 7), Glasgow (2018)
7. Saffman, P.G.: Vortex Dynamics. Cambridge University Press, Cambridge (1992)
8. Shcheglov, G.A., Dergachev, S.A.: Hydrodynamic loads simulation for 3D bluff bodies by using the vortex loops based modification of the vortex particle method. In: Wriggers P., et al. (eds.) 5th International Conference on Particle-Based Methods—Fundamentals and Applications, PARTICLES 2017, Hannower (2017)
9. Weissmann, S., Pinkall, U.: Filament-based smoke with vortex shedding and variational reconnection. In: 37th International Conference and Exhibit on Computer Graphics and Interactive Technologies, Los Angeles (2010)

Balancing Domain Decomposition Method on Additive Schwartz Framework for Multi-Level Implementation

Tomonori Yamada and Kazuya Goto

Abstract A new implementation of the balancing domain decomposition (BDD) method on additive Schwartz framework is proposed in this paper. BDD family is one of the most effective approaches for parallel computing of large scale structural finite element analyses. In the balancing domain decomposition by constraints (BDDC), the coarse grid correction procedure is applied in an additive manner, while it is applied multiplicatively in original BDD. Here, BDD on additive Schwartz framework is proposed and its multi-level implementation is discussed. Detailed computing performance is investigated with some numerical examples.

Keywords Large-scale analysis · Domain decomposition method · Additive schwartz framework · Multi-level implementation

1 Introduction

In Domain Decomposition Method (DDM), which is one of the most popular parallelized structural finite element analysis methods, effective parallel computation is made possible by decomposing the analytical domain into multiple subdomains and by assigning the computation of each subdomain to a processor (to a core, in the case of a multi-core processors) [1]. High parallelization efficiency can be achieved especially with iterative substructuring approach [2, 3], where solution for the entire domain is computed by solving the local stiffness equation at each subdomain while conducting a global iterative operation to ensure continuity between subdomains. However, since the number of iterations increases with the number of subdomains,

T. Yamada (✉)
The University of Tokyo, Tokyo, Japan
e-mail: tyamada@sys.t.u-tokyo.ac.jp

K. Goto
PExProCS, LLC., Tokyo, Japan
e-mail: goto@pexprocs.jp

H. van Brummelen et al. (eds.), *Numerical Methods for Flows*,
Lecture Notes in Computational Science and Engineering 132,
https://doi.org/10.1007/978-3-030-30705-9_26

methods that combine iterative substructuring and coarse grid correction, a variant of multigrid method, is widely studied in recent years; such as FETI (Finite Element Tearing and Interconnecting) [2] and BDD (Balancing Domain Decomposition) [4]. In these methods, convergence rate is reported to be constant for a given element count per subdomain regardless of the number of subdomains because of the coarse grid correction [5].

BDD is regarded as a two-level multigrid method where coarse grid correction is added to the Neumann-Neumann preconditioner [6] in a multiplicative manner, to make sure that the residuals, which become right-hand sides of the local problems, are 'balanced' [4]. The degrees of freedom in coarse grid correction are rigid body motion of subdomains, which can be uniquely determined, thus its implementation for general complex structures is straightforward. In large-scale problems, the size of coarse grid problem increases as the number of subdomains increases, and the solution of the coarse grid problem becomes bottleneck due to low parallel efficiency of this procedure [7].

The BDDC (Balancing Domain Decomposition by Constraints) [8, 9] method proposed in 2003 is formulated on the additive Schwartz framework, which makes it relatively simple to implement multi-level variants [10–12], thus it is suitable for large-scale DOF analyses. In fact, scalable parallel implementation [13] and extreme scale computation [14] are already available with BDDC. In BDDC, the degrees of freedom in coarse grid correction are selected from analytical degrees of freedom as primal constraints. Because of this procedure, the coarse grid matrix in BDDC is sparser than that of BDD, but the selection of the primal constraints is an important problem and is not simple to implement for general complex structures.

In this study, we introduce the additive Schwartz framework to BDD and then implement its three-level variant. The proposed methods are evaluated with numerical examples, and applicability to large-scale problems are discussed.

2 Overview of Methods

2.1 Iterative Substructuring Method

Simultaneous linear equations obtained by discretizing the governing equation of structural problems by the finite element method are described as follows:

$$\mathbf{K}\,\mathbf{u} = \mathbf{f}, \tag{1}$$

where $\mathbf{K}$ is the positive-definite symmetrical overall stiffness matrix, $\mathbf{u}$ is the nodal displacement vector, and $\mathbf{f}$ is the nodal load vector. With iterative substructuring taking displacement as unknowns, the entire analytical domain is decomposed into N subdomains without overlapping. By sorting the stiffness matrix, the nodal displacement vector, and the nodal load vector into the inside domain DOF component and the domain boundary DOF component as

$$\mathbf{K}_i = \begin{bmatrix} \mathbf{K}_{IIi} & \mathbf{K}_{IBi} \\ \mathbf{K}_{BIi} & \mathbf{K}_{BBi} \end{bmatrix}, \quad \mathbf{u}_i = \begin{Bmatrix} \mathbf{u}_{Ii} \\ \mathbf{u}_{Bi} \end{Bmatrix}, \quad \mathbf{f}_i = \begin{Bmatrix} \mathbf{f}_{Ii} \\ \mathbf{f}_{Bi} \end{Bmatrix}, \tag{2}$$

the stiffness equation for the inside subdomain DOF, which are independent of each other, can be obtained as follows:

$$\mathbf{K}_{IIi}\,\mathbf{u}_{Ii} + \mathbf{K}_{IBi}\,\mathbf{u}_{Bi} = \mathbf{f}_{Ii}, \tag{3}$$

where subscripts I and B are the DOF within the subdomain and at the domain boundary, respectively, and the subscript i denotes the subdomain number.

On the other hand, the stiffness equation for the DOF on the subdomain boundary can be expressed using the 0-1 matrix $\mathbf{R}_{Bi}$, which maps the DOF at each subdomain boundary onto the DOF for the entire analytical domain boundary.

$$\sum_{i=1}^{N} \mathbf{R}_{Bi}\,\mathbf{K}_{BIi}\,\mathbf{u}_{Ii} + \sum_{i=1}^{N} \mathbf{R}_{Bi}\,\mathbf{K}_{BBi}\,{\mathbf{R}_{Bi}}^{T}\,\mathbf{u}_{B} = \sum_{i=1}^{N} \mathbf{R}_{Bi}\,\mathbf{f}_{Bi} \tag{4}$$

By using Eq. (3) to eliminate each DOF inside the subdomains, Eq. (4) becomes the equilibrium equation for the DOF on the subdomain boundary, as follows:

$$\mathbf{S}\,\mathbf{u}_B = \mathbf{g}, \tag{5}$$

where $\mathbf{S}$ and $\mathbf{g}$ are the statically condensed matrix (hereinafter referred to as the Schur complement matrix) and the nodal load vector, respectively. Equation (5) is referred to as the interface problem in the iterative substructuring method, and it can be solved by iterations via iterative solvers such as the conjugate gradient method with a preconditioner.

2.1.1 Neumann-Neumann Preconditioner

The Neumann-Neumann preconditioner [6] is a preconditioning method introduced to accelerate convergence in iterative substructuring, which takes displacement as unknowns. Typically, to apply the preconditioning matrix $\mathbf{M}$ in the conjugate gradient method, the following equation must be solved.

$$\mathbf{M}\,\mathbf{z} = \mathbf{r}, \tag{6}$$

where $\mathbf{z}$ is the solution vector, and $\mathbf{r}$ is the residual vector at each iterative step. For the statically condensed expression shown in Eq. (5) for the entire analysis domain, the local Schur complement matrix $\mathbf{S}_i$ is considered for each subdomain.

$$\mathbf{S}_i = \mathbf{K}_{BBi} - \mathbf{K}_{BIi}\,{\mathbf{K}_{IIi}}^{-1}\,\mathbf{K}_{IBi} \tag{7}$$

With the Neumann-Neumann preconditioner, the preconditioning matrix can be defined as below using the local Schur complement matrix for each domain.

$$\mathbf{M}_{\text{N-N}}{}^{-1} = \sum_{i=1}^{N} \mathbf{R}_{Bi}\,\mathbf{D}_i\,\mathbf{S}_i{}^{+}\,\mathbf{D}_i{}^{T}\,\mathbf{R}_{Bi}{}^{T}, \tag{8}$$

where $\mathbf{S}_i{}^{+}$ is a generalized inverse matrix, and $\mathbf{D}_i$ is a weighting matrix for domain decomposition, which is introduced so that the DOF evaluation weight sum for each subdomain boundary contained in multiple subdomains would equal unity.

2.1.2 BDD

The BDD method [4] can be implemented by applying coarse grid correction to solutions obtained locally at each subdomain. A coarse problem matrix can be obtained by utilizing the rigid body motion (translation and rotation) at each subdomain as follows:

$$\mathbf{S}_C = \mathbf{R}_C{}^{T}\,\mathbf{S}\,\mathbf{R}_C, \tag{9}$$

where $\mathbf{R}_C{}^{T}$ is a projection matrix from the subdomain boundary onto the coarse problem. The overlap onto the Neumann-Neumann preconditioner is conducted in a multiplicative manner, and the preconditioning matrix can be expressed as follows.

$$\mathbf{M}_{\text{BDD}}{}^{-1} = \mathbf{R}_C\,\mathbf{S}_C{}^{-1}\mathbf{R}_C{}^{T} + (\mathbf{I} - \mathbf{R}_C\,\mathbf{S}_C{}^{-1}\,\mathbf{R}_C{}^{T}\,\mathbf{S})\,\mathbf{M}_{\text{N-N}}{}^{-1}\,(\mathbf{I} - \mathbf{S}\,\mathbf{R}_C\,\mathbf{S}_C{}^{-1}\,\mathbf{R}_C{}^{T}) \tag{10}$$

When the residual vector $\mathbf{r}$ is balanced, or in other words, when the coarse problem component is not included in the residual vector $\mathbf{r}$, Eq. (10) can be simplified as below.

$$\mathbf{M}_{\text{BDD}}{}^{-1} = (\mathbf{I} - \mathbf{R}_C\,\mathbf{S}_C{}^{-1}\,\mathbf{R}_C{}^{T}\,\mathbf{S})\,\mathbf{M}_{\text{N-N}}{}^{-1} \tag{11}$$

In three-dimensional structural analyses, the size of the coarse problem matrix $\mathbf{S}_C$ becomes the product of the number of subdomains and the 6-DOF rigid body motion in each subdomain. The preconditioning matrix in the BDD method shown in Eq. (11) includes the Neumann-Neumann preconditioning matrix $\mathbf{M}_{\text{N-N}}{}^{-1}$, the Schur complement matrix $\mathbf{S}$, and the coarse problem inverse matrix $\mathbf{S}_C{}^{-1}$. Therefore, the main preconditioning calculation load in the BDD method resides in the calculation of products of these three matrices with vectors.

Additionally, BDD-DIAG [15] is proposed as a variant of BDD implementation that does not use the Neumann-Neumann preconditioner, but uses an ordinary diagonal scaling preconditioner.

2.2 BDD on Additive Schwartz Framework

In the original BDD, global coarse grid correction is added to the Neumann-Neumann preconditioner in a multiplicative manner as in Eqs. (10) and (11), to make sure that the residuals are balanced. The BDD preconditioner on additive Schwartz framework is defined by simply adding the global coarse grid correction and the local preconditioner, such as Neumann-Neumann one or diagonal scaling, as follows:

$$\mathbf{M}_{\text{BDD}}^{ADD^{-1}} = \mathbf{R}_C\,\mathbf{S}_C{}^{-1}\,\mathbf{R}_C{}^T + \mathbf{M}_{\text{Local}}{}^{-1} \tag{12}$$

Compared to Eq. (11), a multiplication of matrix **S** is reduced. When the local preconditioner is the Neumann-Neumann one, the convergence rate can be harmed with additive Schwartz framework because of its mathematical requirement of balanced residuals. However, diagonal scaling preconditioner could be useful for BDD on additive Schwartz framework because the balanced residuals are not required in this case. To confirm the performance of BDD on additive Schwartz framework with different local preconditioners, numerical benchmarks were conducted in Sect. 3.

2.3 Multi-Level BDD on Additive Schwartz Framework

The increased analysis scale leads to an increase in computational costs of coarse grid correction in BDD. With the additive framework, multi-level variants can easily be derived, and thus multi-level implementations of BDDC have been proposed so far. Multi-level implementation is also straightforward in case of BDD on additive framework. For instance, a three-level variant can be defined as follows. First, the coarse problem is defined by decomposing the subdomains into multiple subdomain groups without overlapping, and by making the rigid body motion components of each subdomain group the coarse DOF. Second, the middle problem is defined in each subdomain group, treating rigid body motion components of each subdomain as the middle DOF. In this case, the following three-level preconditioner is obtained.

$$\mathbf{M}_{\text{BDD}}^{M\_ADD^{-1}} = \mathbf{R}_C\,\mathbf{S}_C{}^{-1}\,\mathbf{R}_C{}^T + \mathbf{R}_M\,\mathbf{S}_M{}^{-1}\,\mathbf{R}_M{}^T + \mathbf{M}_{\text{N-N}}{}^{-1}, \tag{13}$$

where $\mathbf{S}_M{}^{-1}$ is the inverse matrix of the middle problem and $\mathbf{R}_M{}^T$ is a projection matrix from the subdomain boundary onto the middle problem. Here, $\mathbf{S}_M{}^{-1}$ can be comuted in each subdomain group independently.

3 Numerical Examples

3.1 Cubic Model

We conducted a benchmark study on a cube discretized by hexahedral first-order finite elements. A schematic view of problem setting is illustrated in Fig. 1. The bottom face is fully constrained, and uniform compressive force is applied to the top face. The uniform load is 1 MPa, the Young's modulus is 1 MPa, and the Poisson's ratio is 0.3. Structured grid mesh was generated with 1,404,928 elements, 1,442,897 nodes, and 4,328,691 DOF. With 4096 subdomains, each subdomain contains 343 elements, 512 nodes, and 1536 DOF. Calculations were conducted on a Fujitsu BX900 32 core system. Thread parallelization using OpenMP was not conducted; the calculation was parallelized for 32 processes using MPI only. 128 subdomains were assigned to each core.

Table 1 shows a comparison of the number of iterations and computation time for different combination of local preconditioner (Neumann-Neumann or DIAG) and the BDD framework (multiplicative (two-level), additive (two-level) or additive (three-level)). When Neumann-Neumann preconditioner was used as the local preconditioner (original BDD), number of iterations increased more than five times by introducing additive framework. The reason of this deterioration of convergence rate is that the rigid body motion of each subdomain caused by the Neumann-Neumann preconditioner cannot be eliminated by coarse grid correction

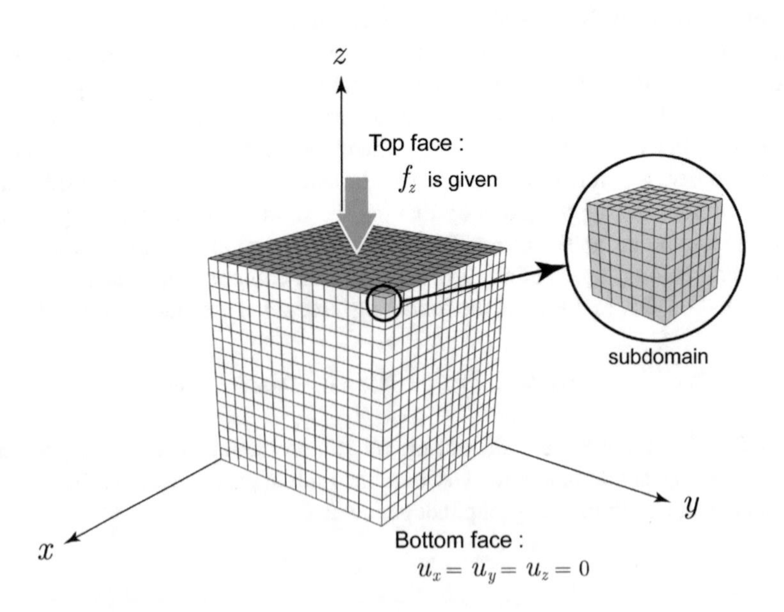

Fig. 1 Domain decomposition of whole analysis domain of cube model

Table 1 Number of CG loops and computation time (total, preprocessing before iteration starts, and entire iterations) of cube model

Local preconditioner	Framework	Num. of CG loops	[Ratio]	Time [s]		
				Total	Pre.	Iter.
Neumann-Neumann	Multiplicative (two-level)	13	[1.00]	73.19	68.11	5.08
	Additive (two-level)	67	[5.15]	90.15	68.06	22.09
	Additive (three-level)	94	[7.23]	32.02	13.19	18.83
DIAG	Multiplicative (two-level)	23	[1.77]	72.06	66.16	5.90
	Additive (two-level)	24	[1.85]	71.24	66.20	5.04
	Additive (three-level)	52	[4.00]	15.15	11.19	3.95

on additive framework without the balanced residuals, thereby negative influence on the convergence rate was observed.

When comparing the multiplicative framework against the additive one for BDD-DIAG, the number of iterations was almost identical, but the computation time for overall iteration decreased because of the reduction in computational volume. And comparing the results of two-level and three-level additive framework with diagonal scaling, the computation time for overall iteration decreased in three-level, because of the reduced computation required for the coarse grid correction, even though the number of iterations increased.

3.2 Plate Model

We conducted another benchmark study on a plate discretized by hexahedral first-order finite elements illustrated in Fig. 2. The bottom face is fully constrained, and uniform shear force is applied to the top face. The uniform load is 1 MPa, the Young's modulus is 1 MPa, and the Poisson's ratio is 0.3. Structured grid mesh was generated with 1,279,648 elements, 1,229,217 nodes, and 3,687,651 DOF. With 4096 subdomains, each subdomain contains 288 elements, 507 nodes, and 1521 DOF. Calculations were conducted again on a Fujitsu BX900 32 core system with the same parallel computing setting as cubic model.

Table 2 shows a comparison of the number of iterations and computation time for different combination of local preconditioner (Neumann-Neumann or DIAG) and the BDD framework (multiplicative (two-level), additive (two-level) or additive (three-level)). The same deterioration of convergence rate with Neumann-Neumann preconditioner on additive framework compared with the multiplicative one was observed. And again the almost identical iteration counts for diagonal scaling preconditioner on additive framework compared with the multiplicative one.

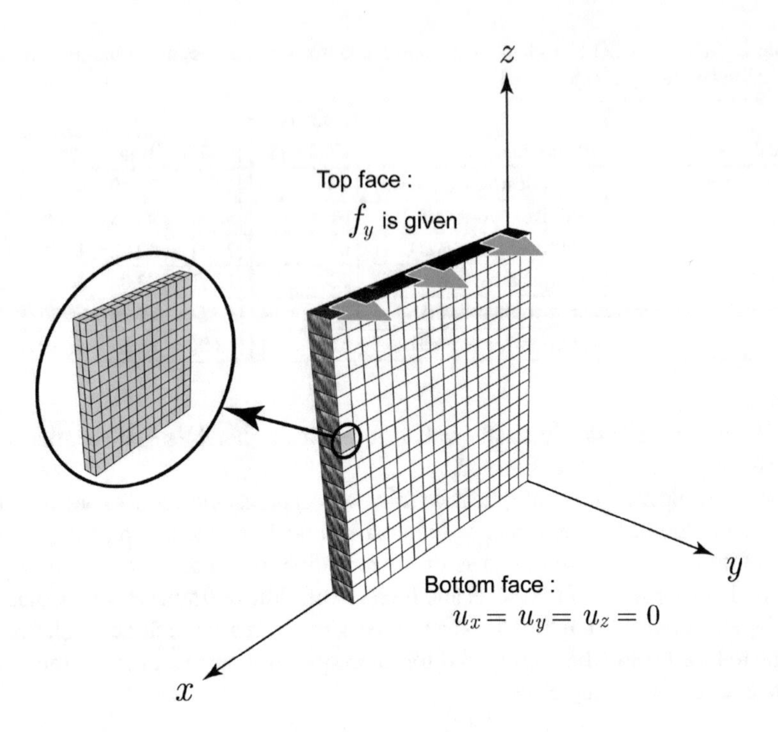

Fig. 2 Domain decomposition of whole analysis domain of plate model

Table 2 Number of CG loops and computation time (total, preprocessing before iteration starts, and entire iterations) of plate model

Local preconditioner	Framework	Num. of CG loops	[Ratio]	Time [s] Total	Pre.	Iter.
Neumann-Neumann	Multiplicative (two-level)	89	[1.00]	76.57	60.42	16.14
	Additive (two-level)	205	[2.30]	96.02	59.95	35.43
	Additive (three-level)	313	[3.52]	21.69	5.98	13.03
DIAG	Multiplicative (two-level)	60	[0.67]	69.25	59.71	9.54
	Additive (two-level)	65	[0.73]	69.26	59.67	9.58
	Additive (three-level)	107	[1.20]	7.02	5.30	1.72

4 Conclusions

BDD on additive Schwartz framework and its three-level variant were implemented and their performance was evaluated. Numerical study indicates the deterioration of convergence rate with Neumann-Neumann preconditioner on additive Schwartz framework compared with the multiplicative one. However, BDD with diagonal scaling preconditioner on additive framework worked well and almost identical number of iterations was observed compared with the multiplicative one. With the three-level variant, although the number of iteration increased, huge reduction in total computation time was obtained.

References

1. Toselli, A., Widlund, O.: Domain decomposition methods-algorithms and theory. In: Springer Series in Computational Mathematics, vol. 34. Springer, Berlin (2004)
2. Farhat, C., Roux, F.X.: A method of finite element tearing and interconnecting and its parallel solution algorithm. Int. J. Numer. Meth. Eng. **32**, 1205–1227 (1991)
3. De Roeck, Y.H., Le Tallec, P.: Analysis and test of a local domain decomposition preconditioner. In: Fourth International Symposium on Domain Decomposition Methods for Partial Differential Equations, pp. 112–128 (1991)
4. Mandel, J.: Balancing domain decomposition. Comm. Numer. Meth. Eng. **9**, 233–241 (1993)
5. Farhat, C., Mandel, J., Roux, F.X.: Optimal convergence properties of the FETI domain decomposition method. Comp. Meth. Appl. Mech. Eng. **115**, 365–385 (1994)
6. Le Tallec, P., De Roeck, Y.H., Vidrascu, M.: Domain decomposition methods for large linearly elliptic three-dimensional problems. J. Comp. Appl. Math. **34**, 93–117 (1991)
7. Badia, S., Martin, A.F., Principe, J.: Implementation and scalability analysis of balancing domain decomposition methods. Arch. Comp. Meth. Eng. **20**, 239–262 (2013)
8. Dohrmann, C.R.: A preconditioner for substructuring based on constrained energy minimization. SIAM J. Sci. Comp. **25**(1), 246–258 (2003)
9. Mandel, J., Dohrmann, C.R.: Convergence of a balancing domain decomposition by constraints and energy minimization. Numer. Linear Algebra Appl. **10**, 639–659 (2003)
10. Tu, X.: Three-level BDDC in two dimensions. Int. J. Numer. Meth. Eng. **69**, 33–59 (2007)
11. Tu, X.: Three-level BDDC in three dimensions. SIAM J. Sci. Comp. **29**(4), 1759–1780 (2007)
12. Mandel, J., Sousedik, B., Dohrmann, C.R.: Multispace and multilevel BDDC. Computing **83**, 55–85 (2008)
13. Badia, S., Martin, A., Principe, J.: A highly scalable parallel implementation of balancing domain decomposition by constraints. SIAM J. Sci. Comp. **36**(2), 190–218 (2014)
14. Badia, S., Martin, A., Principe, J.: Multilevel balancing domain decomposition at extreme scales. SIAM J. Sci. Comp. **38**(1), 22–52 (2016)
15. Ogino, M., Shioya, R., Kanayama, H.: An inexact balancing preconditioner for large-scale structural analysis. J. Comp. Sci. Tech. **2**, 150–161 (2008)

Algebraic Dual Polynomials for the Equivalence of Curl-Curl Problems

Marc Gerritsma, Varun Jain, Yi Zhang, and Artur Palha

Abstract In this paper we will consider two curl-curl equations in two dimensions. One curl-curl problem for a scalar quantity F and one problem for a vector field $\boldsymbol{E}$. For Dirichlet boundary conditions $\boldsymbol{n} \times \boldsymbol{E} = \hat{E}_{\dashv}$ on $\boldsymbol{E}$ and Neumann boundary conditions $\boldsymbol{n} \times \mathbf{curl}\, F = \hat{E}_{\dashv}$, we expect the solutions to satisfy $\boldsymbol{E} = \mathbf{curl}\, F$. When we use algebraic dual polynomial representations, these identities continue to hold at the discrete level. Equivalence will be proved and illustrated with a computational example.

Keywords Spectral element method · Algebraic dual polynomials · Curl-curl problems

1 Introduction

Numerical methods lead invariably to approximations, but a judicious choice of finite dimensional function spaces allows one to preserve, at the discrete level, identities that hold at the continuous level. In this paper we will focus on the finite dimensional representation of the curl operator; or, to put it more correctly, the *curl operators*. This is particularly clear in the two-dimensional setting where one curl operator maps scalar fields to vector fields and the other curl operator is its adjoint and therefore maps vector fields to scalar fields.

M. Gerritsma (✉) · V. Jain · Y. Zhang
Faculty of Aerospace Engineering, TU Delft, Delft, The Netherlands
e-mail: M.I.Gerritsma@tudelft.nl; V.Jain@tudelft.nl; Y.Zhang-14@tudelft.nl

A. Palha
Department of Mechanical Engineering, Eindhoven University of Technology, Eindhoven, The Netherlands
e-mail: A.PalhaDaSilvaClerigo@tudelft.nl

H. van Brummelen et al. (eds.), *Numerical Methods for Flows*,
Lecture Notes in Computational Science and Engineering 132,
https://doi.org/10.1007/978-3-030-30705-9_27

Faraday's and Ampère's law demonstrate the importance of the curl operator in electromagnetism. In fluid mechanics the curl operator appears in the relation between the stream function and the mass fluxes, and the definition of vorticity, [1].

In $\mathbb{R}^2$ we define the curl operators $\mathbf{curl}\, F = (\partial F/\partial y, -\partial F/\partial x)^\top$ for a scalar field $F(x, y)$ and $\mathrm{curl}\, \boldsymbol{E} = \partial E_y/\partial x - \partial E_x/\partial y$ for a vector field $\boldsymbol{E} = \left(E_x, E_y\right)$. We define the functions space

$$H(\mathrm{curl};\, \Omega) = \left\{ \boldsymbol{E} \in \left[L^2(\Omega)\right]^2 \,\middle|\, \mathrm{curl}\, \boldsymbol{E} \in L^2(\Omega) \right\}, \tag{1}$$

with associated inner product

$$(\boldsymbol{D}, \boldsymbol{E})_{H(\mathrm{curl})} := (\boldsymbol{D}, \boldsymbol{E})_{L^2} + (\mathrm{curl}\, \boldsymbol{D}, \mathrm{curl}\, \boldsymbol{E})_{L^2}\,,$$

and the function space

$$H(\mathbf{curl};\, \Omega) = \left\{ F \in L^2(\Omega) \,\middle|\, \mathbf{curl}\, F \in \left[L^2(\Omega)\right]^2 \right\}, \tag{2}$$

with inner product

$$(F, G)_{H(\mathbf{curl})} := (F, G)_{L^2} + (\mathbf{curl}\, F, \mathbf{curl}\, G)_{L^2}\,.$$

The corresponding norms are $\|\boldsymbol{E}\|^2_{H(\mathrm{curl})} = (\boldsymbol{E}, \boldsymbol{E})_{H(\mathrm{curl})}$ and $\|F\|^2_{H(\mathbf{curl})} = (F, F)_{H(\mathbf{curl})}$.

2 The Equivalent Curl-Curl Dual Problems

In [2] the equivalence between several Dirichlet and Neumann problems is introduced. The main question we want to address in this paper is whether we can preserve these equivalences at the discrete level. In [3] this equivalence was already established at the discrete level for the scalar grad-div problem. In this paper we want to focus on the curl-curl equivalence problem: Given $\hat{E}_{\dashv} \in H^{-\frac{1}{2}}(\mathrm{div};\, \partial\Omega)$, [4], find the solution $\boldsymbol{E} \in H(\mathrm{curl};\, \Omega)$ of the Dirichlet problem satisfying

$$\begin{cases} n \times \boldsymbol{E} = \hat{E}_{\dashv} & \text{on } \partial\Omega \\ \mathbf{curl}\,(\mathrm{curl}\, \boldsymbol{E}) + \boldsymbol{E} = \mathbf{0} & \text{in } \Omega \end{cases}, \tag{3}$$

and the associated Neumann problem given by: For $\hat{E}_{\dashv} \in H^{-\frac{1}{2}}(\text{div}; \partial\Omega)$ find the solution $F \in H(\mathbf{curl}; \Omega)$ such that

$$\begin{cases} n \times \mathbf{curl}\, F = \hat{E}_{\dashv} & \text{on } \partial\Omega \\ \text{curl}\,(\mathbf{curl}\, F) + F = 0 & \text{in } \Omega \end{cases}, \tag{4}$$

where $n \times \boldsymbol{E} = n_x E_y - n_y E_x$ and $n \times \mathbf{curl}\, F = -n_x \partial F/\partial x - n_y \partial F/\partial y$, which can also be written as $n \times \mathbf{curl}\, F = -n \cdot \mathbf{grad}\, F$. At the continuous level we know that these two problems are equivalent in the sense that $\boldsymbol{E}$ solves (3) and F solves (4) if and only if $\boldsymbol{E} = \mathbf{curl}\, F$. In addition, we have that

$$\|\boldsymbol{E}\|_{H(\text{curl};\Omega)} = \|F\|_{H(\mathbf{curl};\Omega)} = \|\hat{E}_{\dashv}\|_{H^{-\frac{1}{2}}(\text{div};\partial\Omega)}. \tag{5}$$

In this paper we want to present a spectral element formulation for both problems (3) and (4) such that the equivalence, $\boldsymbol{E}^h = \mathbf{curl}\, F^h$, continues to hold for the discrete solutions. Moreover, we want to show that we have $\|\boldsymbol{E}^h\|_{H(\text{curl};\Omega)} = \|F^h\|_{H(\mathbf{curl};\Omega)}$.

Methods which preserve the equivalence of the two Dirichlet-Neumann problems in the discrete setting such that (5) continues to hold, are of importance in DG methods, such as [5], and hybrid finite element formulations, see [6], for instance.

3 Primal Spectral Element Formulation

Consider the partitioning $-1 = \xi_0 < \xi_1 <, \ldots, < \xi_{N-1} < \xi_N = 1$ of the interval $I = [-1, 1]$, where ξ_i are the roots of $(1 - \xi^2)L'_N(\xi)$, with $L'_N(\xi)$ the derivative of the Legendre polynomial of degree N. With these nodes ξ_i we associate the Lagrange polynomials, $h_i(\xi)$, of degree N which satisfy $h_i(\xi_j) = \delta_{ij}$. Any polynomial p of degree N defined on I can be written as

$$p(\xi) = \sum_{i=0}^{N} p_i h_i(\xi). \tag{6}$$

Since the Lagrange polynomials are linearly independent, $p(\xi) = 0$ iff all $p_i = 0$.

The derivative of (6) is given by

$$p'(\xi) = \sum_{i=0}^{N} p_i \frac{dh_i}{d\xi}(\xi) = \sum_{i=1}^{N} (p_i - p_{i-1})\, e_i(\xi), \tag{7}$$

where, [7],

$$e_i(\xi) = -\sum_{k=1}^{i-1} \frac{dh_k}{d\xi}(\xi).$$

Note that the $dh_i/d\xi$, $i = 0, \ldots, N$, do not form a basis, while the functions $e_i(\xi)$, $i = 1, \ldots, N$ do form a basis. Therefore, $p'(\xi) = 0$ iff $p_i = p_{i-1}$ for $i = 1, \ldots, N$, which in turn means that $p(\xi) = const$, as required. Note that differentiation of the Lagrange expansion (6) amounts to a linear combination of the expansion coefficients, $(p_i - p_{i-1})$ in (7) and a representation in a different basis, $e_i(\xi)$ in (7). An important property of the edge polynomials, $e_i(\xi)$, is

$$\int_{\xi_{j-1}}^{\xi_j} e_i(\xi)\, \mathrm{d}\xi = \begin{cases} 1 & \text{if } i = j \\ 0 & \text{if } i \neq j \end{cases} . \tag{8}$$

In the two dimensional case we consider $(\xi, \eta) \in \hat{\Omega} = [-1, 1]^2 \subset \mathbb{R}^2$ and the partitioning ξ_i in the ξ-direction as given in the one dimensional case and we choose the same partitioning in the η-direction, see Fig. 1. Here Ω is a contractible domain with Lipschitz continuous boundary. For the representation of F we use the tensor product of the nodal representation

$$F^h(\xi, \eta) = \sum_{i=0}^{N} \sum_{j=0}^{N} F_{i,j} h_i(\xi) h_j(\eta) = \Psi^{(0)}(\xi, \eta)\mathsf{F}. \tag{9}$$

Here $\Psi^{(0)}(\xi, \eta)$ is a row vector with the basis functions and F is a column vector with the nodal degrees of freedom

$$\Psi^{(0)}(\xi, \eta) = \left[h_0(\xi)h_0(\eta) \ldots h_N(\xi)h_N(\eta) \right], \quad \mathsf{F} = \begin{bmatrix} F_{0,0} \\ F_{1,0} \\ \vdots \\ F_{N-1,N} \\ F_{N,N} \end{bmatrix}, \tag{10}$$

where $F_{i,j} = F^h(\xi_i, \eta_j)$. The curl of F^h is then given by

$$\mathbf{curl}\, F^h = \begin{pmatrix} \sum_{i=0}^{N} \sum_{j=1}^{N} (F_{i,j} - F_{i,j-1}) h_i(\xi) e_j(\eta) \\ \sum_{i=1}^{N} \sum_{j=0}^{N} (-F_{i,j} + F_{i-1,j}) e_i(\xi) h_j(\eta) \end{pmatrix} = \Psi^{(1)}(\xi, \eta)\mathbb{E}^{1,0}\mathsf{F}. \tag{11}$$

In (11) we used (7) to represent the curl in a different basis. The basis $\Psi^{(1)}(\xi,\eta)$ and the incidence matrix $\mathbb{E}^{1,0}$ are given by

$$\Psi^{(1)}(\xi,\eta) = \begin{bmatrix} h_0(\xi)e_1(\eta) \ldots h_N(\xi)e_N(\eta) & 0 & \ldots & 0 \\ 0 & \ldots & 0 & e_1(\xi)h_0(\eta) \ldots e_N(\xi)h_N(\eta) \end{bmatrix}, \tag{12}$$

$$\mathbb{E}^{1,0} = \begin{pmatrix}
-1 & 0 & 0 & 0 & 1 & 0 & 0 & 0 & 0 & 0 & 0 & 0 & 0 & 0 & 0 & 0 \\
0 & -1 & 0 & 0 & 0 & 1 & 0 & 0 & 0 & 0 & 0 & 0 & 0 & 0 & 0 & 0 \\
0 & 0 & -1 & 0 & 0 & 0 & 1 & 0 & 0 & 0 & 0 & 0 & 0 & 0 & 0 & 0 \\
0 & 0 & 0 & -1 & 0 & 0 & 0 & 1 & 0 & 0 & 0 & 0 & 0 & 0 & 0 & 0 \\
0 & 0 & 0 & 0 & -1 & 0 & 0 & 0 & 1 & 0 & 0 & 0 & 0 & 0 & 0 & 0 \\
0 & 0 & 0 & 0 & 0 & -1 & 0 & 0 & 0 & 1 & 0 & 0 & 0 & 0 & 0 & 0 \\
0 & 0 & 0 & 0 & 0 & 0 & -1 & 0 & 0 & 0 & 1 & 0 & 0 & 0 & 0 & 0 \\
0 & 0 & 0 & 0 & 0 & 0 & 0 & -1 & 0 & 0 & 0 & 1 & 0 & 0 & 0 & 0 \\
0 & 0 & 0 & 0 & 0 & 0 & 0 & 0 & -1 & 0 & 0 & 0 & 1 & 0 & 0 & 0 \\
0 & 0 & 0 & 0 & 0 & 0 & 0 & 0 & 0 & -1 & 0 & 0 & 0 & 1 & 0 & 0 \\
0 & 0 & 0 & 0 & 0 & 0 & 0 & 0 & 0 & 0 & -1 & 0 & 0 & 0 & 1 & 0 \\
0 & 0 & 0 & 0 & 0 & 0 & 0 & 0 & 0 & 0 & 0 & -1 & 0 & 0 & 0 & 1 \\
1 & -1 & 0 & 0 & 0 & 0 & 0 & 0 & 0 & 0 & 0 & 0 & 0 & 0 & 0 & 0 \\
0 & 1 & -1 & 0 & 0 & 0 & 0 & 0 & 0 & 0 & 0 & 0 & 0 & 0 & 0 & 0 \\
0 & 0 & 1 & -1 & 0 & 0 & 0 & 0 & 0 & 0 & 0 & 0 & 0 & 0 & 0 & 0 \\
0 & 0 & 0 & 0 & 1 & -1 & 0 & 0 & 0 & 0 & 0 & 0 & 0 & 0 & 0 & 0 \\
0 & 0 & 0 & 0 & 0 & 1 & -1 & 0 & 0 & 0 & 0 & 0 & 0 & 0 & 0 & 0 \\
0 & 0 & 0 & 0 & 0 & 0 & 1 & -1 & 0 & 0 & 0 & 0 & 0 & 0 & 0 & 0 \\
0 & 0 & 0 & 0 & 0 & 0 & 0 & 0 & 1 & -1 & 0 & 0 & 0 & 0 & 0 & 0 \\
0 & 0 & 0 & 0 & 0 & 0 & 0 & 0 & 0 & 1 & -1 & 0 & 0 & 0 & 0 & 0 \\
0 & 0 & 0 & 0 & 0 & 0 & 0 & 0 & 0 & 0 & 1 & -1 & 0 & 0 & 0 & 0 \\
0 & 0 & 0 & 0 & 0 & 0 & 0 & 0 & 0 & 0 & 0 & 0 & 1 & -1 & 0 & 0 \\
0 & 0 & 0 & 0 & 0 & 0 & 0 & 0 & 0 & 0 & 0 & 0 & 0 & 1 & -1 & 0 \\
0 & 0 & 0 & 0 & 0 & 0 & 0 & 0 & 0 & 0 & 0 & 0 & 0 & 0 & 1 & -1
\end{pmatrix},$$

where this incidence matrix corresponds to the layout depicted in Fig. 1, i.e. $N = 3$. Note that this incidence matrix only contains the entries -1, 0 and 1 and that the matrix is extremely sparse. The important thing to note is that this incidence matrix remains unchanged if we map the standard element $\hat{\Omega}$ to an arbitrary curved element. The basis functions $\Psi^{(1)}(\xi,\eta)$ do change, but the incidence matrix remains invariant. This is another reason to decompose a derivative into a part that acts on the degrees of freedom and new basis functions, as was done in (7).

The mass matrix $\mathbb{M}^{(0)}$ associated with the basis functions (10) is given by

$$\mathbb{M}^{(0)} = \int_{\hat{\Omega}} \Psi^{(0)\top} \Psi^{(0)} \, \mathrm{d}\Omega. \tag{13}$$

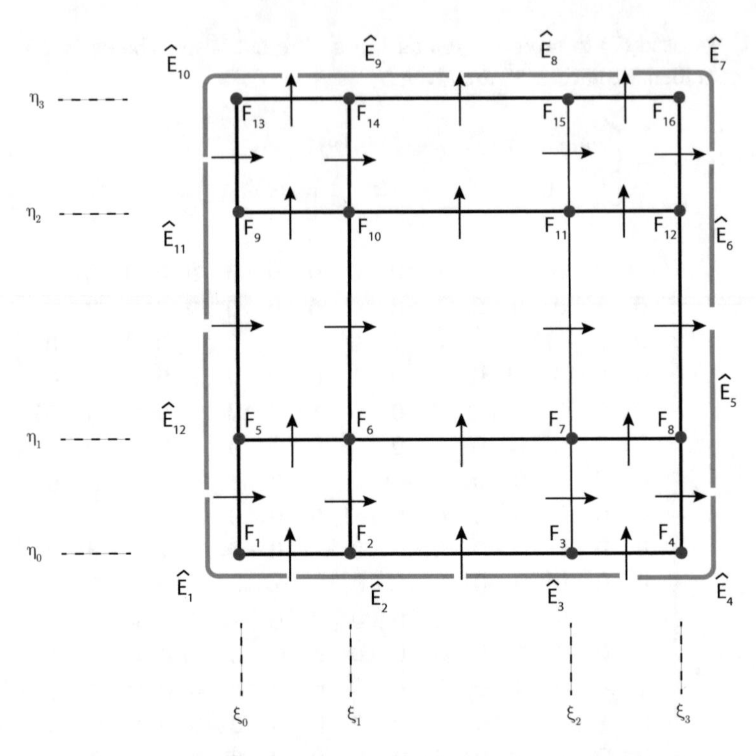

Fig. 1 Layout of one spectral element: the degrees of freedom for F indicated by the blue points. The curl of F is represented on the mesh line segments by the arrows crossing the line segments in the mesh. The boundary condition $\boldsymbol{n} \times \mathbf{curl}\, F = \hat{E}_{\dashv}$ is represented by the red outer line segments

Likewise, the mass matrix $\mathbb{M}^{(1)}$ associated with the basis functions (12) is given by

$$\mathbb{M}^{(1)} = \int_{\hat{\Omega}} \Psi^{(1)^{\top}} \Psi^{(1)} \,\mathrm{d}\Omega. \tag{14}$$

4 Dual Spectral Element Formulation

4.1 Duality in the Interior of the Domain

In the previous section we expanded the discrete solution in terms of basis functions $h_i(\xi)h_j(\eta)$ for F^h and $h_i(\xi)e_j(\eta) \otimes e_i(\xi)h_j(\eta)$ for the curl of F^h, respectively. With every linear vector space, $\mathcal{V}$, we can associate the space of linear functionals acting on that space $\mathcal{L}(\mathcal{V}, \mathbb{R}) = \mathcal{V}'$, called the *algebraic dual space*. Let $\alpha \in \mathcal{V}'$ and $u \in \mathcal{V}$, then $\alpha(u) \in \mathbb{R}$. Because we work in a Hilbert space, the Riesz representation theorem tells us that for every $\alpha \in \mathcal{V}'$ there exists a unique $v_\alpha \in \mathcal{V}$ such that

$$\alpha(u) = (v_\alpha, u), \quad \forall u \in \mathcal{V}, \tag{15}$$

where $(\cdot, \cdot)$ denotes the inner product in $\mathcal{V}$, [8, 9]. We first apply these ideas to the degrees of freedom (the expansion coefficients) which also form a linear vector space. Let F^h and G^h be expanded as in (9)

$$F^h(\xi, \eta) = \Psi^{(0)}(\xi, \eta)\mathsf{F} \quad \text{and} \quad G^h(\xi, \eta) = \Psi^{(0)}(\xi, \eta)\mathsf{G}.$$

Then we define *the dual degrees of freedom* $\widetilde{\mathsf{F}}$ analogous to (15) by, [3, 10]

$$\widetilde{\mathsf{F}}^\top \mathsf{G} := \mathsf{F}^\top \mathbb{M}^{(0)} \mathsf{G}, \quad \forall \mathsf{G} \in \mathbb{R}^{(N+1)^2}. \tag{16}$$

Therefore, the dual degrees of freedom are related to the primal degrees of freedom by $\widetilde{\mathsf{F}} = \mathbb{M}^{(0)}\mathsf{F}$. The canonical dual basis functions are then given by

$$\widetilde{\Psi}^{(2)} := \Psi^{(0)} {\mathbb{M}^{(0)}}^{-1}, \tag{17}$$

such that

$$\int_\Omega \widetilde{\Psi}^{(2)^\top} \Psi^{(0)} \, \mathrm{d}\Omega = \mathbb{I}, \tag{18}$$

where $\mathbb{I}$ is the identity matrix on $\mathbb{R}^{(N+1)^2}$. The relation (18) is analogous to the canonical basis $e_i^* \in \mathcal{V}'$ with the property $e_i^*(e_j) = \delta_{ij}$, when e_i form a basis for $\mathcal{V}$. If the basis functions change under a transformation, then the dual basis functions also change and (18) continues to hold.

Let the vector field $\boldsymbol{E}^h$ be expanded as in (11)

$$\boldsymbol{E}^h(\xi, \eta) = \Psi^{(1)}(\xi, \eta)\mathsf{E}.$$

The corresponding dual degrees of freedom are then given by $\widetilde{\mathsf{E}} = \mathbb{M}^{(1)}\mathsf{E}$ and the associated dual basis is related to the primal basis by $\widetilde{\Psi}^{(1)}(\xi, \eta) = \Psi^{(1)}(\xi, \eta){\mathbb{M}^{(1)}}^{-1}$.

4.2 Duality in the Boundary

The construction of a primal and a dual representation in the interior of the domain $\hat{\Omega}$ can also be applied along the boundary of the domain $\partial\hat{\Omega}$. We can restrict F^h to the boundary of the domain using (9), which gives

$$F^h\Big|_{\partial\hat{\Omega}} = \begin{cases} F^h\big|_{\partial\hat{\Omega}_E} = F_{N,j}h_j(\eta) \\ F^h\big|_{\partial\hat{\Omega}_N} = F_{i,N}h_i(\xi) \\ F^h\big|_{\partial\hat{\Omega}_W} = F_{0,j}h_j(\eta) \\ F^h\big|_{\partial\hat{\Omega}_S} = F_{i,0}h_j(\xi) \end{cases}. \tag{19}$$

This boundary expansion is essentially a one-dimensional expansion, (6), in terms of the four 1D elements which make up the boundary of a single spectral element. From this expansion we can compute the associated mass matrix, which for this boundary integral we will denote by $\mathbb{B}^{(0)}$. With the nodal degrees of freedom on the boundary, F_b, we can now define the dual degrees of freedom, $\widetilde{\mathsf{E}}_b$ by setting $\widetilde{\mathsf{E}}_b = \mathbb{B}^{(0)}\mathsf{F}_b$.

We introduce the matrix $\mathbb{T}$, given by

$$\mathbb{T} = \begin{pmatrix} 1&0&0&0&0&0&0&0&0&0&0&0&0&0&0&0 \\ 0&1&0&0&0&0&0&0&0&0&0&0&0&0&0&0 \\ 0&0&1&0&0&0&0&0&0&0&0&0&0&0&0&0 \\ 0&0&0&1&0&0&0&0&0&0&0&0&0&0&0&0 \\ 0&0&0&0&0&0&0&1&0&0&0&0&0&0&0&0 \\ 0&0&0&0&0&0&0&0&0&0&0&1&0&0&0&0 \\ 0&0&0&0&0&0&0&0&0&0&0&0&0&0&0&1 \\ 0&0&0&0&0&0&0&0&0&0&0&0&0&0&1&0 \\ 0&0&0&0&0&0&0&0&0&0&0&0&0&1&0&0 \\ 0&0&0&0&0&0&0&0&0&0&0&0&1&0&0&0 \\ 0&0&0&0&0&0&0&0&1&0&0&0&0&0&0&0 \\ 0&0&0&0&1&0&0&0&0&0&0&0&0&0&0&0 \end{pmatrix}. \tag{20}$$

The matrix $\mathbb{T}$ restricts the field F^h to the boundary of the domain. The curl of the representation $\widetilde{\boldsymbol{E}}^h$ is defined in the weak sense as

$$\begin{aligned} \int_{\hat{\Omega}} F^h \,\mathrm{curl}\, \widetilde{\boldsymbol{E}}^h \,\mathrm{d}\Omega &= \int_{\hat{\Omega}} \mathbf{curl} F^h \,\widetilde{\boldsymbol{E}}^h \,\mathrm{d}\Omega + \int_{\partial\hat{\Omega}} F^h \,\boldsymbol{n} \times \widetilde{\boldsymbol{E}}^h \,\mathrm{d}\Gamma \\ &= \mathsf{F}^\top \mathbb{E}^{1,0^\top} \widetilde{\mathsf{E}} + \mathsf{F}^\top \mathbb{T}^\top \widetilde{\mathsf{E}}_b, \quad \forall \mathsf{F}, \end{aligned} \tag{21}$$

where $\widetilde{\mathsf{E}}_b$ are the degrees of freedom along the boundary indicated by the red line segments in Fig. 1. Note also that minus signs in $\boldsymbol{n} \times \boldsymbol{E}$ cancel with the minus signs originating from the counter-clockwise evaluated boundary integral in (21). Also, (21) shows that the degrees of freedom for curl $\widetilde{\boldsymbol{E}}^h$ are given by $\mathbb{E}^{1,0^\top}\widetilde{\mathsf{E}} + \mathbb{T}^\top\widetilde{\mathsf{E}}_b$.

5 Discrete Formulation of the Curl-Curl Problem

5.1 The Neumann Problem

The variational formulation of the Neumann problem (4) is given by: Find F^h such that for all G^h we have

$$(G^h, F^h)_{H(\mathbf{curl})} = (\mathbf{curl}\ G^h, \mathbf{curl}\ F^h)_{L^2} + (G^h, F^h)_{L^2} = -\int_{\partial\Omega} G^h E^h_{\dashv}\, \mathrm{d}\Gamma, \tag{22}$$

where F^h is expanded as in (9). Using (11) for $\mathbf{curl}\ F^h$, and (13) and (14) for the mass matrices, we can write the left hand side of (22) as,

$$(\mathbf{curl}\ G^h, \mathbf{curl}\ F^h)_{L^2} + (G^h, F^h)_{L^2} \\ = \mathsf{G}^\top \mathbb{E}^{1,0^\top} \mathbb{M}^{(1)} \mathbb{E}^{1,0} \mathsf{F} + \mathsf{G}^\top \mathbb{M}^{(0)} \mathsf{F}. \tag{23}$$

The boundary conditions are prescribed on the right hand side with the help of duality pairing

$$\int_{\partial\Omega} G^h E_{\dashv}\, \mathrm{d}\Gamma = G^\top \mathbb{T}^\top \hat{\mathsf{E}}^h_{\dashv}. \tag{24}$$

Combining, (23) and (24) in (22), we have,

$$\mathbb{E}^{1,0^\top} \mathbb{M}^{(1)} \mathbb{E}^{1,0} \mathsf{F} + \mathbb{M}^{(0)} \mathsf{F} = -\mathbb{T}^\top \hat{\mathsf{E}}_{\dashv}. \tag{25}$$

5.2 The Dirichlet Problem

The variational formulation for the Dirichlet problem (3) is given by: Find $\widetilde{\boldsymbol{E}}^h$ such that for all $\widetilde{\boldsymbol{G}}^h$ we have,

$$(\widetilde{\boldsymbol{G}}^h, \widetilde{\boldsymbol{E}}^h)_{H(\mathrm{curl})} = (\mathrm{curl}\ \widetilde{\boldsymbol{G}}^h, \mathrm{curl}\ \widetilde{\boldsymbol{E}}^h)_{L^2} + (\widetilde{\boldsymbol{G}}^h, \widetilde{\boldsymbol{E}}^h)_{L^2} = -\int_{\partial\Omega} \mathrm{curl} \widetilde{\boldsymbol{G}}^h \hat{E}_{\dashv}\, \mathrm{d}\Gamma. \tag{26}$$

Here, $\widetilde{\boldsymbol{E}}^h$ is expanded in terms of dual polynomials

$$\widetilde{\boldsymbol{E}}^h(\xi, \eta) = \widetilde{\Psi}^{(1)}(\xi, \eta)\widetilde{\mathsf{E}}. \tag{27}$$

Then the weak formulation (26) can be written as

$$(\text{curl}\,\widetilde{\boldsymbol{G}}^h, \text{curl}\,\widetilde{\boldsymbol{E}}^h)_{L^2} + (\widetilde{\boldsymbol{G}}^h, \widetilde{\boldsymbol{E}}^h)_{L^2} \\ = \widetilde{\mathsf{G}}^T \mathbb{E}^{1,0} \widetilde{\mathbb{M}}^{(2)} \mathbb{E}^{1,0^T} \widetilde{\mathsf{E}} + \widetilde{\mathsf{G}}^T \widetilde{\mathbb{M}}^{(1)} \widetilde{\mathsf{E}}, \tag{28}$$

and

$$\int_{\partial\Omega} \text{curl}\widetilde{\boldsymbol{G}}^h \hat{E}_{\dashv}\, \mathrm{d}\Gamma = \widetilde{\mathsf{G}}^T \mathbb{E}^{1,0} \widetilde{\mathbb{M}}^{(2)} \mathbb{T}^T \hat{\mathsf{E}}_{\dashv}. \tag{29}$$

So the weak formulation (26) can be written as

$$\mathbb{E}^{1,0} \widetilde{\mathbb{M}}^{(2)} \mathbb{E}^{1,0^T} \widetilde{\mathsf{E}} + \widetilde{\mathbb{M}}^{(1)} \widetilde{\mathsf{E}} = -\mathbb{E}^{1,0} \widetilde{\mathbb{M}}^{(2)} \mathbb{T}^T \hat{\mathsf{E}}_{\dashv}. \tag{30}$$

5.3 *The Equivalence Condition*

In this part, we prove that the two discrete formulations (25) and (30) are related by the discrete relation $\boldsymbol{E}^h = \mathbf{curl} F^h$, which, in terms of the degrees of freedom is equivalent to $\widetilde{\mathsf{E}} = \mathbb{M}^{(1)} \mathbb{E}^{1,0} \mathsf{F}$. If we substitute this relation in the left hand side of (30) we get

$$\mathbb{E}^{1,0} \widetilde{\mathbb{M}}^{(2)} \mathbb{E}^{1,0^T} \widetilde{\mathrm{E}} + \widetilde{\mathbb{M}}^{(1)} \widetilde{\mathrm{E}} = \mathbb{E}^{1,0} \widetilde{\mathbb{M}}^{(2)} \mathbb{E}^{1,0^T} \mathbb{M}^{(1)} \mathbb{E}^{1,0} \mathrm{F} + \widetilde{\mathbb{M}}^{(1)} \mathbb{M}^{(1)} \mathbb{E}^{1,0} \mathrm{F}. \tag{31}$$

Then we use (25) in the first term on the right hand side and use the fact that $\widetilde{\mathbb{M}}^{(1)} \mathbb{M}^{(1)} = \mathbb{I}$ for the second term on the right hand side to get,

$$\mathbb{E}^{1,0} \widetilde{\mathbb{M}}^{(2)} \mathbb{E}^{1,0^T} \widetilde{\mathrm{E}} + \widetilde{\mathbb{M}}^{(1)} \widetilde{\mathrm{E}} = \mathbb{E}^{1,0} \widetilde{\mathbb{M}}^{(2)} (-\mathbb{T}^T \hat{\mathrm{E}}_{\dashv} - \mathbb{M}^{(0)} \mathrm{F}) + \mathbb{E}^{1,0} \mathrm{F}. \tag{32}$$

A further simplification of the bracket terms and using, $\widetilde{\mathbb{M}}^{(2)} \mathbb{M}^{(0)} = \mathbb{I}$, we get,

$$\mathbb{E}^{1,0} \widetilde{\mathbb{M}}^{(2)} \mathbb{E}^{1,0^T} \widetilde{\mathrm{E}} + \widetilde{\mathbb{M}}^{(1)} \widetilde{\mathrm{E}} = -\mathbb{E}^{1,0} \widetilde{\mathbb{M}}^{(2)} \mathbb{T}^T \hat{\mathrm{E}}_{\dashv}, \tag{33}$$

which shows that $\widetilde{\mathsf{E}}$ satisfies (30), as required.

5.4 *Equality of Norms*

The degrees of freedom of the curl of $\boldsymbol{E}^h$ are given by (21) and the associated basis functions in which these degrees are expanded are given by $\widetilde{\Psi}^{(2)}$, (17). Therefore we have

$$
\begin{aligned}
\|\boldsymbol{E}^h\|^2_{H(\mathrm{curl})} &= \left(\widetilde{\mathsf{E}}^T\mathbb{E}^{1,0}+\hat{\mathsf{E}}_{\dashv}^T\mathbb{T}\right)\widetilde{\mathbb{M}}^{(2)}\left(\mathbb{T}^T\hat{\mathsf{E}}_{\dashv}+\mathbb{E}^{1,0^T}\widetilde{\mathsf{E}}\right)+\widetilde{\mathsf{E}}^T\widetilde{\mathbb{M}}^{(1)}\widetilde{\mathsf{E}} \\
&\overset{\widetilde{\mathsf{E}}=\mathbb{M}^{(1)}\mathbb{E}^{1,0}\mathsf{F}}{=} \left(\mathsf{F}^T\mathbb{E}^{1,0^T}\mathbb{M}^{(1)}\mathbb{E}^{1,0}+\hat{\mathsf{E}}_{\dashv}^T\mathbb{T}\right)\widetilde{\mathbb{M}}^{(2)}\left(\mathbb{T}^T\hat{\mathsf{E}}_{\dashv}+\mathbb{E}^{1,0^T}\mathbb{M}^{(1)}\mathbb{E}^{1,0}\mathsf{F}\right) \\
&\quad +\mathsf{F}^T\mathbb{E}^{1,0^T}\mathbb{M}^{(1)}\widetilde{\mathbb{M}}^{(1)}\mathbb{M}^{(1)}\mathbb{E}^{1,0}\mathsf{F} \\
&\overset{(25)}{=} \mathsf{F}^T\mathbb{M}^{(0)}\widetilde{\mathbb{M}}^{(2)}\mathbb{M}^{(0)}\mathsf{F}+\mathsf{F}^T\mathbb{E}^{1,0^T}\mathbb{M}^{(1)}\mathbb{E}^{1,0}\mathsf{F} \\
&\overset{\widetilde{\mathbb{M}}^{(2)}\mathbb{M}^{(0)}=\mathbb{I}}{=} \mathsf{F}^T\mathbb{M}^{(0)}\mathsf{F}+\mathsf{F}^T\mathbb{E}^{1,0^T}\mathbb{M}^{(1)}\mathbb{E}^{1,0}\mathsf{F} \\
&= \|F^h\|^2_{H(\mathbf{curl})},
\end{aligned}
$$

which demonstrates that also in the finite dimensional setting the norms of $\boldsymbol{E}^h$ and F^h are the same.

6 Test Case

In this section we show the results for spectral element approximations of (3) and (4) for $\Omega \in [-1,1]^2$, using one spectral element.

We choose an exact solution for scalar F, and the corresponding vector field $\boldsymbol{E}$, given by,

$$
F_{ex} = e^x + e^y \; ; \qquad \boldsymbol{E}_{ex} = (e^y, -e^x). \tag{34}
$$

The problem (4) is discretized using a primal representation where we prescribe the Neumann boundary condition,

$$
\hat{E}_{\dashv} = \boldsymbol{n} \times (\mathbf{curl}\; F_{ex}). \tag{35}
$$

For the problem (3) we use a dual representation where we prescribe the Dirichlet boundary conditions,

$$
\hat{E}_{\dashv} = \boldsymbol{n} \times \boldsymbol{E}_{ex}. \tag{36}
$$

In Fig. 2 we show the difference between the ξ and η components of vector field $\boldsymbol{E}^h$ and **curl** F^h, for a low order spectral element approximation $N = 3$. The case $N = 3$ corresponds to the grid shown in Fig. 1. Here, we choose a very low order approximation to show that the equivalence of the duality relation derived in Sect. 5.3 holds true even for low order approximations.

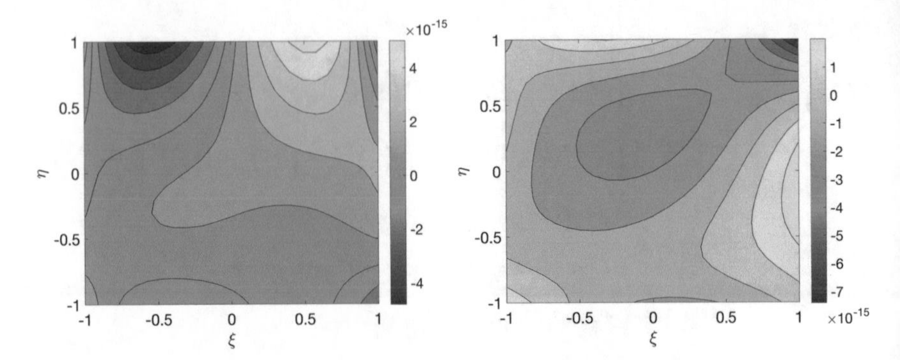

Fig. 2 Difference in vector field $\boldsymbol{E}^h$ and vector field ($\mathbf{curl}\ F^h$) for $N = 3$. *Left*: ξ-component. *Right*: η-component

Table 1 Norms $\|\boldsymbol{E}^h\|_{H(\mathrm{curl})}$ and $\|F^h\|_{H(\mathbf{curl})}$ as a function of the polynomial degree N

N	$\|F^h\|_{\mathrm{H}(\mathbf{curl})}$	$\|\boldsymbol{E}^h\|_{H(\mathrm{curl})}$
1	5.62334036	5.62334036
2	6.28815932	6.28815932
3	6.32851719	6.32851719
4	6.32957061	6.32957061
5	6.32958640	6.32958640
6	6.32958655	6.32958655
7	6.32958656	6.32958656
8	6.32958656	6.32958656
9	6.32958656	6.32958656

The difference observed in Fig. 2 is of the order $O(10^{-15})$; the two discrete vector fields thus agree up to machine precision.

From Sect. 5.4 we know that in the continuous setting, the $H(\mathrm{curl})$-norm of vector field $\boldsymbol{E}$ is equal to the $H(\mathbf{curl})$-norm of scalar field F. For this test case with exact solution given by (34) we have

$$\| F_{ex} \|_{H(\mathrm{curl})} = \sqrt{\| F_{ex} \|_{L^2}^2 + \| \mathbf{curl}\ F_{ex} \|_{L^2}^2} = \sqrt{8(\sinh(2) + \sinh^2(1))} = 6.32958656. \tag{37}$$

In Table 1 we show the calculated value of these discrete norms for increasing polynomial degree N of the basis functions. We observe that the discrete norms are exactly equal to each other and they converge to the theoretical value, (37). From Sect. 2 we saw that at the continuous level we have $\|\boldsymbol{E}\|_{H(\mathrm{curl};\Omega)} = \|F\|_{H(\mathbf{curl};\Omega)} = \|\hat{E}_{\dashv}\|_{H^{-\frac{1}{2}}(\mathrm{div};\partial\Omega)}$. Table 1 reveals that in the discrete setting, we can now also unambiguously define a norm $\|\hat{E}_{\dashv}\|_{H^{-\frac{1}{2}}(\mathrm{div};\partial\Omega)}$ on the trace space.

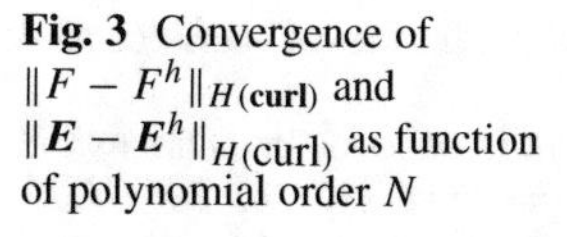

Fig. 3 Convergence of $\|F - F^h\|_{H(\mathbf{curl})}$ and $\|\boldsymbol{E} - \boldsymbol{E}^h\|_{H(\mathrm{curl})}$ as function of polynomial order N

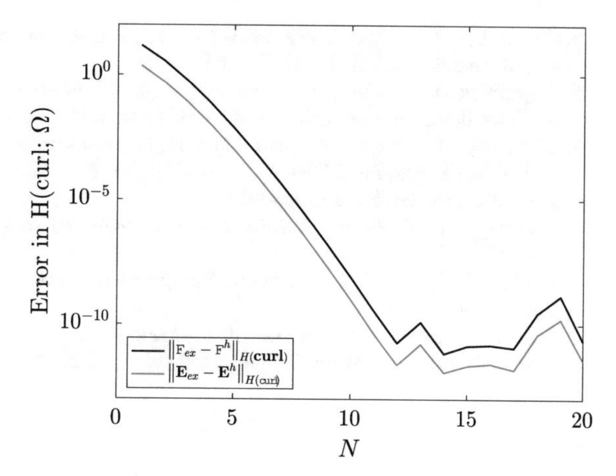

In Fig. 3 we show the convergence of error of F in $H(\mathbf{curl}; \Omega)$ norm, and the convergence of error of $\boldsymbol{E}$ in H(curl; Ω) norm. Both the errors converge exponentially to machine precision level.

7 Conclusions

The two curl-curl problems introduced in Sect. 2 are equivalent in the sense that $F = \mathrm{curl}\, \boldsymbol{E}$ and the norms of F and $\boldsymbol{E}$ are the same. In this paper it is proved that this equivalence continues to hold in finite dimensional spaces, if one of the degrees of freedom is expressed in terms of primal unknowns, F^h, and the other in dual degrees of freedom, $\boldsymbol{E}^h$. Equivalence of the approximate solutions and their norms is shown in Sect. 5, while in Sect. 6 this was illustrated for a specific test case.

References

1. Palha, A., Gerritsma, M.I.: A mass, energy, enstrophy and vorticity conserving (MEEVC) mimetic spectral element discretization for the 2D incompressible Navier-Stokes equations. J. Comput. Phys. **328**, 200–220 (2017)
2. Carstensen, C., Demkowicz, L., Gopalakrishnan, J.: Breaking spaces and form for the DPG method and applications including the Maxwell equations. Comput. Math. Appl. **72**, 494–522 (2016)
3. Jain, V., Zhang, Y., Palha, A., Gerritsma, M.I.: Construction and application of algebraic dual polynomial representations for finite element methods. Comput. Methods Appl. Math. (2018, submitted), arXiv: 1712.09472v2
4. Buffa, A., Ciarlet Jr., P.: On traces for functional spaces related to Maxwell's equations Part I: an integration by parts formula in Lipschitz polyhedra. Math. Methods Appl. Sci. **24**, 9–30 (2001)

5. Chung, E.T., Lee, C.S.: A staggred discontinuous Galerkin method for the curl-curl operator. IMA J. Numer. Anal. **32**(3), 1241–1265 (2012)
6. Degerfeldt, D., Rylander, T.: A brick-tetrahedron finite-element interface with stable hybrid explicit-implicit time-stepping for Maxwell's equations. J. Comput. Phys. 220, 383–393 (2006)
7. Gerritsma, M.I.: Edge functions for spectral element methods. In: Hesthaven, J.S., Rønquist, E.M. (eds.) Spectral and High Order Methods for Partial Differential Equations, vol. 76, pp. 199–207. Springer, Heidelberg (2011)
8. Kreyszig, E.: Introductory functional analysis with applications, 2nd edn. Wiley, New York (1978)
9. Oden, J.T., Demkowicz, L.: Applied Functional Analysis, 2nd edn. Chapman & Hall/CRC, Boca Raton (2010)
10. Zhang, Y., Jain, V., Palha, A., Gerritsma, M.I.: The discrete Steklov–Poincaré operator using algebraic dual polynomials. Comput. Methods Appl. Math. **19**(3), 645–661 (2019)

Multiple-Precision Iterative Methods for Solving Complex Symmetric Electromagnetic Systems

Koki Masui, Masao Ogino, and Lijun Liu

Abstract This paper deals with multiple-precision iterative methods for solving the complex symmetric linear equation derived from high-frequency electromagnetic field analysis using the edge finite element method. Double-precision iterative methods for solving this problem are slow to converge. Therefore, we implement a multiple-precision calculation, specifically, double-double (DD) precision numbers. The DD-precision complex numbers and arithmetic operations are implemented using an error-free transformation based on Knuth's and Dekker's algorithm. Moreover, some iterative methods with mixed precision are proposed to reduce computational cost. Using developed systems and the QD library that supports DD and quad-double arithmetics, some numerical methods are demonstrated experimentally and their performances are evaluated.

Keywords High precision calculation · Complex symmetric matrices · Iterative methods · Electromagnetic field analysis

1 Introduction

In solving the complex symmetric linear equations derived from high-frequency electromagnetic field analysis [1] using the edge finite element method [2], iterative methods such as the conjugate orthogonal conjugate gradient (COCG) method [3]

K. Masui (✉)
Graduate School of Informatics, Nagoya University, Nagoya, Japan
e-mail: masui@hpc.itc.nagoya-u.ac.jp

M. Ogino
Faculty of Informatics, Daido University, Nagoya, Japan
e-mail: m-ogino@daido-it.ac.jp

L. Liu
Graduate School of Engineering, Osaka University, Suita, Japan
e-mail: liu@mech.eng.osaka-u.ac.jp

H. van Brummelen et al. (eds.), *Numerical Methods for Flows*,
Lecture Notes in Computational Science and Engineering 132,
https://doi.org/10.1007/978-3-030-30705-9_28

suffer from slow convergence rates. Therefore, to perform electromagnetic field analysis with high efficiency, there is an urgent need for an effective iterative method for the complex symmetric linear equation. In addition, software libraries of multiple-precision arithmetic have been developed and the double-double (DD) precision number using two IEEE double precision numbers has been widely researched. Moreover, an optimized implementation of DD arithmetic for modern computer architecture has been reported [4]. Multiple-precision calculation is considered to be a useful strategy for improving the convergence behavior of iterative methods for linear systems. In flow problems, a mixed-precision calculation Krylov subspace method has been applied to improve the convergence by reducing round-off error [5].

However, there are few studies on multiple-precision complex number calculation for electromagnetic field analysis. In this study, we apply multiple-precision arithmetic in large-scale electromagnetic field analysis and evaluate the convergence performance and total calculation time through numerical experiments. Specifically, we develop multiple-precision and mixed-precision iterative methods based on DD-precision numbers and compare performances using the COCG, conjugate orthogonal conjugate residual (COCR) [6], and MINRES-like_CS [7] methods.

2 Multiple-Precision Complex Number Calculation

In this study, multiple precision is a precision that is higher than double precision. The DD precision used in this paper was proposed by Bailey [8] and Briggs [9], which is based on an error-free transformation based on Knuth's [10] and Dekker's [11] algorithm and uses two double-precision numbers to realize approximately quad-precision arithmetic. Double-precision floating-point operations are supported in hardware in most modern computers, but DD-precision operations are implemented in software. Therefore, DD-precision arithmetic requires about 10–20 times the number of operations required for double-precision arithmetic.

Software libraries that support multiple-precision arithmetic have been developed; however, there are few libraries that implement multiple-precision complex number arithmetic. Most libraries handle complex numbers by utilizing a C++ complex class template. This study focuses on the QD library (http://crd-legacy.lbl.gov/~dhbailey/mpdist/), which is an open source DD calculation library. It provides DD and quad-double (QD) precision arithmetic, which provides pseudo octuple-precision floating-point operations using four double-precision numbers. We use the *dd_real* and *qd_real* classes for DD- and QD-precision real numbers, respectively. Moreover, we can use the *complex<dd_real>* and *complex<qd_real>* classes for DD- and QD-precision complex numbers, respectively. In the C++ language and QD library, arithmetic operators are overloaded. However, it is difficult to write efficient code with this approach while considering mixed-precision complex number arithmetic such as double-precision and DD-precision complex numbers.

Therefore, in this study, we develop in-house code in the C language for mixed-precision complex number calculations of double-precision and DD-precision complex numbers. In our in-house code, a double-precision complex data type is defined by two double-precision members for the real and imaginary parts. Moreover, a DD-precision complex data type is defined by four double-precision members for the upper and lower bits of the real and imaginary parts.

There are some calculation methods at DD precision for the four arithmetic operations. To implement DD-precision complex number calculation, we choose *cray_add*, which has the same algorithm as *sloppy_add*, which is implemented in the QD library, as DD-precision addition and *fast_div* [12] as DD-precision division. Two advantages of these methods are that *cray_add* satisfies the error limit of the cray format and should yield high performance calculation, and *fast_div* does not use DD multiplication; therefore, the calculation cost is less than those of *sloppy_div* and *accurate_div*, which are implemented in the QD library.

3 Mixed-Precision Iterative Methods

For electromagnetic field analysis using the edge finite element method, the convergence rate and calculation time of the iterative method used to solve the system of linear equations are of great interest. In this study, a finite element equation is assumed to be constructed using double-precision arithmetic operations. In contrast, we use multiple-precision arithmetic operation in the iterative method for solving the system. In other words, to solve a linear equation $A\mathbf{x} = \mathbf{b}$, the coefficient matrix A and right-hand-side vector $\mathbf{b}$ (i.e., the input to the linear solver) are double precision, and other vectors are DD precision. For instance, when calculating sparse matrix-vector multiplication $\mathbf{q} = A\mathbf{p}$, matrix A is made up of double-precision complex numbers and vectors $\mathbf{q}$ and $\mathbf{p}$ are DD-precision complex numbers. Moreover, preconditioning matrix M, derived from coefficient matrix A, is calculated using double-precision complex numbers. Consequently, we develop mixed-precision iterative methods, specifically, the COCG, COCR, and MINRES-like_CS methods are implemented to solve complex symmetric linear systems.

4 Numerical Experiments

Our experiments were carried out on a 64-bit Linux system with a 3.60 GHz Intel Core i7-7700 CPU and the gcc/g++ ver 6.3.0 compiler. Note that the "-mfma" compiler option should be used to enable fused multiply-add (FMA) instructions in DD-precision calculation. We chose TEAM Workshop Problem 29 [13], shown in Fig. 1. This problem is a standard problem of high-frequency electromagnetic field

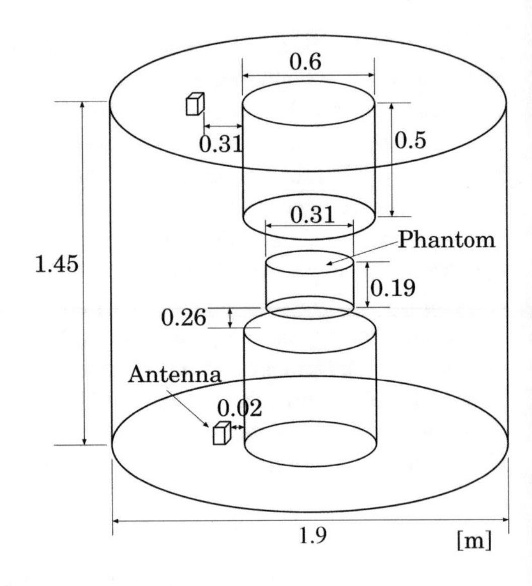

Fig. 1 The model of TEAM Workshop Problem 29

analysis. The finite element equation and using edge elements are given as follows [14]:

$$\int_{\Omega} \mathrm{rot} E_h \cdot \mu^{-1} \mathrm{rot} E_h^* d\Omega - \int_{\Omega} (\omega^2 \varepsilon' - i\omega\sigma) E_h \cdot E_h^* d\Omega = i\omega \int_{\Omega} J_h \cdot E_h^* d\Omega \quad (1)$$

where Ω is a finite element mesh of an analysis domain with boundary $\partial\Omega$, E_h [V/m] is the finite element approximation of the electric field, E_h^* [V/m] the test function satisfying $E_h^* \times n = 0$ on $\partial\Omega$, J_h [A/m^2] is the finite element approximation of current density, μ [H/m] is the permeability, ε' (=80.0 F/m) is the real part of the complex permittivity, σ (=0.52 S/m) is the electric conductivity, $\omega = 2\pi f$ [rad/s] is the single angular frequency, f [Hz] is the frequency, n is the outward normal vector on the boundary, and i is the imaginary unit. To construct the coefficient matrix of Eq. (1), we used ADVENTURE_Magnetic (http://adventure.sys.t.u-tokyo.ac.jp/), developed and published by the ADVENTURE project. Equation (1) leads to a complex symmetric linear system. The number of unknowns of the equation is 134,573 and the number of non-zero elements in the coefficient matrix is 1,129,211. Let $A\mathbf{x} = \mathbf{b}$ be the linear equation to be solved. Then, $\mathbf{b}$ is given by computing $\mathbf{b} = A\mathbf{x}$ with $\mathbf{x} = (1.0 + 1.0i, \ldots, 1.0 + 1.0i)^T$, where the initial guess $\mathbf{x}^0$ is zero. To determine convergence, the stopping condition $\|\mathbf{r}^n\|_2/\|\mathbf{r}^0\|_2 \leq 10^{-6}$ is used.

To solve this problem, we used three kinds of iterative methods, the COCG, COCR, and MINRES-like_CS methods, and two kinds of preconditioning, IC(0) decomposition with an acceleration factor [15] and SSOR preconditioning [16]. The acceleration factor of the IC(0) preconditioning and the relaxation coefficient of the SSOR preconditioning are 1.1 and 0.4, respectively. These values are optimal values for double-precision determined by numerical experiments. We evaluated the performances of the iterative methods with either double, mixed, DD, or QD precision arithmetic. The results of the iteration counts and calculation times are shown in Table 1 for SSOR preconditioning and Table 2 for IC(0) preconditioning. We note that, except for the QD calculation, which was done by the QD library, all other calculations were performed by our in-house code.

As can be seen in Tables 1 and 2, our proposed mixed-precision iterative methods based on the DD arithmetic reduces the number of iterations to about two-thirds compared with the double precision calculation. Moreover, our methods converge within almost the same number of steps as that of DD-precision calculation but in less computational time. Figures 2, 3, and 4 plot the convergence histories of IC(0) preconditioning. In all cases, we found that IC(0) preconditioning is more effective than SSOR preconditioning without the need for precision in the floating-point operations. In addition, although the COCR and MINRES-like_CS methods converge faster than the COCG method, the changes in the convergence and calculation times of the three methods using multiple-precision arithmetic tend

Table 1 Computational performance in solving the TEAM 29 (SSOR preconditioning)

	COCG		COCR		MINRES-like_CS	
	Iteration counts	Time [s]	Iteration counts	Time [s]	Iteration counts	Time [s]
Double	21,025	775	11,482	593	12,203	640
Mixed	14,730	1078	8615	808	8960	954
DD	14,691	1248	8245	1144	9003	1643
QD	9721	28,000	6192	40,486	7684	73,106

Table 2 Computational performance in solving the TEAM 29 (IC(0) preconditioning)

	COCG		COCR		MINRES-like_CS	
	Iteration counts	Time [s]	Iteration counts	Time [s]	Iteration counts	Time [s]
Double	8421	431	5932	214	6592	519
Mixed	5420	734	3939	533	4117	787
DD	5416	770	3940	565	4180	826
QD	3624	13,633	2559	18,018	2609	26,075

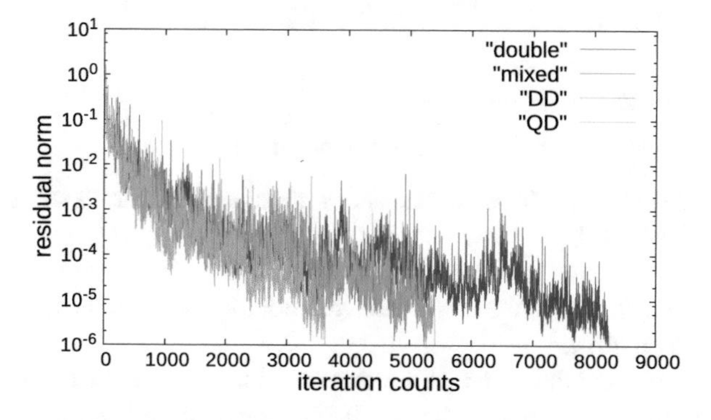

Fig. 2 Convergence history of IC(0) preconditioning for the COCG method

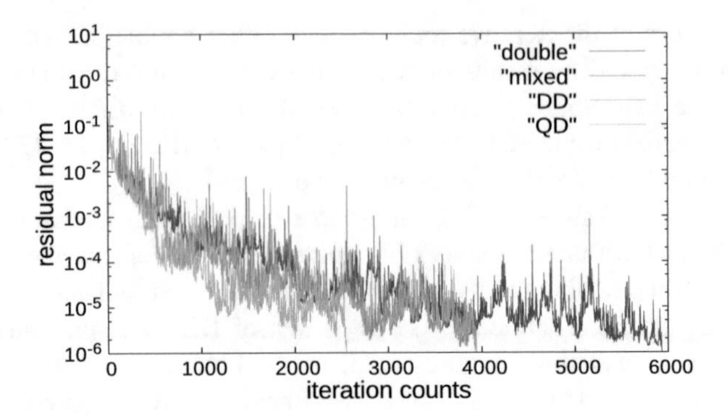

Fig. 3 Convergence history of IC(0) preconditioning for the COCR method

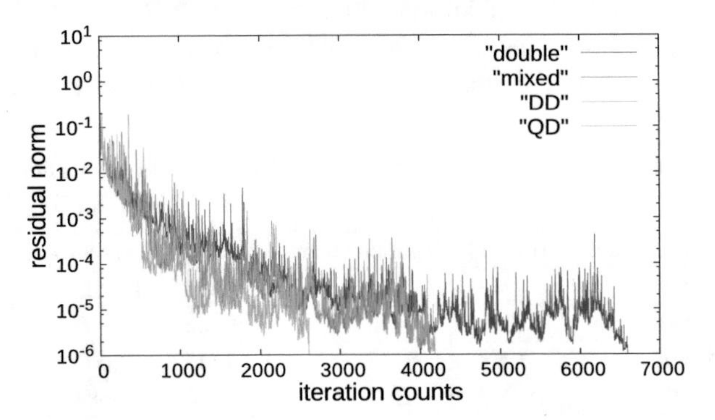

Fig. 4 Convergence history of IC(0) preconditioning for the MINRES-like_CS method

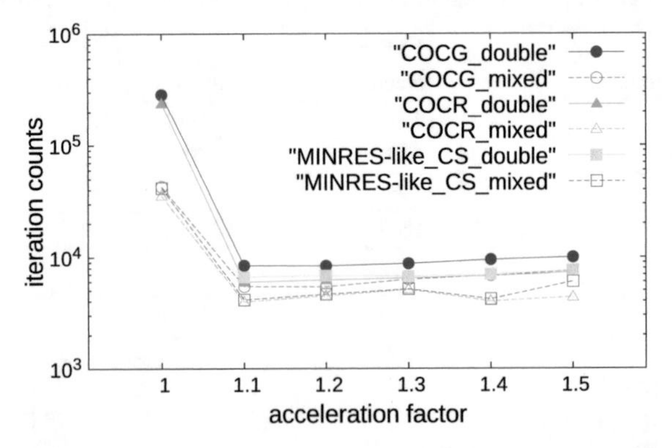

Fig. 5 Acceleration factor of IC preconditioning vs. iteration counts for the COCG, COCR, and MINRES-like_CS method

to be the same. Moreover, we confirmed that double-precision is sufficient for the preconditioning matrix.

Next, we evaluated the effect of the preconditioning parameters and calculation precision on the iteration count. For IC(0) preconditioning, the value of the acceleration factor is often chosen between 1.0 and 1.5. Figure 5 shows the number of iterations with different acceleration factors. The results of DD became almost the same as mixed. As these figures show, when the acceleration factor is 1.1–1.5, the double-precision calculation requires about 40% more steps than multiple-precision calculation. However, when the acceleration factor is 1.0, which indicates no acceleration, the iteration takes about 572% more steps to converge in double-precision arithmetic than in multiple-precision calculation. Especially, the MINRES-like_CS method with double-precision and no acceleration shows no convergence behavior in practical computation time. Therefore, applying multiple-precision calculation could be very helpful for improving the convergence behavior if an optimum acceleration factor for the IC(0) preconditioning is uncertain.

The SSOR preconditioning has a relaxation coefficient ω as a parameter and it is necessary to use an optimum value within $0 < \omega < 2$. Figure 6 shows the iteration counts for different values of relaxation coefficient. Similarly, the results of DD became almost the same as mixed. These figures show that using multiple-precision calculation instead of double-precision arithmetic, we succeeded in suppressing the increase of the iteration counts, even with different relaxation coefficient values. Moreover, the change in iteration counts is much less using multiple-precision calculation than when using double-precision calculation. In terms of computation time, if we choose the optimum relaxation factor of 0.4 in our experiments, mixed-precision arithmetic is 1.5 times slower than double-precision operation in all iterative methods. However, in some cases, the iteration counts in double-precision calculation are substantially increased. In these situations, mixed-precision calculation is faster than double-precision calculation. That is, our mixed-precision iterative methods are less influenced by the relaxation coefficient and provide stable convergence behavior. For both cases of IC(0) and SSOR preconditioning, by using multiple-precision arithmetic, we succeeded in suppressing the influence of the relaxation coefficient parameter and obtained a convergent solution for cases that could not be solved at double precision in practical calculation time. Consequently, it can be said that the multiple-precision iterative method is a robust method.

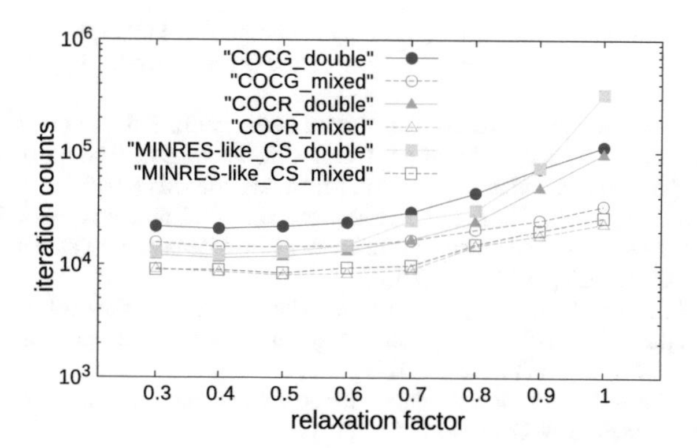

Fig. 6 Relaxation factor of SSOR preconditioning vs. iteration counts for the COCG, COCR, MINRES-like_CS method

5 Conclusion

In this research, we focused on multiple-precision calculation to speed up the solution of complex symmetric linear systems of equations derived from high-frequency electromagnetic field analysis using the edge finite element method. We developed mixed-precision iterative methods in which the precision of the coefficient matrix and preconditioning matrix are double-precision and the vectors are DD precision. The results of our numerical experiments yield the following findings.

- Multiple-precision calculation such as DD-precision and mixed-precision calculations are more effective than double-precision calculation for reducing the iteration counts of iterative methods for complex symmetric linear equations in electromagnetic field analysis.
- From the viewpoint of calculation time, (pseudo) octuple-precision calculation is unnecessary.
- When the optimum value of the parameter in the preconditioning is uncertain, it is possible to reduce the influence of this parameter using multiple-precision arithmetic operation and obtain a convergent solution more quickly.
- When using the same precision, the COCR and MINRES-like_CS methods need fewer iteration counts than the COCG method.

Acknowledgements This research was partially supported by Grant-in-Aid for Scientific Research (B) No. 17H02829 and JST-CREST "Development of Hierarchical Split Type Numerical Solution Library for Post Petascale Simulation." One of the authors was supported by ACT-I, JST. We thank Kim Moravec, PhD, from Edanz Group (www.edanzediting.com/ac) for editing a draft of this manuscript.

References

1. Takei, A., Yoshimura, S., Kanayama, H.: Large-scale parallel finite element analyses of high frequency electromagnetic field in commuter trains. Comput. Model. Eng. Sci. **31**(1), 13–24 (2008)
2. Nédélec, J.C.: Mixed finite elements in 3. Numer. Math. **35**(3), 315–341 (1980)
3. Van der Vorst, H.A., Melissen, J.B.M.: A Petrov-Galerkin type method for solving Ax=b and where A is symmetric complex. IEEE Trans. Magn. **26**, 706–708 (1990)
4. Hishinuma, T., Fujii, A., Tanaka, T., Hasegawa, H.: AVX acceleration of DD arithmetic between a sparse matrix and vector. In: International Conference on Parallel Processing and Applied Mathematics, pp. 622–631 (2013)
5. Furuichi, M., Dave, A.M., Paul, J.T: Development of a Stokes flow solver robust to large viscosity jumps using a Schur complement approach with mixed precision arithmetic. J. Comput. Phys. **230**(24), 8835–8851 (2011)
6. Sogabe, T., Zhang, S.L.: A COCR method for solving complex symmetric linear systems. J. Comput. Appl. Math. **199**(2), 297–303 (2007)
7. Ogino, M., Takei, A., Sugimoto, S., Yoshimura, S.: A numerical study of iterative substructuring method for finite element analysis of high frequency electromagnetic fields. Comput. Math. Appl. **72**(8), 2020–2027 (2016)

8. Bailey, D.H.: A fortran-90 double-double library. http://crd-legacy.lbl.gov/~dhbailey/mpdist/
9. Briggs, K.: Double double floating point arithmetic. http://keithbriggs.info/doubledouble.html
10. Knuth, D.E.: The Art of Computer Programming: Seminumerical Algorithms, vol. 2. Addison Wesley, Reading (1981)
11. Dekker, T.J.: A foating-point technique for extending the available precision. Numer. Math. **18**, 224–242 (1971)
12. Yamanaka, N., Oishi, S.: Error of quadruple precision arithmetic method based on double precision and its application (in Japanese). RIMS Kokyuroku **1791**, 216–225 (2012)
13. Kanai, Y.: Description of TEAM Workshop Problem 29: whole body cavity resonator. In: TEAM Workshop in Tucson (1998)
14. Takei, A., Sugimoto, S., Ogino, M., Yoshimura, S., Kanayama, H.: Full wave analyses of electromagnetic fields with an iterative domain decomposition method. IEEE Trans. Magn. **46**(8), 2860–2863 (2010)
15. Fujiwara, K., Nakata, T., Fusayasu, H.: Acceleration of convergence characteristic of the ICCG method. IEEE Trans. Magn. **29**(2), 1958–1961 (1993)
16. Young, D.M.: Iterative Solution of Large Linear Systems. Academic, New York (1971)

Gradient-Based Limiting and Stabilization of Continuous Galerkin Methods

Dmitri Kuzmin

Abstract In this paper, we stabilize and limit continuous Galerkin discretizations of a linear transport equation using an algebraic approach to derivation of artificial diffusion operators. Building on recent advances in the analysis and design of edge-based algebraic flux correction schemes for singularly perturbed convection-diffusion problems, we derive algebraic stabilization operators that generate nonlinear high-order stabilization in smooth regions and enforce discrete maximum principles everywhere. The correction factors for antidiffusive element or edge contributions are defined in terms of nodal gradients that vanish at local extrema. The proposed limiting strategy is linearity-preserving and provides Lipschitz continuity of constrained terms. Numerical examples are presented for two-dimensional test problems.

Keywords Hyperbolic conservation laws · Finite element methods · Discrete maximum principles · Algebraic flux correction · Linearity preservation

1 Introduction

Bound-preserving discretizations of hyperbolic conservation laws and convection-dominated transport problems use limiting techniques to enforce discrete maximum principles. Recent years have witnessed an increased interest of the finite element community in algebraic flux correction (AFC) schemes [9] based on various generalizations of flux-corrected transport (FCT) algorithms and total variation diminishing (TVD) methods. A major breakthrough in the theoretical analysis of AFC for continuous finite elements was achieved by Barrenechea et al. [3, 5] whose recent work has provided a set of design principles for derivation of limiters that lead to well-posed nonlinear problems in the context of stationary convection-

D. Kuzmin (✉)
TU Dortmund University, Institute of Applied Mathematics (LS III), Dortmund, Germany
e-mail: kuzmin@math.uni-dortmund.de

H. van Brummelen et al. (eds.), *Numerical Methods for Flows*,
Lecture Notes in Computational Science and Engineering 132,
https://doi.org/10.1007/978-3-030-30705-9_29

diffusion equations. Limiting techniques for continuous Galerkin discretizations of hyperbolic problems were proposed in [2, 7, 12]. As shown in [7, 12], the use of the standard Galerkin method as the AFC target for hyperbolic conservation laws may give rise to bounded ripples and nonphysical weak solutions. In fact, the Galerkin discretization may even produce singular matrices on criss-cross (Union Jack) meshes [13]. The use of limiters restricts the range of possible solution values but does not rule out spurious oscillations within this range. In this paper, we design artificial diffusion operators that introduce high-order stabilization in smooth regions and enforce preservation of local bounds in the vicinity of steep fronts. The element or edge contributions to the residual of the nonlinear system are constrained using limiters defined in terms of nodal gradients rather than nodal correction factors. This approach leads to a limiting procedure that satisfies all essential design criteria.

2 Artificial Diffusion Operators

To make the presentation self-contained, we begin with an outline of the basic AFC methodology [9] for C^0 finite element discretizations of the hyperbolic equation

$$\frac{\partial u}{\partial t} + \nabla \cdot (\mathbf{v}u) = 0 \qquad \text{in } \Omega \tag{1}$$

to be solved in a bounded domain Ω with a Lipschitz boundary Γ. The velocity field $\mathbf{v}$ is assumed to be known. At the inlet $\Gamma_{\text{in}} = \{\mathbf{x} \in \Gamma \,:\, \mathbf{v} \cdot \mathbf{n} < 0\}$, we impose the boundary condition $u = 0$ in a weak sense by using the variational formulation

$$\int_\Omega w \frac{\partial u}{\partial t} \mathrm{d}\mathbf{x} + \int_\Gamma wu \max\{0, \mathbf{v} \cdot \mathbf{n}\} \mathrm{d}\mathbf{s} - \int_\Omega \nabla w \cdot (\mathbf{v}u) \mathrm{d}\mathbf{x} = 0, \tag{2}$$

where $\mathbf{n}$ is the unit outward normal and $w \in H^1(\Omega)$ is an admissible test function.

The numerical solution $u_h = \sum_{j=1}^N u_j \varphi_j$ is defined in terms of continuous piecewise-linear or multilinear Lagrange basis functions φ_j associated with vertices $\mathbf{x}_j$ of a mesh (alias *triangulation*) $\mathscr{T}_h$. The standard Galerkin discretization leads to

$$M_C \frac{\mathrm{d}u}{\mathrm{d}t} + Au = 0, \tag{3}$$

where $M_C = \{m_{ij}\}$ is the consistent mass matrix, $A = \{a_{ij}\}$ is the discrete transport operator, and $u = \{u_i\}$ is the vector of time-dependent nodal values.

Introducing the lumped mass matrix $M_L = \{\delta_{ij} \sum_j m_{ij}\}$ and a symmetric artificial diffusion operator $D = \{d_{ij}\}$, we construct the low-order approximation

$$M_L \frac{\mathrm{d}u}{\mathrm{d}t} + (A - D)u = 0 \tag{4}$$

which is provably bound-preserving if $\sum_j d_{ij} = 0$ and $d_{ij} \geq \max\{a_{ij}, 0, a_{ji}\}$ for all $j \neq i$ [3, 9]. The original Galerkin discretization (3) can be written as

$$M_L \frac{\mathrm{d}u}{\mathrm{d}t} + (A - D)u = f\left(u, \frac{\mathrm{d}u}{\mathrm{d}t}\right), \qquad f_i = \sum_m f_i^m, \tag{5}$$

where $f = \{f_i\}$ is the antidiffusive part that requires limiting. In edge-based AFC schemes, f_i^m is the contribution of edge m to node i, and there exists a neighbor node $j \neq i$ such that $f_j^m = -f_i^m$ [3, 7, 9]. In element-based versions, f_i^m is the contribution of element m to node i and $\sum_i f_i^m = 0$ by definition [10, 12]. In the 1D case, the decompositions of f into edge and element contributions are equivalent.

In the process of limiting, each component f_i^m is multiplied by a solution-dependent correction factor $\alpha^m \in [0, 1]$. This modification leads to the nonlinear system

$$M_L \frac{\mathrm{d}u}{\mathrm{d}t} + (A - D)u = \bar{f}\left(u, \frac{\mathrm{d}u}{\mathrm{d}t}\right), \qquad \bar{f}_i = \sum_m \alpha^m f_i^m. \tag{6}$$

We define f_i^m and α^m in the next section. The discretization in time can be performed using a strong stability preserving (SSP) Runge-Kutta method [6]. Note that only the backward Euler method is SSP without any restrictions on the time step.

3 Limiting of Antidiffusive Terms

First and foremost, the definition of correction factors α^m should guarantee that the limited antidiffusive term $\bar{f}_i$ be *local extremum diminishing* (LED), i.e.,

$$u_i^{\max} := \max_{j \in \mathcal{N}_i} u_j = u_i \quad \Rightarrow \quad \bar{f}_i \leq 0, \tag{7}$$

$$u_i^{\min} := \min_{j \in \mathcal{N}_i} u_j = u_i \quad \Rightarrow \quad \bar{f}_i \geq 0, \tag{8}$$

where $\mathcal{N}_i = \{j \in \{1, \dots, N\} : m_{ij} \neq 0\}$ is the computational stencil of node i.

Obviously, the LED property (7), (8) holds for α^m satisfying (cf. [4, 5])

$$\alpha^m \leq \alpha_i := \min\left\{1, \frac{\gamma_i \min\{u_i^{\max} - u_i, u_i - u_i^{\min}\}}{\max\{u_i^{\max} - u_i, u_i - u_i^{\min}\} + \epsilon h}\right\} \quad \forall i \in \mathcal{N}^m, \tag{9}$$

where $\mathcal{N}^m$ is the set of nodes belonging to the element or edge, $\gamma_i > 0$ is a parameter to be defined in Sect. 3.2, h is the mesh size, and ϵ is a small positive constant.

Theoretical and numerical studies of AFC schemes indicate that the use of linearity-preserving limiters is an essential prerequisite for achieving optimal accuracy on general meshes [4, 5, 9, 10]. The bound α_i in formula (9) is linearity preserving if $\alpha_i = 1$ whenever u_h is linear on the patch $\bar{\Omega}_i = \{K \in \mathcal{T}_h : \mathbf{x}_i \in K\}$ of elements containing an internal node $\mathbf{x}_i \in \Omega$. According to the analysis of Barrenechea et al. [4, 5], the nodal correction factor α_i defined by (9) possesses this property if

$$\gamma_i \geq \gamma_i^{\min} := \frac{\max_{\mathbf{x}_j \in \partial\Omega_i} |\mathbf{x}_i - \mathbf{x}_j|}{\operatorname{dist}\{\mathbf{x}_i, \partial\Omega_i^{\text{conv}}\}}, \tag{10}$$

where $|\cdot|$ is the Euclidean norm and Ω_i^{conv} is the convex hull of points $\mathbf{x}_j \in \bar{\Omega}_i$.

Implicit time integration requires iterative solution of nonlinear systems and only converged solutions are guaranteed to be bound preserving. Therefore, convergence behavior of iterative solvers should also be taken into account. It is essential to guarantee that each product $\alpha^m f_i^m$ be a Lipschitz-continuous function of nodal values. This property is used in proofs of existence and uniqueness for the nonlinear discrete problem [3, 5] and secures convergence of fixed-point iterations based on deferred correction methods (see [1, Proposition 4.3]). Faster convergence can be achieved, e.g., using Anderson acceleration [9] or differentiable LED limiters [2].

3.1 Nonlinear High-Order Stabilization

The straightforward choice $\alpha^m = \min_{i \in \mathcal{N}^m} \alpha_i$ of the correction factor α^m for f_i^m corresponds to using the oscillatory Galerkin scheme (3) as the limiting target. In this section, we construct a stabilized AFC scheme using a definition of α^m in terms of limited nodal gradients $\mathbf{g}_i^*$ such that $\mathbf{g}_i^* = 0$ if $u_i = u_i^{\max}$ or $u_i = u_i^{\min}$ at an internal node $\mathbf{x}_i \in \Omega$. Additionally, the gradient recovery procedure should be exact for linear functions. In Sect. 3.2, we use nodal correction factors α_i of the form (9) to correct a linearity-preserving gradient reconstruction $\mathbf{g}_i$ and obtain a Lipschitz-continuous approximation $\mathbf{g}_i^* = \alpha_i \mathbf{g}_i$ satisfying the above requirements.

An element-based AFC scheme with a stabilized high-order target is defined by

$$\alpha^m = \left(\frac{\min\{\| |\nabla u_h^m| \|_{C(K^m)}, p \min_{i \in \mathcal{N}^m} |\mathbf{g}_i^*|\}}{\| |\nabla u_h^m| \|_{C(K^m)} + \epsilon} \right)^q, \tag{11}$$

where $\|\cdot\|_{C(K)}$ is the maximum norm. The parameters $p \geq 1$ and $q \in \mathbb{N}$ act as steepeners that make the limiter less diffusive. By default, we use $p = q = 2$.

If a local extremum is attained at node i for any $i \in \mathcal{N}^m$, then $|\mathbf{g}_i^*| = 0$ and, therefore, $\alpha^m = 0$ in accordance with the LED criterion. If u_h is linear on $\bar{\Omega}_i$ and the parameter γ_i is defined by (16), then $\alpha_i = 1$ and, therefore, $\mathbf{g}_i^* = \mathbf{g}_i = \nabla u_h^m$. In general, our formula (11) will produce $\alpha^m = 1$ if the magnitude of ∇u_h^m does not exceed that of the smallest nodal gradient by more than a factor of p. Lipschitz continuity of $\alpha^m f_i^m$ can be shown following Lohmann's [11] proofs for edge-based tensor limiters.

An edge-based counterpart of our gradient-based limiter (11) is defined by

$$\alpha^m = \left(\frac{\min\{p|\mathbf{g}_i^* \cdot (\mathbf{x}_i - \mathbf{x}_j)|, |u_i - u_j|, p|\mathbf{g}_j^* \cdot (\mathbf{x}_i - \mathbf{x}_j)|\}}{|u_i - u_j| + \epsilon} \right)^q. \tag{12}$$

A proof of Lipschitz continuity for $q \in \mathbb{N}$ follows from Lohmann's analysis [11].

If the gradient is nonsmooth, our method may produce $\alpha^m < 1$ even in the case when $\mathbf{g}_i^* = \mathbf{g}_i$ for all i. In contrast to limiters designed to recover the standard Galerkin discretization whenever it satisfies the LED constraints, our definition of α^m generates nonlinear high-order dissipation in elements free of local extrema. On a uniform mesh of 1D linear elements both (11) and (12) lead to a *symmetric limited positive* (SLIP) scheme [8] that switches between second- and fourth-order dissipation.

In predictor-corrector algorithms of FCT type, high-order dissipation can also be generated by adding nonlinear entropy viscosity [7] or linear gradient-based stabilization [10, 12]. However, the use of such artificial diffusion operators in iterative AFC schemes inhibits convergence due to the lack of Lipschitz continuity.

3.2 *Recovery of Nodal Gradients*

If u_h is linear on $\bar{\Omega}_i$, then $\mathbf{g}_i = \nabla u_h(\mathbf{x}_i)$ holds for any weighted average $\mathbf{g}_i$ of the one-sided element gradients $\nabla u_h|_K(\mathbf{x}_i)$, $K \in \bar{\Omega}_i$. For example, the global lumped-mass L^2 projection yields the averaged nodal gradient [9]

$$\mathbf{g}_i = \frac{1}{m_i} \sum_{j \in \mathcal{N}_i} \mathbf{c}_{ij} u_j = \frac{1}{m_i} \sum_{j \in \mathcal{N}_i \setminus \{i\}} \mathbf{c}_{ij}(u_j - u_i), \tag{13}$$

$$m_i = \int_\Omega \varphi_i \mathrm{d}\mathbf{x}, \qquad \mathbf{c}_{ij} = \int_\Omega \varphi_i \nabla \varphi_j \mathrm{d}\mathbf{x}. \tag{14}$$

However, the so-defined $\mathbf{g}_i$ does not necessarily vanish if a local maximum or minimum is attained at $\mathbf{x}_i \in \Omega$. To rectify this, we consider the limited gradient

$$\mathbf{g}_i^* = \alpha_i \mathbf{g}_i = \frac{1}{m_i} \sum_{j \in \mathcal{N}_i \setminus \{i\}} \mathbf{c}_{ij} \alpha_i (u_j - u_i), \tag{15}$$

where α_i is the nodal correction factor defined by (9). The limited gradient reconstruction $\mathbf{g}_i^*$ does vanish at local extrema. Lipschitz continuity of $\alpha_i(u_j - u_i)$ can be shown using Lemma 6 in [3]. Linearity preservation is guaranteed under condition (10) since $\alpha_i = 1$ if u_h is linear on $\bar{\Omega}_i$. The use of the sharp bound $\gamma_i := \gamma_i^{\min}$ defined by (10) requires calculation of the distance to the convex hull and leads to rather diffusive *minmod* limiters like the one proposed in [4]. To simplify the formula for γ_i and make the LED constraints less restrictive, we define γ_i as follows.

The anisotropy of a mesh element $K \in \mathscr{T}_h$ can be characterized by the ratio of the local mesh size $h_K = \mathrm{diam}(K)$ and the diameter ρ_K of the largest ball that fits into K. A family of triangulations $\{\mathscr{T}_h\}$ is called regular if there exists a constant $\sigma > 0$ such that $\frac{h_K}{\rho_K} \le \sigma$ for all $K \in \mathscr{T}_h$ and all h. For triangular elements, ρ_K is the diameter of the inscribed circle. Since $h_K \ge \max_{\mathbf{x}_j \in K} |\mathbf{x}_i - \mathbf{x}_j|$ and $\rho_K \le \mathrm{dist}\{\mathbf{x}_i, \partial K \cap \partial \Omega_i^{\mathrm{conv}}\}$ for all $K \in \bar{\Omega}_i$, condition (10) holds for parameters $\gamma_i = \gamma_i(s)$, $s \ge 1$ defined by

$$\gamma_i = s \frac{\max_{K \in \bar{\Omega}_i} h_K}{\min_{K \in \bar{\Omega}_i} \rho_K}. \tag{16}$$

A reasonable default setting for iterative AFC schemes is $s = 2$. The limiter becomes less diffusive as s is increased but the use of $\gamma_i \gg \gamma_i^{\min}$ may cause convergence problems when it comes to iterative solution of nonlinear systems.

3.3 Recovery of Nodal Time Derivatives

By (3) and (6), the term to be limited is given by $f(u, \dot{u}) = (M_L - M_C)\dot{u} - Du$. Using the Neumann series approximation [7] to M_C^{-1}, we obtain

$$\dot{u} = M_L^{-1}(I + (M_L - M_C)M_L^{-1})(\bar{f}(u, 0) - (A - D)u). \tag{17}$$

This definition of $\dot{u}$ makes it possible to determine $\bar{f}(0, \dot{u})$ without recalculating $\bar{f}(u, 0)$. Moreover, the correct steady state behavior is preserved for $\dot{u} = 0$.

In our AFC scheme for time-dependent problems, we limit $f_i^m(u, 0)$ using α^m defined in Sect. 3.1. To provide continuity, the contribution of $f_i^m(0, \dot{u})$ is limited using a correction factor $\dot{\alpha}^m$ such that $\dot{\alpha}^m = 0$ if $\alpha^m = 0$ (see below). In the element-based version of (6), the limited antidiffusive components are given by

$$\bar{f}_i^m(u, \dot{u}) = \sum_{j \in \mathcal{N}^m \setminus \{i\}} \left[m_{ij}^m \dot{\alpha}^m (\dot{u}_i - \dot{u}_j) + d_{ij}^m \alpha^m (u_i - u_j) \right], \qquad i \in \mathcal{N}^m. \tag{18}$$

The coefficients m_{ij}^m and d_{ij}^m represent the contribution of $K^m \in \mathscr{T}_h$ to the global matrix entries m_{ij} and d_{ij}, respectively. Algebraic residual correction schemes based on such decompositions into element contributions can be found in [10, 12].

The evolutionary part $f_i^m(0, \dot{u})$ is constrained using correction factors of the form

$$\dot{\alpha}^m = \min\left\{1, \beta^m \frac{\alpha^m \| |\nabla u_h^m| \|_{C(K^m)}}{\| |\nabla \dot{u}_h^m| \|_{C(K^m)} + \epsilon}\right\} \tag{19}$$

such that $(\dot{\alpha}^m \| |\nabla \dot{u}_h^m| \|_{C(K^m)})/(\alpha^m \| |\nabla u_h^m| \|_{C(K^m)}) \leq \beta^m$, where $\beta^m > 0$ should have units of the reciprocal second s^{-1}. In this work, we use $\beta^m = \|\mathbf{v}\|_{C(K^m)}/(2h_{K^m})$.

In the edge-based version of (6), we limit $f_i^m = f_{ij}$ and $f_j^m = -f_{ij}$ as follows:

$$\bar{f}_i^m(u, \dot{u}) = m_{ij}\dot{\alpha}^m(\dot{u}_i - \dot{u}_j) + d_{ij}\alpha^m(u_i - u_j), \qquad \{i, j\} =: \mathcal{N}^m, \tag{20}$$

$$\dot{\alpha}^m = \min\left\{1, \beta^m \frac{\alpha^m |u_i - u_j|}{|\dot{u}_i - \dot{u}_j| + \epsilon}\right\}. \tag{21}$$

In pseudo-time-stepping schemes for steady-state computations, we use $\dot{\alpha}^m = 0$.

4 Numerical Examples

In Figs. 1 and 2, we present the AFC results for the time-dependent solid body rotation benchmark and a stationary circular convection test. For the formulation of the corresponding (initial-)boundary value problems, we refer the reader to [9, 10]. In this numerical study, we use the element-based version of (6). The stationary problem is solved using implicit pseudo-time-stepping and a fixed-point iteration method [9]. The employed parameter settings and discretization parameters are summarized in the captions. The constrained Galerkin solutions satisfy local

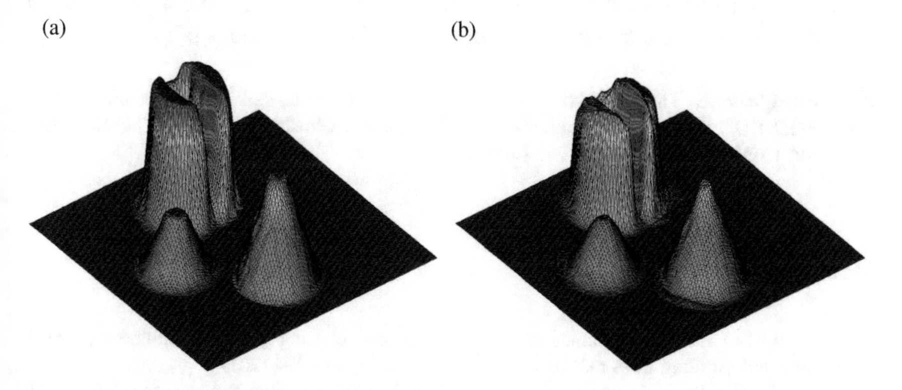

Fig. 1 Solid body rotation: uniform triangular mesh, 2×128^2 P_1 elements, AFC scheme based on (11), explicit second-order SSP-RK time-stepping, $\Delta t = 10^{-3}$, solutions at $T = 2\pi$. **(a)** $\alpha^m(\mathbf{g}_h^*, s = 5)$, $u_h \in [0.0, 0.992]$. **(b)** $\alpha^m(\mathbf{g}_h) =: \dot{\alpha}^m$, $u_h \in [-0.029, 1.08]$

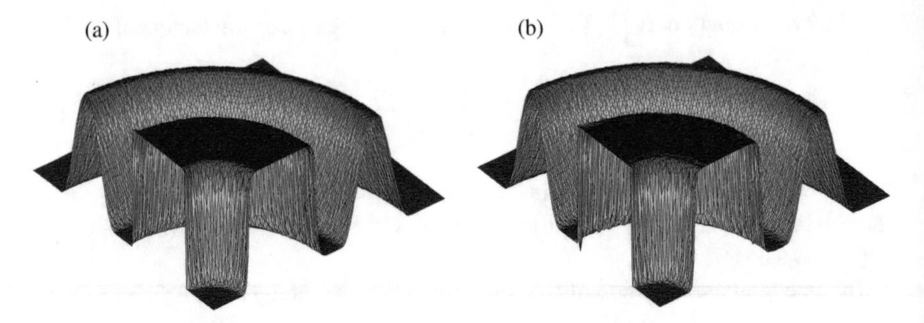

Fig. 2 Circular convection: perturbed triangular mesh, 2×128^2 P_1 elements, AFC scheme based on (11), backward Euler pseudo-time-stepping, steady-state solutions. (**a**) $\alpha^m(\mathbf{g}_h^*, s = 2)$, $u_h \in [0.0, 1.0]$. (**b**) $\alpha^m(\mathbf{g}_h)$, $u_h \in [-0.082.0, 1.026]$

Table 1 Circular convection: L^1 convergence history for triangular meshes consisting of $2/h^2$ cells, P_1 approximation, AFC scheme based on (11), $s = 2$, smooth inflow profile

h	Uniform, $\alpha^m(\mathbf{g}_h^*)$	EOC	Perturbed, $\alpha^m(\mathbf{g}_h^*)$	EOC	Perturbed, $\alpha^m(\mathbf{g}_h)$	EOC
1/32	0.185E−01		0.150E−01		0.141E−01	
1/64	0.511E−02	1.85	0.473E−02	1.67	0.446E−02	1.66
1/128	0.117E−02	2.14	0.155E−02	1.61	0.133E−02	1.75
1/256	0.256E−03	2.19	0.499E−03	1.64	0.438E−03	1.60

maximum principles if $\alpha^m = \alpha^m(\mathbf{g}_h^*)$ is defined by (11). To assess the amount of intrinsic high-order stabilization, we also present the numerical solutions obtained using the target scheme ($\alpha^m = \alpha^m(\mathbf{g}_h)$ and $\dot{\alpha}^m := \alpha^m$ in the unsteady case). The L^1 convergence rates for the circular convection test without the discontinuous portion of the inflow profile are shown in Table 1. The AFC scheme based on $\alpha^m = \alpha^m(\mathbf{g}_h^*)$ exhibits second-order superconvergence on uniform meshes. The convergence rates on perturbed meshes are comparable to those for $\alpha^m = \alpha^m(\mathbf{g}_h)$ and higher than the optimal order 1.5 for continuous P_1 finite element discretizations of (1).

Acknowledgements This research was supported by the German Research Association (DFG) under grant KU 1530/23-1. The author would like to thank Christoph Lohmann (TU Dortmund University) for helpful discussions and suggestions.

References

1. Abgrall, R.: High order schemes for hyperbolic problems using globally continuous approximation and avoiding mass matrices. J. Sci. Comput. **73**, 461–494 (2017)
2. Badia, S., Bonilla, J.: Monotonicity-preserving finite element schemes based on differentiable nonlinear stabilization. Comput. Methods Appl. Mech. Eng. **313**, 133–158 (2017)
3. Barrenechea, G., John, V., Knobloch, P.: Analysis of algebraic flux correction schemes. SIAM J. Numer. Anal. **54**, 2427–2451 (2016)

4. Barrenechea, G., John, V., Knobloch, P.: A linearity preserving algebraic flux correction scheme satisfying the discrete maximum principle on general meshes. Math. Models Methods Appl. Sci. **27**, 525–548 (2017)
5. Barrenechea, G., John, V., Knobloch, P., Rankin, R.: A unified analysis of algebraic flux correction schemes for convection-diffusion equations. SeMA (2018). https://doi.org/10.1007/s40324-018-0160-6
6. Gottlieb, S., Shu, C.-W., Tadmor, E.: Strong stability-preserving high-order time discretization methods. SIAM Rev. **43**, 89–112 (2001)
7. Guermond, J.-L., Nazarov, M., Popov, B., Yang, Y.: A second-order maximum principle preserving Lagrange finite element technique for nonlinear scalar conservation equations. SIAM J. Numer. Anal. **52**, 2163–2182 (2014)
8. Jameson, A.: Positive schemes and shock modelling for compressible flows. Int. J. Numer. Methods Fluids **20**, 743–776 (1995)
9. Kuzmin, D.: Algebraic flux correction I. Scalar conservation laws. In: Kuzmin, D., Löhner, R., Turek, S. (eds.) Flux-Corrected Transport: Principles, Algorithms, and Applications, 2nd edn., pp. 145–192. Springer, Dordrecht (2012)
10. Kuzmin, D., Basting, S., Shadid, J.N.: Linearity-preserving monotone local projection stabilization schemes for continuous finite elements. Comput. Methods Appl. Mech. Eng. **322**, 23–41 (2017)
11. Lohmann, C.: Algebraic flux correction schemes preserving the eigenvalue range of symmetric tensor fields. Ergebnisber. Inst. Angew. Math. 584, TU Dortmund University (2018)
12. Lohmann, C., Kuzmin, D., Shadid, J.N., Mabuza, S.: Flux-corrected transport algorithms for continuous Galerkin methods based on high order Bernstein finite elements. J. Comput. Phys. **344**, 151–186 (2017)
13. Neta, B., Williams, R.T.: Stability and phase speed for various finite element formulations of the advection equation. Comput. Fluids **14**, 393–410 (1986)

High Order CG Schemes for KdV and Saint-Venant Flows

Julie Llobell, Sebastian Minjeaud, and Richard Pasquetti

Abstract Strategies we have recently proposed to efficiently address dispersive equations and hyperbolic systems with high order continuous Galerkin schemes are first recalled. Using the Spectral Element Method (SEM), we especially consider the Korteweg-De Vries equation to explain how to handle the third order derivative term with an only C^0-continuous approximation. Moreover, we focus on the preservation of two invariants, namely the mass and momentum invariants. With a stabilized SEM, we then address the Saint-Venant system to show how a strongly non linear viscous stabilization, namely the entropy viscosity method (EVM), can allow to support the presence of dry-wet transitions and shocks. The new contribution of the paper is a sensitivity study to the EVM parameters, for a shallow water problem involving many interactions and shocks. A comparison with a computation carried out with a second order Finite Volume scheme that implements a shock capturing technique is also presented.

Keywords Korteweg-de Vries equation · Shallow water equations · Dispersive problems · Hyperbolic problems · Spectral element method

1 Introduction

The Spectral Element Method (SEM) allows a high order approximation of partial differential equations (PDEs) and combines the advantages of spectral methods, that is accuracy and rapid convergence, with those of the finite element method (FEM), that is geometrical flexibility. The SEM has proved for a long time to be efficient for the highly accurate resolution of elliptic or parabolic problems, but hyperbolic problems and dispersive equations still remain challenging. As relevant examples

J. Llobell · S. Minjeaud · R. Pasquetti (✉)
Université Côte d'Azur, CNRS, INRIA, Lab. J.A. Dieudonné, Nice, France
e-mail: julie.llobell@unice.fr; sebastian.minjeaud@unice.fr; richard.pasquetti@unice.fr

H. van Brummelen et al. (eds.), *Numerical Methods for Flows*,
Lecture Notes in Computational Science and Engineering 132,
https://doi.org/10.1007/978-3-030-30705-9_30

of such problems, here we consider the Korteweg-De Vries (KdV) and the shallow water equations, and develop some strategies to address them.

The SEM is based on a nodal Continuous Galerkin (CG) approach, such that the approximation space contains all C^0 functions whose restriction in each element is associated to a polynomial of degree N. More precisely, in the master element $(-1, 1)^d$, with d for the space dimension, the basis functions are Lagrange polynomials associated to the $(N+1)^d$ Gauss-Lobatto-Legendre (GLL) points, which are also used as quadrature points to evaluate the integrals obtained when using a weak form of the problem. The fact that interpolation and quadrature points coincide implies that the mass matrix is diagonal. The SEM algorithms that we describe hereafter make strongly use of this property both (1) to address evolution problems with explicit (or implicit-explicit) time schemes and (2) to define high order differentiation operators in the frame of C^0-continuous approximations.

We describe in Sect. 2 the algorithms that we have developed for the KdV equation, which is a well known example of dispersive equation. In Sect. 3 we consider an hyperbolic system of PDEs, namely the Saint-Venant system, using for stabilization the Entropy Viscosity Method (EVM). In Sect. 4 we address an academic but complex Saint-Venant problem to carry out a sensitivity study to the EVM control parameters. A comparison with results obtained using a Finite Volume (FV) scheme with shock capturing strategy is presented in Sect. 5, and we conclude in Sect. 6.

2 SEM Approximation of the KdV Equation

Here we summarize the SEM method that we have developed for the KdV equation. Details and references may be found in [7].

The KdV problem writes: Find $u(x,t)$, $x \in \Omega$ and $t \in \mathbb{R}^+$, such that

$$\partial_t u + u\partial_x u + \beta\partial_{xxx} u = 0 \tag{1}$$

with the initial condition $u(x, t = 0) = u_0(x)$ and, e.g., periodic boundary conditions (β given parameter). With KdV equation, the main difficulties are (1) the approximation of the dispersive term $\beta\partial_{xxx}u$ and (2) the preservation of at least two invariants: mass and energy

$$I_1 = \int_\Omega u\,dx\,, \quad I_2 = \int_\Omega u^2\,dx\,, \tag{2}$$

which is required to get correct results for long time computations. Due to the presence of the third order derivative term, the standard FEM approximation does not apply. Indeed, after integration by parts a second order derivative remains on the unknown function u or on the test function, say w. To overcome such a difficulty, one generally makes use of a C^1-continuous FEM or a Petrov-Galerkin approach with C^1 test functions. Such approaches generally yield less efficient algorithms, due to the increase of the bandwidth of the resulting algebraic systems, and are often not easy to implement, especially in the multidimensional case or when non trivial boundary conditions are involved. Moreover, the C^1-continuity is not sufficient for PDEs involving higher order derivative terms, since e.g. the C^3-continuity would be required for a fifth order derivative term.

Alternatively, one can introduce new variables. Thus, in the frame of C^0-continuous FEM it is natural to set $f = \partial_{xx} u$, this is the so-called "natural approach" mentioned hereafter. Then, if the convection term is treated explicitly, in such a way it can be assimilated at each time-step to a source term, one obtains the semi-discrete problem:

$$\begin{aligned} M \partial_t \boldsymbol{u} + \beta D \boldsymbol{f} &= \boldsymbol{S} \\ M \boldsymbol{f} + B \boldsymbol{u} &= 0 \end{aligned}$$

with M: mass matrix, B: stiffness matrix and D: Differentiation matrix, and where the vectors of the grid-point values are denoted in bold. By elimination of $\boldsymbol{f}$ one obtains:

$$M \partial_t \boldsymbol{u} - \beta D M^{-1} B \boldsymbol{u} = \boldsymbol{S} \,. \tag{3}$$

At this point the problem is that an inversion of the mass matrix is required. Such an inversion is however trivial if using the SEM, because matrix M is diagonal. Moreover, the $DM^{-1}B$ algebraic operator is sparse.

In the spirit of Discontinuous Galerkin (DG) methods, one can also use the following strategy: Set $g = \partial_x u$ and $f = \partial_x g$, then a C^0-continuous FEM approximation yields:

$$\begin{aligned} M \partial_t \boldsymbol{u} + \beta D \boldsymbol{f} &= \boldsymbol{S} \\ M \boldsymbol{f} &= D \boldsymbol{g} \\ M \boldsymbol{g} &= D \boldsymbol{u} \,. \end{aligned}$$

By elimination of $\boldsymbol{f}$ and $\boldsymbol{g}$ one obtains:

$$M \partial_t \boldsymbol{u} + \beta D (M^{-1} D)^2 \boldsymbol{u} = \boldsymbol{S} \,. \tag{4}$$

If using the SEM this new differentiation operator can be easily set up. Its bandwidth is larger than for the previous natural approach, but one can check that its spectral properties are similar.

Using the natural approach or the DG like one, the present definitions of the high order differentiation operator are of course not restricted to 1D problems. When using quadrangular or parrallelipipedic elements, the SEM mass matrices are also diagonal, since the master element is defined by tensorial product. Moreover, a diagonal mass matrix can also be obtained with triangular elements, if using cubature points of the triangle for both the interpolation and quadrature points, see e.g. [10] and references herein.

In time, we suggest using high order implicit-explicit (IMEX) Runge-Kutta (RK) schemes. Then, for stability reasons the dispersive term is handled implicitly whereas the non linear convective term is handled explicitly, under the usual Courant-Friedrich-Lewy (CFL) condition.

Concerning the non linear term, $\int u\partial_x u\, w\, dx$, where w is the test function, it is of interest to exactly compute it by using the GLL quadrature rule associated to polynomials of degree M such that $2M-1=3N-1$, i.e. $M=3N/2$. Indeed, our numerical experiments have shown that this allows to get satisfactory results without introducing any stabilization term, see [7]. This is interesting since the stabilizing effect results from an improvement in the computation of the convective term and not from the introduction of an artificial dissipation term. The same stabilizing effect is observed in other contexts, see e.g. [5, 8] where the use of $M=N+1$ or $M=N+2$ allows to avoid the spurious oscillations.

As mentioned previously, for KdV equation the preservation of at least two invariants is important. Indeed, from a physical point of view it gives sense to the numerical solution since it ensures mass conservation and energy conservation, and from a mathematical point of view it ensures in some sense the stability of the method since here the L^2 norm of the discrete solution is preserved. As a direct consequence of the weak formulation together with the accuracy of the GLL quadrature rule, preserving the mass invariant is natural in the frame of the SEM. Concerning the energy invariant two approaches have been investigated. First, one can take into account the two invariants as constraints and introduce Lagrange multipliers. Second, one can make use of two IMEX schemes, yielding two slightly different solutions, say at time t_n, u_1^n and u_2^n, and write u^n as a linear combination of them: $u^n=(1-\lambda)u_1^n+\lambda u_2^n$. The mass invariant is then preserved and one must look for λ such that $I_2=Constant$, see (2). It turns out that λ solves

$$S[(u_2-u_1)^2]\lambda^2+2S[u_1(u_2-u_1)]\lambda+S[u_1^2]-I_2=0$$

where $S[.]$ stands for a quadrature formula on the grid-points.

The computational price of such an approach is a priory twice greater, since it is needed to compute u_1^n and u_2^n to get u^n, but this is not true if using embedded IMEX schemes, that only differ by the final recombination of the intermediate values. The second IMEX scheme (giving u_2^n) is then generally only first order accurate, but one can demonstrate that the accuracy of the leading RK scheme (giving u_1^n) is generally preserved. All details are given in [7].

To conclude this section we consider the KdV equation with $\beta=0.022^2$ in the periodic domain $(0,2)$ and assume the initial condition $u_0(x)=\cos(\pi x)$, see e.g. [3, 13]. The numerical solution is computed with $K=160$ elements, a polynomial approximation degree $N=5$ and a time step $\tau=2.5\times10^{-4}$. The contour levels of the numerical solution in the (x,t)-plane are plotted in Fig. 1 between times 0 and t_R at left, and between times $19t_R$ and $20t_R$ at right, where $t_R\approx 9.68$ is the so

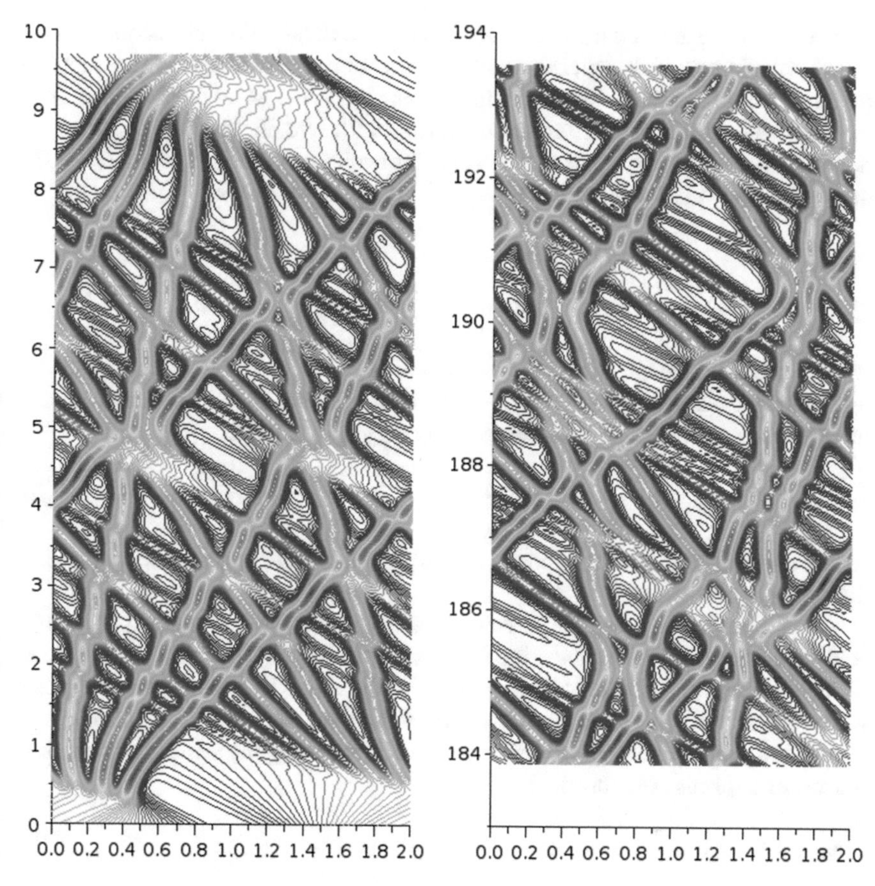

Fig. 1 Contour levels of the numerical solution in the (x, t)-plane, between times 0 and t_R at left, between times $19t_R$ and $20t_R$ at right

called recurrence time, at which one expects to (approximately) recover the initial condition.

Additional test-cases, that e.g. show accuracy results in both periodic and non periodic domains, are provided in [7].

3 EVM-Stabilized SEM of the Saint-Venant System

We consider now a more involved fluid flow model that also constitutes a challenging problem for high order CG approaches, namely the shallow water equations. For the paper to be self contained, we give here some details of our EVM-stabilized approximation of the Saint-Venant system, see [9, 11] for details, references and examples of applications.

The Saint-Venant system results from an approximation of the incompressible Euler equations which assumes that the pressure is hydrostatic and that the perturbations of the free surface are small compared to the water height. Then, from the mass and momentum conservation laws and with $\Omega \subset \mathbb{R}^2$ for the computational domain, one obtains equations that describe the evolution of the height $h : \Omega \to \mathbb{R}^+$ and of the horizontal velocity $\boldsymbol{u} : \Omega \to \mathbb{R}^2$: For $(\boldsymbol{x}, t) \in \Omega \times \mathbb{R}^+$:

$$\begin{aligned} &\partial_t h + \nabla \cdot (h\boldsymbol{u}) = 0 \\ &\partial_t (h\boldsymbol{u}) + \nabla \cdot (h\boldsymbol{u} \otimes \boldsymbol{u} + gh^2\mathbb{I}/2) + gh\nabla z = 0 \end{aligned} \tag{5}$$

with $\mathbb{I}$, identity tensor, g, gravity acceleration, and where $z(\boldsymbol{x})$ describes the topography, assumed such that $\nabla z \ll 1$. Moreover, for the saint-Venant system there exists a convex entropy (actually the energy E) such that

$$\partial_t E + \nabla \cdot ((E + gh^2/2)\boldsymbol{u}) \leq 0, \quad E = h\boldsymbol{u}^2/2 + gh^2/2 + ghz. \tag{6}$$

so that one may think to implement the EVM for stabilization of the following SEM discrete approximation.

Set $\boldsymbol{q} = h\boldsymbol{u}$ and let $h_N(t)$ (resp. $\boldsymbol{q}_N(t)$) to be the piecewise polynomial continuous approximation of degree N of $h(t)$ (resp. $\boldsymbol{q}(t)$). The proposed stabilized SEM relies on the Galerkin approximation of the Saint-Venant system completed with viscous terms for both the mass and momentum equations. For any w_N, $\boldsymbol{w}_N$ (scalar and vector valued functions, respectively) spanning the same approximation spaces, in semi-discrete form:

$$\begin{aligned} &(\partial_t h_N + \nabla \cdot \boldsymbol{q}_N, w_N)_N = -(\nu_h \nabla h_N, \nabla w_N)_N \\ &(\partial_t \boldsymbol{q}_N + \nabla \cdot I_N(\boldsymbol{q}_N \otimes \boldsymbol{q}_N / h_N) + gh_N \nabla (h_N + z_N), \boldsymbol{w}_N)_N = -(\nu_q \nabla \boldsymbol{q}_N, \nabla \boldsymbol{w}_N)_N \end{aligned} \tag{7}$$

where $\nu_h \propto \nu_q = \nu$, with ν: entropy viscosity (in the rest of the paper we simply use $\nu_h = \nu_q$). The usual SEM approach is used here: I_N is the piecewise polynomial interpolation operator, based for each element on the tensorial product of Gauss-Lobatto-Legendre (GLL) points, and $(.,.)_N$ stands for the SEM approximation of the $L^2(\Omega)$ inner product, using for each element the GLL quadrature formula which is exact for polynomials of degree less than $2N-1$ in each variable. Note that thanks to using $\nabla \cdot I_N(gh_N^2\mathbb{I}/2) \approx gh_N \nabla h_N$ (while h_N^2 is generally piecewise polynomial of degree greater than N), and thus grouping in (7) the pressure and topography terms, a well balanced scheme is obtained by construction: If $\boldsymbol{q}_N \equiv 0$ and $h_N \neq 0$, then $h_N + z_N = Constant$. Of course, a difficulty comes from the required positivity of h_N, as discussed at the end of the present section.

It remains to define the entropy viscosity ν. To this end we make use of an entropy that does not depend on z but on ∇z, which is of interest, at the discrete level, to get free of the choice of the coordinate system. Taking into account the mass conservation equation (into the entropy equation) one obtains:

$$\partial_t \tilde{E} + \nabla \cdot ((\tilde{E} + gh^2/2)\boldsymbol{u}) + gh\boldsymbol{u} \cdot \nabla z \leq 0, \quad \tilde{E} = h\boldsymbol{u}^2/2 + gh^2/2. \tag{8}$$

At each time-step, we then compute the entropy viscosity $\nu(\boldsymbol{x})$ at the GLL grid points, using the following three steps procedure:

- Assuming all variables given at time t_n, compute the entropy residual, using a Backward Difference Formula, e.g. the BDF2 scheme, to approximate $\partial_t \tilde{E}_N$

$$r_E = \partial_t \tilde{E}_N + \nabla \cdot I_N((\tilde{E}_N + gh_N^2/2)\boldsymbol{q}_N/h_N) + g\boldsymbol{q}_N \cdot \nabla z_N$$

  where $\tilde{E}_N = \boldsymbol{q}_N^2/(2h_N) + gh_N^2/2$. Then set up a viscosity ν_E such that:

$$\nu_E = \beta |r_E| \delta x^2 / \Delta E \, ,$$

  where ΔE is a reference entropy, β a user defined control parameter and δx the local GLL grid-size, defined such that δx^2 equals the surface of the quadrilateral cell (of the dual GLL mesh) surrounding the GLL point, and using symmetry assumptions for the points at the edges and vertices of the element.
- Define a viscosity upper bound based on the wave speeds: $\lambda_{\pm} = u \pm \sqrt{gh}$:

$$\nu_{max} = \alpha \max_{\Omega}(|\boldsymbol{q}_N/h_N| + \sqrt{gh_N})\delta x$$

  where α is a $O(1)$ user defined parameter (recall that for the advection equation $\alpha = 1/2$ is well suited).
- Compute the entropy viscosity:

$$\nu = \min(\nu_{max}, \nu_E)$$

  and smooth: (1) locally (in each element), e.g. in 1D: $(\nu_{i-1} + 2\nu_i + \nu_{i+1})/4 \to \nu_i$; (2) globally, by projection onto the space of the C^0 piecewise polynomials of degree N. Note that operation (2) is cheap because the SEM mass matrix is diagonal.

The positivity of h_N is difficult to enforce as soon as $N > 1$, so that for problems involving dry-wet transitions the present EVM methodology must be completed. The algorithm that we propose is the following: In dry zones, i.e. for any element Q_{dry} such that at one GLL point $\min h_N < h_{thresh}$, where h_{thresh} is a user defined threshold value (typically a thousandth of the reference height):

- Modify the entropy viscosity technique, by using in Q_{dry} the upper bound first order viscosity:

$$\nu = \nu_{max} \quad \text{in} \quad Q_{dry}$$

- In the momentum equation assume that:

$$h_N g \nabla(h_N + z_N) \equiv 0 \quad \text{in} \quad Q_{dry}$$

- Moreover, notice that the upper bound viscosity $\nu_{\max}$ is not local but global, and that the entropy scaling ΔE used in the definition of ν_E is time independent. This has improved the robustness of the general approach described in [4].

Simulations with dry-wet transitions and comparisons to exact solutions are given in [11] and [9], for 1D and 2D flows, respectively.

4 Sensitivity Study to the EVM Control Parameters

We address a shallow water problem, the "falling columns" test proposed in [1], whose solution is characterized by many interactions and shocks. Thus, it constitutes a good benchmark to check the sensitivity of our SEM model to the control parameters of the EVM. The flow is governed by the Saint-Venant system (5), in which the dimensionless gravity acceleration is taken equal to 2. The computational domain is the square $(-1, 1)^2$, with free slip condition at the boundary. At the initial time the fluid is at rest, $\boldsymbol{u}(t = 0) = 0$, and the height is given by:

$$h(t = 0) = 3 + 1_{(\boldsymbol{x}-\boldsymbol{x}_1)^2<0.15^2} + 1_{(\boldsymbol{x}-\boldsymbol{x}_2)^2<0.15^2} + 2\, 1_{\boldsymbol{x}^2<0.2^2}$$

with $\boldsymbol{x}_1 = (0.5, 0.5)$ and $\boldsymbol{x}_2 = (-0.5, -0.5)$, and where 1_ω is the indicator function of subdomain ω.

A first computation has been carried out without the EVM stabilization. As expected, in this case the computation crashes, since a stabilization is needed when shocks develop. Computations have been done for the following values of the pair (α, β): $(0.5, \infty)$, $(1, \infty)$, $(0.5, 1)$, $(0.5, 2)$, $(0.5, 3)$, $(1, 3)$ and $(1, 5)$. Mentioning $\beta = \infty$ means that we simply use a first order viscosity everywhere. Note that choosing $\alpha = 0.5$ is very natural, since for an advection equation it yields a $O(h)$ diffusion term equivalent to the implicit one of the upwind scheme. The three pairs such that $1 \leq \beta \leq 3$ show the influence of β, while keeping $\alpha = 0.5$. In the two last tests the stabilization is strengthened, by increasing α up to 1 and β up to 5.

One uses a polynomial approximation of degree $N = 5$ in each quadrangle of a regular $K = 100 \times 100$ mesh. This yields 255,001 interpolation points in Ω, with 91,001 of them at the quadrangle boundaries. All computations have been made with a time step $\tau = 10^{-4}$. Such time and space discretizations allow a fair comparison with FV results in Sect. 5.

The height of the flow at the final time, $t_f = 1.035$, is visualized for the different simulations in Fig. 2. As desired, the result obtained without EVM but only a first order viscosity is very smooth, but clearly completely false. If implementing the EVM, then the correct solution is captured. One observes that strengthening the

Fig. 2 Visualizations of the height at the final time $t_f = 1.035$ for the (α, β) pairs (0.5, ∞), (0.5, 1), (0.5, 2) , (0.5, 3), (1, 3) and (1, 5), from up to down and left to right. For all graphics the color bar ranges from 2.7 to 4 and the extrema are (2.821, 3.611), (2.690, 4.010), (2.695, 3.967), (2.699, 3.933), (2.701, 3.924) and (2.701, 3.924), respectively

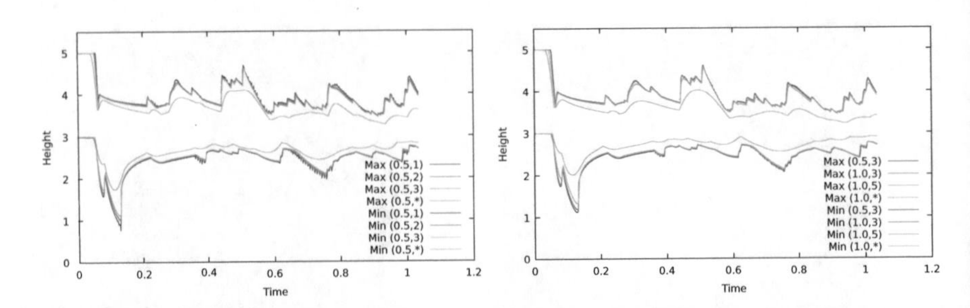

Fig. 3 Evolution of the maximum and minimum of the height when using the first order ($\beta = *$) and the EVM stabilization for different (α, β) pairs. The influence of β is mainly shown in the left panel and the influence of α in the right one

stabilization allows to filter some spurious oscillations. Note that the presence of such oscillations is not surprising, since the discontinuities of the initial height enforces the Gibbs phenomenon. The present study of the influence of the EVM control parameters is of course very qualitative, and moreover only based of the height at the final time.

In order to complete such a qualitative study, we show in Fig. 3 the evolutions of the extrema of the height during the simulation. Clearly, (1) the first order viscosity result is not correct and (2) the stronger is the EVM-stabilization, the smoother are the extrema evolutions. Additionally, one observes the EVM-stabilization becomes too strong for ($\alpha = 1, \beta = 5$), since the corresponding curve no longer coincides with the other EVM ones.

5 Comparison with a Second Order FV Computation

For comparison purposes, we provide in this section the results obtained using a first order and a second order FV scheme, that can be viewed as an extension to the 2D and to the second order accuracy of the scheme presented in [2]. These schemes work on staggered Cartesian grids and, in contrast to the colocalized approach for conservative system, it make use of a discretization of the physical variables, the height and the velocity separately, instead of a discretization of the conservative variables. The height is stored at the cell centers whereas the horizontal (resp. vertical) component of the velocity is stored at the vertical (resp. horizontal) edges like in the well-known MAC (Marker-and-Cell) scheme. The numerical fluxes are derived using the framework of the so-called (kinetic) Boltzmann schemes. In the spirit of hydrodynamic limits which allow to derive the Euler equations from Boltzmann equation, the Saint Venant system is seen as the limit of a vector BGK (Bhatnagar-Gross-Krook) equation, see e.g. [12]. This is a transport equation for a kinetic variable f (i.e. a variable which depends on $(\boldsymbol{x}, t)$ but also on an auxiliary "ghost" velocity variable ξ) with a relaxation term towards a given equilibrium state which depends only on ξ and on the zeroth moment of f. This equilibrium state is especially designed to ensure that, at least formally, the zeroth moment of f satisfies the Saint-Venant equation when the relaxation parameter goes to zero. A numerical scheme for the BGK equation is obtained by decoupling into two successive steps the transport and the relaxation process. A basic upwind scheme is then used for the (linear) transport step. Finally, we get rid of the "ghost" velocity variable by integrating the formula with respect to ξ: it provides formula of fluxes for updating the height and the momentum (see [2]). Note that this formula, which may be written explicitly, can be viewed as an upwinding of the transported variables (height and momentum) with respect to the sign of the characteristic velocities, the pressure being centered. The second order accuracy is reached thanks to a MUSCL-like (Monotonic Upwind Scheme for Conservation Laws) procedure using the MinMod limiter. The first order FV scheme is coupled with an explicit Euler time discretization whereas a second order ERK (explicit Runge-Kutta) scheme is used with the second order space discretization. All the details can be found in [6].

The results obtained for the height at the final time $t_f = 1.035$ are presented in Fig. 4. The grid is a 512×512 Cartesian mesh and the time step is $\tau = 10^{-4}$, so that the number of degrees of freedom for the height is 262,144 allowing a fair comparison with the results obtained using the SEM in Fig. 2. As expected, the result obtained with the first order FV scheme is smooth and close to the one obtained with the SEM when adding a first order viscosity whereas using the second FV scheme allows to recover the correct solution (free of spurious oscillations) very close to the one obtained with the EVM.

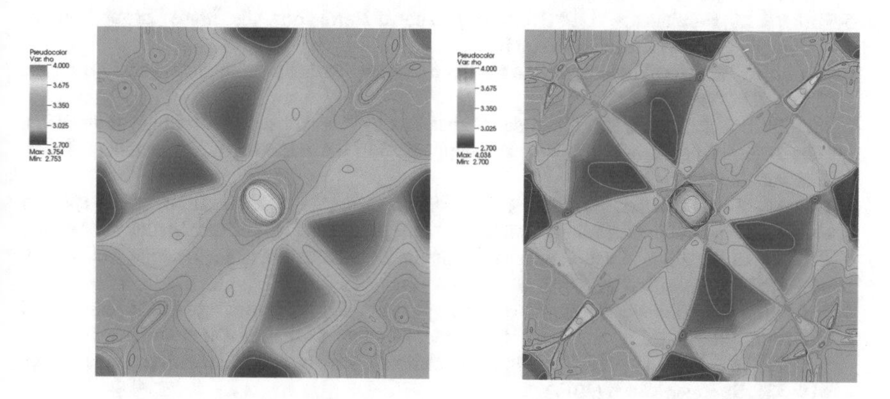

Fig. 4 Visualizations of the height at the final time $t_f = 1.035$ using a first order FV scheme (at left) and a second order FV scheme (at right). The color bar is the same as in Fig. 2, from 2.7 to 4, and the extrema are (2.753, 3.754) and (2.700, 4.038), respectively

6 Conclusion

A lot of numerical methods have been developed in the past, and are still developed, to address the KdV and Saint-Venant problems. In this spirit, but in contrast with studies based on the celebrated FV or DG methods, here we have proposed to use a high order CG method, namely the SEM. For KdV the main advantage of the SEM is the diagonal structure of the mass matrix. This indeed allows to simply eliminate intermediate variables and thus set up efficient algorithms. For hyperbolic problems a stabilization technique is however required. For Saint-Venant flows, we have investigated the EVM capabilities and additionally provided a sensitivity study to the EVM parameters as well as a comparison with FV results. Additional tests and comparisons for less academical problems will be focused on in our future works.

References

1. Aguillon, N.: Problèmes d'interfaces et couplages singuliers dans les systèmes hyperboliques: analyse et analyse numérique. PhD thesis, University of Paris Sud (2014)
2. Berthelin, F., Goudon, T., Minjeaud, S.: Kinetic schemes on staggered grids for barotropic Euler models: entropy-stability analysis. Math. Comput. **84**, 2221–2262 (2015)
3. Cui, Y., Mao, D.K.: Numerical method satisfying the first two conservation laws for the Korteweg-de Vries equation. J. Comput. Phys. **227**(1), 376–399 (2007)
4. Guermond, J.L., Pasquetti, R., Popov, B.: Entropy viscosity method for non-linear conservation laws. J. Comput. Phys. **230** (11), 4248–4267 (2011)
5. Kirby, R.M., Karniadakis, G.E.: De-aliasing on non uniform grids: algorithms and applications. J. Comput. Phys. **191**, 249–264 (2003)
6. Llobell, J.: Schémas Volumes Finis à mailles décalées pour la dynamique des gaz. PhD thesis, Université Côte d'Azur (2018)
7. Minjeaud, S., Pasquetti, R.: High order C^0-continuous Galerkin schemes for high order PDEs, conservation of quadratic invariants and application to the Korteweg-De Vries model. J. Sci. Comput. **74**, 491–518 (2018)
8. Ohlsson, J.P., Schlatter, P., Fischer, P.F., Henningson, D.S.: Stabilization of the spectral element method in turbulent flow simulations. In: Spectral and High Order Methods for Partial Differential Equations. Lecture Notes in Computational Science and Engineering, vol. 76, pp. 449–458. Springer, Berlin (2011)
9. Pasquetti, R.: Viscous stabilizations for high order approximations of Saint-Venant and Boussinesq flows. In: Spectral and High Order Methods for Partial Differential Equations – ICOSAHOM 2016. Lecture Notes in Computational Science and Engineering, vol. 119, pp. 519–531. Springer, Cham (2017)
10. Pasquetti, R., Rapetti, F.: Cubature versus Fekete-Gauss nodes for spectral element methods on simplicial meshes. J. Comput. Phys. **347**, 463–466 (2017)
11. Pasquetti, R., Guermond, J.L., Popov, B.: Stabilized spectral element approximation of the Saint-Venant system using the entropy viscosity technique. In: Spectral and High Order Methods for Partial Differential Equations – ICOSAHOM 2014. Lecture Notes in computational Science and Engineering, vol. 106, pp. 397–404. Springer, Cham (2015)
12. Saint Raymond, L.: Hydrodynamic Limits of the Boltzmann Equation. Lecture Notes in Mathematics, vol. 1971. Springer, Berlin (2009)
13. Zabusky, N.J., Kruskal, M.D.: Interaction of solitons in a collisionless plasma and the recurrence of initial states. Phys. Rev. Lett. **15**(6), 240–243 (1965)

Editorial Policy

1. Volumes in the following three categories will be published in LNCSE:

i) Research monographs
ii) Tutorials
iii) Conference proceedings

Those considering a book which might be suitable for the series are strongly advised to contact the publisher or the series editors at an early stage.

2. Categories i) and ii). Tutorials are lecture notes typically arising via summer schools or similar events, which are used to teach graduate students. These categories will be emphasized by Lecture Notes in Computational Science and Engineering. **Submissions by interdisciplinary teams of authors are encouraged.** The goal is to report new developments – quickly, informally, and in a way that will make them accessible to non-specialists. In the evaluation of submissions timeliness of the work is an important criterion. Texts should be well-rounded, well-written and reasonably self-contained. In most cases the work will contain results of others as well as those of the author(s). In each case the author(s) should provide sufficient motivation, examples, and applications. In this respect, Ph.D. theses will usually be deemed unsuitable for the Lecture Notes series. Proposals for volumes in these categories should be submitted either to one of the series editors or to Springer-Verlag, Heidelberg, and will be refereed. A provisional judgement on the acceptability of a project can be based on partial information about the work: a detailed outline describing the contents of each chapter, the estimated length, a bibliography, and one or two sample chapters – or a first draft. A final decision whether to accept will rest on an evaluation of the completed work which should include

- at least 100 pages of text;
- a table of contents;
- an informative introduction perhaps with some historical remarks which should be accessible to readers unfamiliar with the topic treated;
- a subject index.

3. Category iii). Conference proceedings will be considered for publication provided that they are both of exceptional interest and devoted to a single topic. One (or more) expert participants will act as the scientific editor(s) of the volume. They select the papers which are suitable for inclusion and have them individually refereed as for a journal. Papers not closely related to the central topic are to be excluded. Organizers should contact the Editor for CSE at Springer at the planning stage, see *Addresses* below.

In exceptional cases some other multi-author-volumes may be considered in this category.

4. Only works in English will be considered. For evaluation purposes, manuscripts may be submitted in print or electronic form, in the latter case, preferably as pdf- or zipped ps-files. Authors are requested to use the LaTeX style files available from Springer at http://www.springer.com/gp/authors-editors/book-authors-editors/manuscript-preparation/5636 (Click on LaTeX Template → monographs or contributed books).

For categories ii) and iii) we strongly recommend that all contributions in a volume be written in the same LaTeX version, preferably LaTeX2e. Electronic material can be included if appropriate. Please contact the publisher.

Careful preparation of the manuscripts will help keep production time short besides ensuring satisfactory appearance of the finished book in print and online.

5. The following terms and conditions hold. Categories i), ii) and iii):

Authors receive 50 free copies of their book. No royalty is paid.
Volume editors receive a total of 50 free copies of their volume to be shared with authors, but no royalties.

Authors and volume editors are entitled to a discount of 40 % on the price of Springer books purchased for their personal use, if ordering directly from Springer.

6. Springer secures the copyright for each volume.

Addresses:

Timothy J. Barth
NASA Ames Research Center
NAS Division
Moffett Field, CA 94035, USA
barth@nas.nasa.gov

Michael Griebel
Institut für Numerische Simulation
der Universität Bonn
Wegelerstr. 6
53115 Bonn, Germany
griebel@ins.uni-bonn.de

David E. Keyes
Mathematical and Computer Sciences
and Engineering
King Abdullah University of Science
and Technology
P.O. Box 55455
Jeddah 21534, Saudi Arabia
david.keyes@kaust.edu.sa

and

Department of Applied Physics
and Applied Mathematics
Columbia University
500 W. 120 th Street
New York, NY 10027, USA
kd2112@columbia.edu

Risto M. Nieminen
Department of Applied Physics
Aalto University School of Science
and Technology
00076 Aalto, Finland
risto.nieminen@aalto.fi

Dirk Roose
Department of Computer Science
Katholieke Universiteit Leuven
Celestijnenlaan 200A
3001 Leuven-Heverlee, Belgium
dirk.roose@cs.kuleuven.be

Tamar Schlick
Department of Chemistry
and Courant Institute
of Mathematical Sciences
New York University
251 Mercer Street
New York, NY 10012, USA
schlick@nyu.edu

Editor for Computational Science
and Engineering at Springer:

Martin Peters
Springer-Verlag
Mathematics Editorial IV
Tiergartenstrasse 17
69121 Heidelberg, Germany
martin.peters@springer.com

Lecture Notes in Computational Science and Engineering

1. D. Funaro, *Spectral Elements for Transport-Dominated Equations.*
2. H.P. Langtangen, *Computational Partial Differential Equations.* Numerical Methods and Diffpack Programming.
3. W. Hackbusch, G. Wittum (eds.), *Multigrid Methods V.*
4. P. Deuflhard, J. Hermans, B. Leimkuhler, A.E. Mark, S. Reich, R.D. Skeel (eds.), *Computational Molecular Dynamics: Challenges, Methods, Ideas.*
5. D. Kröner, M. Ohlberger, C. Rohde (eds.), *An Introduction to Recent Developments in Theory and Numerics for Conservation Laws.*
6. S. Turek, *Efficient Solvers for Incompressible Flow Problems.* An Algorithmic and Computational Approach.
7. R. von Schwerin, ***Multi Body System SIMulation.*** Numerical Methods, Algorithms, and Software.
8. H.-J. Bungartz, F. Durst, C. Zenger (eds.), *High Performance Scientific and Engineering Computing.*
9. T.J. Barth, H. Deconinck (eds.), *High-Order Methods for Computational Physics.*
10. H.P. Langtangen, A.M. Bruaset, E. Quak (eds.), *Advances in Software Tools for Scientific Computing.*
11. B. Cockburn, G.E. Karniadakis, C.-W. Shu (eds.), *Discontinuous Galerkin Methods.* Theory, Computation and Applications.
12. U. van Rienen, *Numerical Methods in Computational Electrodynamics.* Linear Systems in Practical Applications.
13. B. Engquist, L. Johnsson, M. Hammill, F. Short (eds.), *Simulation and Visualization on the Grid.*
14. E. Dick, K. Riemslagh, J. Vierendeels (eds.), *Multigrid Methods VI.*
15. A. Frommer, T. Lippert, B. Medeke, K. Schilling (eds.), *Numerical Challenges in Lattice Quantum Chromodynamics.*
16. J. Lang, *Adaptive Multilevel Solution of Nonlinear Parabolic PDE Systems.* Theory, Algorithm, and Applications.
17. B.I. Wohlmuth, *Discretization Methods and Iterative Solvers Based on Domain Decomposition.*
18. U. van Rienen, M. Günther, D. Hecht (eds.), *Scientific Computing in Electrical Engineering.*
19. I. Babuška, P.G. Ciarlet, T. Miyoshi (eds.), *Mathematical Modeling and Numerical Simulation in Continuum Mechanics.*
20. T.J. Barth, T. Chan, R. Haimes (eds.), *Multiscale and Multiresolution Methods.* Theory and Applications.
21. M. Breuer, F. Durst, C. Zenger (eds.), *High Performance Scientific and Engineering Computing.*
22. K. Urban, *Wavelets in Numerical Simulation.* Problem Adapted Construction and Applications.
23. L.F. Pavarino, A. Toselli (eds.), *Recent Developments in Domain Decomposition Methods.*

24. T. Schlick, H.H. Gan (eds.), *Computational Methods for Macromolecules: Challenges and Applications.*

25. T.J. Barth, H. Deconinck (eds.), *Error Estimation and Adaptive Discretization Methods in Computational Fluid Dynamics.*

26. M. Griebel, M.A. Schweitzer (eds.), *Meshfree Methods for Partial Differential Equations.*

27. S. Müller, *Adaptive Multiscale Schemes for Conservation Laws.*

28. C. Carstensen, S. Funken, W. Hackbusch, R.H.W. Hoppe, P. Monk (eds.), *Computational Electromagnetics.*

29. M.A. Schweitzer, *A Parallel Multilevel Partition of Unity Method for Elliptic Partial Differential Equations.*

30. T. Biegler, O. Ghattas, M. Heinkenschloss, B. van Bloemen Waanders (eds.), *Large-Scale PDE-Constrained Optimization.*

31. M. Ainsworth, P. Davies, D. Duncan, P. Martin, B. Rynne (eds.), *Topics in Computational Wave Propagation.* Direct and Inverse Problems.

32. H. Emmerich, B. Nestler, M. Schreckenberg (eds.), *Interface and Transport Dynamics.* Computational Modelling.

33. H.P. Langtangen, A. Tveito (eds.), *Advanced Topics in Computational Partial Differential Equations.* Numerical Methods and Diffpack Programming.

34. V. John, *Large Eddy Simulation of Turbulent Incompressible Flows.* Analytical and Numerical Results for a Class of LES Models.

35. E. Bänsch (ed.), *Challenges in Scientific Computing - CISC 2002.*

36. B.N. Khoromskij, G. Wittum, *Numerical Solution of Elliptic Differential Equations by Reduction to the Interface.*

37. A. Iske, *Multiresolution Methods in Scattered Data Modelling.*

38. S.-I. Niculescu, K. Gu (eds.), *Advances in Time-Delay Systems.*

39. S. Attinger, P. Koumoutsakos (eds.), *Multiscale Modelling and Simulation.*

40. R. Kornhuber, R. Hoppe, J. Périaux, O. Pironneau, O. Wildlund, J. Xu (eds.), *Domain Decomposition Methods in Science and Engineering.*

41. T. Plewa, T. Linde, V.G. Weirs (eds.), *Adaptive Mesh Refinement – Theory and Applications.*

42. A. Schmidt, K.G. Siebert, *Design of Adaptive Finite Element Software.* The Finite Element Toolbox ALBERTA.

43. M. Griebel, M.A. Schweitzer (eds.), *Meshfree Methods for Partial Differential Equations II.*

44. B. Engquist, P. Lötstedt, O. Runborg (eds.), *Multiscale Methods in Science and Engineering.*

45. P. Benner, V. Mehrmann, D.C. Sorensen (eds.), *Dimension Reduction of Large-Scale Systems.*

46. D. Kressner, *Numerical Methods for General and Structured Eigenvalue Problems.*

47. A. Boriçi, A. Frommer, B. Joó, A. Kennedy, B. Pendleton (eds.), *QCD and Numerical Analysis III.*

48. F. Graziani (ed.), *Computational Methods in Transport.*

49. B. Leimkuhler, C. Chipot, R. Elber, A. Laaksonen, A. Mark, T. Schlick, C. Schütte, R. Skeel (eds.), *New Algorithms for Macromolecular Simulation.*

50. M. Bücker, G. Corliss, P. Hovland, U. Naumann, B. Norris (eds.), *Automatic Differentiation: Applications, Theory, and Implementations.*

51. A.M. Bruaset, A. Tveito (eds.), *Numerical Solution of Partial Differential Equations on Parallel Computers.*

52. K.H. Hoffmann, A. Meyer (eds.), *Parallel Algorithms and Cluster Computing.*

53. H.-J. Bungartz, M. Schäfer (eds.), *Fluid-Structure Interaction.*

54. J. Behrens, *Adaptive Atmospheric Modeling.*

55. O. Widlund, D. Keyes (eds.), *Domain Decomposition Methods in Science and Engineering XVI.*

56. S. Kassinos, C. Langer, G. Iaccarino, P. Moin (eds.), *Complex Effects in Large Eddy Simulations.*

57. M. Griebel, M.A Schweitzer (eds.), *Meshfree Methods for Partial Differential Equations III.*

58. A.N. Gorban, B. Kégl, D.C. Wunsch, A. Zinovyev (eds.), *Principal Manifolds for Data Visualization and Dimension Reduction.*

59. H. Ammari (ed.), *Modeling and Computations in Electromagnetics: A Volume Dedicated to Jean-Claude Nédélec.*

60. U. Langer, M. Discacciati, D. Keyes, O. Widlund, W. Zulehner (eds.), *Domain Decomposition Methods in Science and Engineering XVII.*

61. T. Mathew, *Domain Decomposition Methods for the Numerical Solution of Partial Differential Equations.*

62. F. Graziani (ed.), *Computational Methods in Transport: Verification and Validation.*

63. M. Bebendorf, *Hierarchical Matrices.* A Means to Efficiently Solve Elliptic Boundary Value Problems.

64. C.H. Bischof, H.M. Bücker, P. Hovland, U. Naumann, J. Utke (eds.), *Advances in Automatic Differentiation.*

65. M. Griebel, M.A. Schweitzer (eds.), *Meshfree Methods for Partial Differential Equations IV.*

66. B. Engquist, P. Lötstedt, O. Runborg (eds.), *Multiscale Modeling and Simulation in Science.*

67. I.H. Tuncer, Ü. Gülcat, D.R. Emerson, K. Matsuno (eds.), *Parallel Computational Fluid Dynamics 2007.*

68. S. Yip, T. Diaz de la Rubia (eds.), *Scientific Modeling and Simulations.*

69. A. Hegarty, N. Kopteva, E. O'Riordan, M. Stynes (eds.), *BAIL* 2008 – *Boundary and Interior Layers.*

70. M. Bercovier, M.J. Gander, R. Kornhuber, O. Widlund (eds.), *Domain Decomposition Methods in Science and Engineering XVIII.*

71. B. Koren, C. Vuik (eds.), *Advanced Computational Methods in Science and Engineering.*

72. M. Peters (ed.), *Computational Fluid Dynamics for Sport Simulation.*

73. H.-J. Bungartz, M. Mehl, M. Schäfer (eds.), *Fluid Structure Interaction II - Modelling, Simulation, Optimization.*

74. D. Tromeur-Dervout, G. Brenner, D.R. Emerson, J. Erhel (eds.), *Parallel Computational Fluid Dynamics 2008.*

75. A.N. Gorban, D. Roose (eds.), *Coping with Complexity: Model Reduction and Data Analysis.*

76. J.S. Hesthaven, E.M. Rønquist (eds.), *Spectral and High Order Methods for Partial Differential Equations.*

77. M. Holtz, *Sparse Grid Quadrature in High Dimensions with Applications in Finance and Insurance.*

78. Y. Huang, R. Kornhuber, O.Widlund, J. Xu (eds.), *Domain Decomposition Methods in Science and Engineering XIX.*

79. M. Griebel, M.A. Schweitzer (eds.), *Meshfree Methods for Partial Differential Equations V.*

80. P.H. Lauritzen, C. Jablonowski, M.A. Taylor, R.D. Nair (eds.), *Numerical Techniques for Global Atmospheric Models.*

81. C. Clavero, J.L. Gracia, F.J. Lisbona (eds.), *BAIL 2010 – Boundary and Interior Layers, Computational and Asymptotic Methods.*

82. B. Engquist, O. Runborg, Y.R. Tsai (eds.), *Numerical Analysis and Multiscale Computations.*

83. I.G. Graham, T.Y. Hou, O. Lakkis, R. Scheichl (eds.), *Numerical Analysis of Multiscale Problems.*

84. A. Logg, K.-A. Mardal, G. Wells (eds.), *Automated Solution of Differential Equations by the Finite Element Method.*

85. J. Blowey, M. Jensen (eds.), *Frontiers in Numerical Analysis - Durham 2010.*

86. O. Kolditz, U.-J. Gorke, H. Shao, W. Wang (eds.), *Thermo-Hydro-Mechanical-Chemical Processes in Fractured Porous Media - Benchmarks and Examples.*

87. S. Forth, P. Hovland, E. Phipps, J. Utke, A. Walther (eds.), *Recent Advances in Algorithmic Differentiation.*

88. J. Garcke, M. Griebel (eds.), *Sparse Grids and Applications.*

89. M. Griebel, M.A. Schweitzer (eds.), *Meshfree Methods for Partial Differential Equations VI.*

90. C. Pechstein, *Finite and Boundary Element Tearing and Interconnecting Solvers for Multiscale Problems.*

91. R. Bank, M. Holst, O. Widlund, J. Xu (eds.), *Domain Decomposition Methods in Science and Engineering XX.*

92. H. Bijl, D. Lucor, S. Mishra, C. Schwab (eds.), *Uncertainty Quantification in Computational Fluid Dynamics.*

93. M. Bader, H.-J. Bungartz, T. Weinzierl (eds.), *Advanced Computing.*

94. M. Ehrhardt, T. Koprucki (eds.), *Advanced Mathematical Models and Numerical Techniques for Multi-Band Effective Mass Approximations.*

95. M. Azaïez, H. El Fekih, J.S. Hesthaven (eds.), *Spectral and High Order Methods for Partial Differential Equations ICOSAHOM 2012.*

96. F. Graziani, M.P. Desjarlais, R. Redmer, S.B. Trickey (eds.), *Frontiers and Challenges in Warm Dense Matter.*

97. J. Garcke, D. Pflüger (eds.), *Sparse Grids and Applications – Munich 2012.*

98. J. Erhel, M. Gander, L. Halpern, G. Pichot, T. Sassi, O. Widlund (eds.), *Domain Decomposition Methods in Science and Engineering XXI.*

99. R. Abgrall, H. Beaugendre, P.M. Congedo, C. Dobrzynski, V. Perrier, M. Ricchiuto (eds.), *High Order Nonlinear Numerical Methods for Evolutionary PDEs - HONOM 2013.*

100. M. Griebel, M.A. Schweitzer (eds.), *Meshfree Methods for Partial Differential Equations VII.*

101. R. Hoppe (ed.), *Optimization with PDE Constraints - OPTPDE 2014.*

102. S. Dahlke, W. Dahmen, M. Griebel, W. Hackbusch, K. Ritter, R. Schneider, C. Schwab, H. Yserentant (eds.), *Extraction of Quantifiable Information from Complex Systems.*

103. A. Abdulle, S. Deparis, D. Kressner, F. Nobile, M. Picasso (eds.), *Numerical Mathematics and Advanced Applications - ENUMATH 2013.*

104. T. Dickopf, M.J. Gander, L. Halpern, R. Krause, L.F. Pavarino (eds.), *Domain Decomposition Methods in Science and Engineering XXII.*

105. M. Mehl, M. Bischoff, M. Schäfer (eds.), *Recent Trends in Computational Engineering - CE2014.* Optimization, Uncertainty, Parallel Algorithms, Coupled and Complex Problems.

106. R.M. Kirby, M. Berzins, J.S. Hesthaven (eds.), *Spectral and High Order Methods for Partial Differential Equations - ICOSAHOM'14.*

107. B. Jüttler, B. Simeon (eds.), *Isogeometric Analysis and Applications 2014.*

108. P. Knobloch (ed.), *Boundary and Interior Layers, Computational and Asymptotic Methods – BAIL 2014.*

109. J. Garcke, D. Pflüger (eds.), *Sparse Grids and Applications – Stuttgart 2014.*

110. H. P. Langtangen, *Finite Difference Computing with Exponential Decay Models.*

111. A. Tveito, G.T. Lines, *Computing Characterizations of Drugs for Ion Channels and Receptors Using Markov Models.*

112. B. Karazösen, M. Manguoğlu, M. Tezer-Sezgin, S. Göktepe, Ö. Uğur (eds.), *Numerical Mathematics and Advanced Applications - ENUMATH 2015.*

113. H.-J. Bungartz, P. Neumann, W.E. Nagel (eds.), *Software for Exascale Computing - SPPEXA 2013-2015.*

114. G.R. Barrenechea, F. Brezzi, A. Cangiani, E.H. Georgoulis (eds.), *Building Bridges: Connections and Challenges in Modern Approaches to Numerical Partial Differential Equations.*

115. M. Griebel, M.A. Schweitzer (eds.), *Meshfree Methods for Partial Differential Equations VIII.*

116. C.-O. Lee, X.-C. Cai, D.E. Keyes, H.H. Kim, A. Klawonn, E.-J. Park, O.B. Widlund (eds.), *Domain Decomposition Methods in Science and Engineering XXIII.*

117. T. Sakurai, S.-L. Zhang, T. Imamura, Y. Yamamoto, Y. Kuramashi, T. Hoshi (eds.), *Eigenvalue Problems: Algorithms, Software and Applications in Petascale Computing*. EPASA 2015, Tsukuba, Japan, September 2015.

118. T. Richter (ed.), *Fluid-structure Interactions*. Models, Analysis and Finite Elements.

119. M.L. Bittencourt, N.A. Dumont, J.S. Hesthaven (eds.), *Spectral and High Order Methods for Partial Differential Equations ICOSAHOM 2016.* Selected Papers from the ICOSAHOM Conference, June 27-July 1, 2016, Rio de Janeiro, Brazil.

120. Z. Huang, M. Stynes, Z. Zhang (eds.), *Boundary and Interior Layers, Computational and Asymptotic Methods BAIL 2016.*

121. S.P.A. Bordas, E.N. Burman, M.G. Larson, M.A. Olshanskii (eds.), *Geometrically Unfitted Finite Element Methods and Applications*. Proceedings of the UCL Workshop 2016.

122. A. Gerisch, R. Penta, J. Lang (eds.), *Multiscale Models in Mechano and Tumor Biology*. Modeling, Homogenization, and Applications.

123. J. Garcke, D. Pflüger, C.G. Webster, G. Zhang (eds.), *Sparse Grids and Applications - Miami 2016*.

124. M. Schäfer, M. Behr, M. Mehl, B. Wohlmuth (eds.), *Recent Advances in Computational Engineering*. Proceedings of the 4th International Conference on Computational Engineering (ICCE 2017) in Darmstadt.

125. P.E. Bjørstad, S.C. Brenner, L. Halpern, R. Kornhuber, H.H. Kim, T. Rahman, O.B. Widlund (eds.), *Domain Decomposition Methods in Science and Engineering XXIV*. 24th International Conference on Domain Decomposition Methods, Svalbard, Norway, February 6–10, 2017.

126. F.A. Radu, K. Kumar, I. Berre, J.M. Nordbotten, I.S. Pop (eds.), *Numerical Mathematics and Advanced Applications – ENUMATH 2017*.

127. X. Roca, A. Loseille (eds.), *27th International Meshing Roundtable*.

128. Th. Apel, U. Langer, A. Meyer, O. Steinbach (eds.), *Advanced Finite Element Methods with Applications*. Selected Papers from the 30th Chemnitz Finite Element Symposium 2017.

129. M. Griebel, M. A. Schweitzer (eds.), *Meshfree Methods for Partial Differencial Equations IX*.

130. S. Weißer, BEM-based Finite Element *Approaches on Polytopal Meshes*.

131. V. A. Garanzha, L. Kamenski, H. Si (eds.), *Numerical Geometry, Grid Generation and Scientific Computing*. Proceedings of the 9th International Conference, NUMGRID2018/Voronoi 150, Celebrating the 150th Anniversary of G. F. Voronoi, Moscow, Russia, December 2018.

132. H. van Brummelen, A. Corsini, S. Perotto, G. Rozza (eds.), *Numerical Methods for Flows*.

For further information on these books please have a look at our mathematics catalogue at the following URL: `www.springer.com/series/3527`

Monographs in Computational Science and Engineering

1. J. Sundnes, G.T. Lines, X. Cai, B.F. Nielsen, K.-A. Mardal, A. Tveito, *Computing the Electrical Activity in the Heart.*

For further information on this book, please have a look at our mathematics catalogue at the following URL: www.springer.com/series/7417

Texts in Computational Science and Engineering

1. H. P. Langtangen, *Computational Partial Differential Equations.* Numerical Methods and Diffpack Programming. 2nd Edition
2. A. Quarteroni, F. Saleri, P. Gervasio, *Scientific Computing with MATLAB and Octave.* 4th Edition
3. H. P. Langtangen, *Python Scripting for Computational Science.* 3rd Edition
4. H. Gardner, G. Manduchi, *Design Patterns for e-Science.*
5. M. Griebel, S. Knapek, G. Zumbusch, *Numerical Simulation in Molecular Dynamics.*
6. H. P. Langtangen, *A Primer on Scientific Programming with Python.* 5th Edition
7. A. Tveito, H. P. Langtangen, B. F. Nielsen, X. Cai, *Elements of Scientific Computing.*
8. B. Gustafsson, *Fundamentals of Scientific Computing.*
9. M. Bader, *Space-Filling Curves.*
10. M. Larson, F. Bengzon, *The Finite Element Method: Theory, Implementation and Applications.*
11. W. Gander, M. Gander, F. Kwok, *Scientific Computing: An Introduction using Maple and MATLAB.*
12. P. Deuflhard, S. Röblitz, *A Guide to Numerical Modelling in Systems Biology.*
13. M. H. Holmes, *Introduction to Scientific Computing and Data Analysis.*
14. S. Linge, H. P. Langtangen, *Programming for Computations* - A Gentle Introduction to Numerical Simulations with MATLAB/Octave.
15. S. Linge, H. P. Langtangen, *Programming for Computations* - A Gentle Introduction to Numerical Simulations with Python.
16. H.P. Langtangen, S. Linge, *Finite Difference Computing with PDEs* - A Modern Software Approach.
17. B. Gustafsson, *Scientific Computing from a Historical Perspective.*
18. J. A. Trangenstein, *Scientific Computing.* Volume I - Linear and Nonlinear Equations.

19. J. A. Trangenstein, *Scientific Computing*. Volume II - Eigenvalues and Optimization.

20. J. A. Trangenstein, *Scientific Computing*. Volume III - Approximation and Integration.

For further information on these books please have a look at our mathematics catalogue at the following URL: www.springer.com/series/5151

Printed in the United States
by Baker & Taylor Publisher Services